SOLUTION KEY

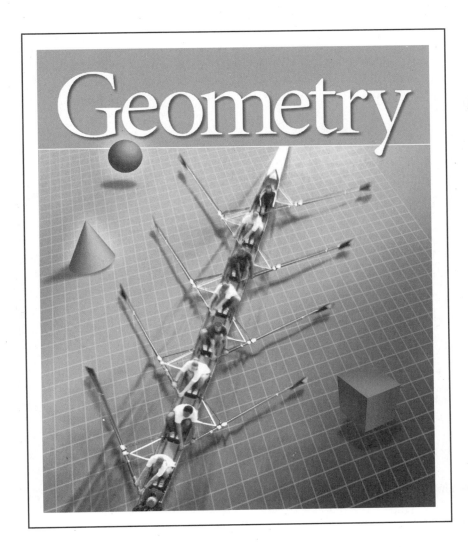

Geometry

HOLT, RINEHART AND WINSTON

A Harcourt Classroom Education Company

Austin • New York • Orlando • Atlanta • San Francisco • Boston • Dallas • Toronto • London

To the Teacher

Geometry Solution Key contains the worked-out solutions for the exercises in the Guided Skills practice, Practice and Apply, Look Back, Look Beyond, Chapter Review and Assessment, Chapter Test, and Cumulative Assessment sections in *Geometry*. Answers for the Activities and Communicate questions are found in the Additional Answers section in *Geometry Teacher's Edition*.

Requests for permission to make copies of any part of the work should be mailed to the following address: Permissions Department, Holt, Rinehart and Winston, 10801 N. MoPac Expressway, Building 3, Austin, Texas 78759.

Photo Credit
Front Cover: Dale Sanders/Masterfile.

Printed in the United States of America

ISBN 0-03-066379-2

1 2 3 4 5 6 7 066 05 04 03 02 01

Table of Contents

CHAPTER 1

Exploring Geometry

6. Sample answer: point A; $\overleftrightarrow{AB}$, $\overline{AB}$, $\overrightarrow{AB}$

7. $\angle 1$, $\angle Q$, $\angle PQR$, or $\angle RQP$

8. Sample answer: plane MNO, plane NOM, or plane $\mathcal{R}$.

PAGES 14–16, PRACTICE AND APPLY

9. $\overline{AB}$, $\overline{BC}$, $\overline{AC}$

10. $\angle A$, $\angle 1$, $\angle BAC$; $\angle B$, $\angle 2$, $\angle ABC$; $\angle C$, $\angle 3$, $\angle ACB$

11. $\angle A$: $\overrightarrow{AB}$, $\overrightarrow{AC}$; $\angle B$: $\overrightarrow{BA}$, $\overrightarrow{BC}$; $\angle C$: $\overrightarrow{CA}$, $\overrightarrow{CB}$

12. plane ABC

13. line

14. point

15. plane

16. plane

17. point

18. False. Lines are infinite, never ending in both directions.

19. False. Planes are infinite, extending without bound and having no edges.

20. False. Two intersecting lines are contained in exactly one plane. Three intersecting lines may be contained in two planes.

21. True. For example, two opposite sides of a box intersect the bottom plane of a box, but they don't intersect each other.

22. True. For example, the side, front, and bottom of a box intersect at a point, the corner of the box.

23. False. There are an infinite number of planes through any two points.

24. False. Through any three *noncollinear* points there is exactly one plane. If the points are collinear the line is contained in an infinite number of planes.

25. False. Four collinear points are contained in an infinite number of planes.

26. Sample answer: line m, $\overleftrightarrow{AE}$, $\overleftrightarrow{EA}$, or $\overleftrightarrow{CE}$

27. point E, point A, or point C

28. point A

29. Sample answer: $\angle DAC$; A; $\overrightarrow{AD}$; $\overrightarrow{AC}$

30. No. There are 4 angles with a vertex at A.

31. line ℓ

32. line ℓ

33. point M

34. point K, point M, or point F

35. point M

36. Sample answer: point I, point L, and point N

37. Sample answer: $\overline{KP}$ and $\overline{MU}$

38. point A

39. line m

40. Sample answer: point M, point A, and point R

41. 1; $\overline{AB}$

42. $2 + 1 = 3$; $\overline{AB}$, $\overline{AC}$, $\overline{BC}$

43. $3 + 2 + 1 = 6$; $\overline{AB}$, $\overline{AC}$, $\overline{AD}$, $\overline{BC}$, $\overline{BD}$, $\overline{CD}$

44. If there are n points, there are $\frac{n(n-1)}{2}$ segments. From each of the n points, $n - 1$ segments may be drawn, giving $n(n - 1)$ segments. But each segment is counted twice, so divide by 2 to get the formula.

45. 1; $\angle AVB$

46. $1 + 2 = 3$; $\angle AVC$, $\angle AVB$, and $\angle BVC$

47. $1 + 2 + 3 = 6$; $\angle AVD$, $\angle AVC$, $\angle BVD$, $\angle AVB$, $\angle BVC$, and $\angle CVD$

48. Each of the n points is contained in $n - 1$ angles. By reasoning as in Exercise 44, there are $\frac{n(n-1)}{2}$ distinct angles.

PAGE 16, LOOK BACK

49. $\begin{aligned} 22 + (-6) &= 22 - 6 \\ &= 16 \end{aligned}$

50. $7 + 15 = 22$

51. $\begin{aligned} 11 - (-4) &= 11 + 4 \\ &= 15 \end{aligned}$

52. $\begin{aligned} -81 - (-30) &= -81 + 30 \\ &= -51 \end{aligned}$

53. $\begin{aligned} \left|-14 + (-35)\right| &= \left|-14 - 35\right| \\ &= \left|-49\right| \\ &= 49 \end{aligned}$

54. $\begin{aligned} \left|13 - 10\right| &= \left|3\right| \\ &= 3 \end{aligned}$

55. $\begin{aligned} -123 - 41 &= -(123 + 41) \\ &= -164 \end{aligned}$

56. $\begin{aligned} \left|21 + (-35)\right| &= \left|21 - 35\right| \\ &= \left|-14\right| \\ &= 14 \end{aligned}$

57. $\begin{aligned} \left|-54 + (-20)\right| &= \left|-54 - 20\right| \\ &= \left|-74\right| \\ &= 74 \end{aligned}$

PAGE 16, LOOK BEYOND

58. Twelve cards are exchanged; six lines can be drawn; each lines represents two cards. Therefore, there are six exchanges.

59. Twenty cards are exchanged with five people. Therefore, there are ten exchanges. When n people exchange cards, there are $n(n - 1)$ cards exchanged and $\frac{n(n-1)}{2}$ exchanges.

1.2 PAGE 21, GUIDED SKILLS PRACTICE

6. MN (or $\mathrm{m}\overline{MN}$) $= \left|-3 - 2\right| = \left|-5\right| = 5$

7. NP (or $\mathrm{m}\overline{NP}$) $= \left|2 - 6\right| = \left|-4\right| = 4$

8. MP (or $\mathrm{m}\overline{MP}$) $= \left|-3 - 6\right| = \left|-9\right| = 9$

9. a. If $\overline{AB} \cong \overline{CD}$, then $AB = CD$.

  b. If $AB = CD$, then $\overline{AB} \cong \overline{CD}$.

10. Let x be the distance in miles from Forsyth to Decator (FD). Then $7x$ is the distance from Bloomington to Forsyth (BF).

$$7x + x = 40$$
$$8x = 40$$
$$x = 5$$
$$7x = 7(5) = 35$$

$FD = 5$ miles and $BF = 35$ miles

PAGES 22–23, PRACTICE AND APPLY

11. $AB = \left|4 - 0\right| = \left|4\right| = 4$

12. $AB = \left|-2 - 0\right| = \left|-2\right| = 2$

13. $AB = \left|5 - 2\right| = \left|3\right| = 3$

14. $AB = \left|-7 - (-2)\right| = \left|-5\right| = 5$

15. $AB = \left|-3 - 2\right| = \left|-5\right| = 5$

16. $AB = \left|-1 - 1\right| = \left|-2\right| = 2$

17. $AB = |-3 - (-1)| = |-1 - (-3)| = 2$
$BC = |-1 - 3| = |3 - (-1)| = 4$
$AC = |-3 - 3| = |3 - (-3)| = 6$

18. $\overline{AC} \cong \overline{CE} \cong \overline{BD} \cong \overline{DF}; \overline{AB} \cong \overline{CD} \cong \overline{EF}$

19. $\overline{AF} \cong \overline{ED}; \overline{FG} \cong \overline{EG}; \overline{BG} \cong \overline{CG}; \overline{FC} \cong \overline{BE}$

20.

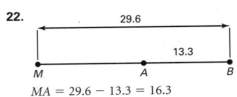

$MB = 30 + 15 = 45$

21.

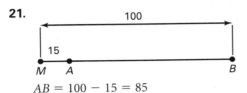

$AB = 100 - 15 = 85$

22.

$MA = 29.6 - 13.3 = 16.3$

23. $2x + 3x = 25$
$5x = 25$
$x = 5$

24. $PQ = 25$
$5x = 25$
$x = 5$
$3x = 15$
$PR = 25 + 15 = 40$

25. $x = 12$
$PR = PQ + QR = 25 + 12 = 37$

26. $AB = x, BC = 2 + 2x$, then
$x + (2 + 2x) = 41$
$3x + 2 = 41$
$3x = 39$
$x = 13$
$AB = 13$ miles; $BC = 2 + 2(13) = 28$ miles

27. $XC = 41 - 6 = 35$ miles

28. The statement $XY = 5000$ yd makes sense because both XY and 5000 are numbers.

29. The statement $\overline{PQ} = 32$ in. does not make sense because $\overline{PQ}$ is not a number.

30. The statement m$ST = 6$ cm does not make sense because ST is a number and numbers do not have measures.

31. The statement $XY + XZ = 32$ cm makes sense because XY, XZ, and 32 are all numbers.

32. The statement m$\overline{PR} = 46$ cm makes sense because both m$\overline{PR}$ and 46 cm are numbers.

33. The statement $\overline{XY} - \overline{XZ} = 12$ cm does not make sense because $\overline{XY}$ and $\overline{XZ}$ are not numbers.

34. Check students' rulers.

35. Sample answer: the Egyptian royal cubit is longer.

36. $\frac{1}{4}$

37. Check students' rulers.

38. Sample answers: $\frac{1}{3}, \frac{5}{6}, \frac{3}{5}, \frac{4}{7}$

PAGE 24, LOOK BACK

39. positive

40. Positive. The result is a positive number.

41. Negative. The result is a positive number.

42. The absolute value is the same, regardless of the order in which the numbers are subtracted.

43. negative

44. Positive. The result is a positive number.

45. Negative. The result is a positive number.

46. The absolute values are the same, regardless of the order in which the numbers are subtracted.

PAGE 24, LOOK BEYOND

47. Check students' rulers. 1000 mm; 10,000 mm

48. $\frac{10^8}{10^6} = 10^2$, 100 times

49. $\frac{32^8}{32^6} = 32^2$, 1024 times

8. $m\angle WVX = 60°$

9. $m\angle QPR = |90 - 65| = |25| = 25°$

10. $m\angle RPS = |65 - 45| = |20| = 20°$

11. $m\angle QPR + m\angle RPS = 25° + 20° = 45°$

12. a. If $m\angle UVW = m\angle XYZ$, then $\angle UVW \cong \angle XYZ$.

   b. If $\angle UVW \cong \angle XYZ$, then $m\angle UVW = m\angle XYZ$.

13. Complementary angle pair: $\angle CAD$ and $\angle DAE$. Supplementary angle pairs: $\angle BAD$ and $\angle DAE$ or $\angle BAC$ and $\angle CAE$.

PAGES 31–33, PRACTICE AND APPLY

14. $45°$

15. $85°$

16. $120°$

17. $85 - 45 = 40°$

18. $120° - 45° = 75°$

19. $120 - 85 = 35°$

20. $m\angle X = 35°$

21. $m\angle Y = 120°$

22. $90°$

23. $m\angle 2 = 30°$

24. $m\angle RST = 93°$

25. $m\angle SRT = 32°$

26. $m\angle STR = 55°$

27. $m\angle BEC = m\angle CED = 25°$

28. $m\angle AEB = 180 - m\angle BED = 180 - 50 = 130°$

29. $m\angle AEC = m\angle AEB + m\angle BEC = 130° + 25° = 155°$

30. $m\angle 2 = m\angle 3$ and $m\angle 2 + m\angle 3 = 60°$, so $m\angle 2 = 30°$

31. $m\angle 3 = m\angle 2 = 30°$

32. $m\angle 1 = m\angle EDG - m\angle 2 = 70 - 30 = 40°$

33. $m\angle 4 = m\angle 1 = 40°$

34. $m\angle EDJ = m\angle 1 + m\angle 2 + m\angle 3 + m\angle 4 = 40 + 30 + 30 + 40 = 140°$

35. $m\angle SLA = 90°$; $\angle SLE$ and $\angle ALE$ are complementary angles.

36. $m\angle WZY = 80 + 34 = 114°$

37. $m\angle XZY = 43 - 21 = 22°$

38. $m\angle WZX = 52 - 34 = 18°$

39. $87° = (2x - 8)° + (x + 50)°$
$87° = (3x + 42)°$
$45° = (3x)°$
$15 = x$

40. $x = 15$
$m\angle KNL = (x + 50)° = (15 + 50)° = 65°$

41. $x = 15$
$m\angle LNM = (2x - 8)° = (30 - 8)° = 22°$

42. $(43 + x)° = (6x - 6)° + (3x + 1)°$
$(43 + x)° = (9x - 5)°$
$(8x)° = (48)°$
$x = 6$
$(43 + x)° = (43 + 6)°$
$m\angle ADC = 49°$

43. $(6x - 6)° = (6(6) - 6)°$
$= (36 - 6)°$
$= 30°$
$m\angle ADB = 30°$

44. $(3x + 1)° = (3(6) + 1)°$
$= (18 + 1)°$
$= 19°$
$m\angle BDC = 19°$

45. $\frac{360}{12} = 30$, There are 30° between each pair of consecutive numbers. $3(30) = 90$. At 3:00 the angle is 90°.

46. $\frac{360}{12} = 30$, There are 30° between each pair of consecutive numbers. $5(30) = 150$. At 5:00 the angle is 150°.

47. A gradian is smaller than 1 degree since one gradian is $\frac{9}{10}$ of one degree.

48. m∠*TVM* and m∠*OVB* = 20°; m∠*BVM*, m∠*RVT*, and m∠*TVO* = 45°; m∠*BVT* and m∠*MVR* = 65°

49. 000

50. 090

51. 180

52. 270

53. $\frac{90°}{2}$ = 45°; 045

54. 180° + 45° = 225°; 225

55. $\frac{45°}{2}$ = 22.5°; 022.5

56. 180° + 22.5° = 202.5°; 202.5

57. Less than 355. The pilot would have to fly more to the west.

58. 045, 135, 225; The diver must always move the same distance in each direction in order to move in a square.

PAGE 34, LOOK BACK

59. ∠*CAR*

60. Lines *m* and *n*

61. Lines ℓ, *m*, and *n*

62. Points *A*, *B*, and *C*

63. $\left|-9 - (-3)\right| = 6$

64. $\left|-6 - 10\right| = 16$

65. $4x - 3 + x = 27$
$5x = 30$
$x = 6 = BC$
$AC = 4x - 3 = 4(6) - 3 = 24 - 3 = 21$

PAGE 34, LOOK BEYOND

66. The number 360 has many divisors so it would be easy to work with angle measures that are factors of 360.

67. a. $\frac{60 \text{ seconds}}{1 \text{ minute}} \cdot \frac{60 \text{ minutes}}{1 \text{ degree}} = 3600$ seconds/degree
There are 3600 seconds in each degree.

b. $1.5 \text{ minutes} \cdot \frac{60 \text{ seconds}}{1 \text{ minute}} = 90$ seconds
There are 90 seconds in 1.5 minutes.

c. $1.75 \text{ degrees} \cdot \frac{60 \text{ minutes}}{1 \text{ degree}} = 105$ minutes
There are 105 minutes in 1.75 degrees.

1.4 PAGE 39, GUIDED SKILLS PRACTICE

5. Lines ℓ and *m* are parallel.

6. The distance from *P* to ℓ is *PB*.

7. *AC* and *BC* are equal.

8. *BX* and *CX* are equal.

PAGES 40–41, PRACTICE AND APPLY

9.

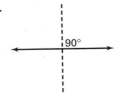

10.

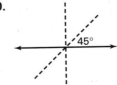

11.

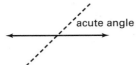
acute angle

12.

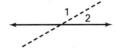
obtuse angle

13. ∠1 and ∠2 are complementary.

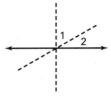

14. ∠1 and ∠2 are supplementary.

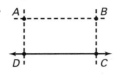

15. Lines ℓ and *m* are parallel.

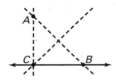

16.

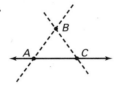

17.

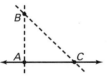

18.

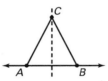

19.

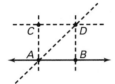

20.

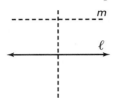

21.

22. The lengths are equal.

23. Triangles that have a vertex located on the perpendicular bisector of one of the sides have at least two sides of equal length.

24. Trace $\overline{AB}$. Construct its perpendicular bisector *m*. Fold a line through *A* so that *B* falls on m. Mark this point *C*. △ABC will have 3 congruent sides.

25. $AC = BC$ and $AC + BC = AB = 10$, so $AC = BC = 5$.

26. $m\angle FEX = m\angle DEX = 15°$
$m\angle DEF = m\angle FEX + m\angle DEX = (15 + 15)° = 30°$

27. The line ℓ is also the bisector of ∠*WVZ*.

28. Lines ℓ and *m* are perpendicular.

29. The bisectors of the angles formed by intersecting lines are perpendicular.

30.

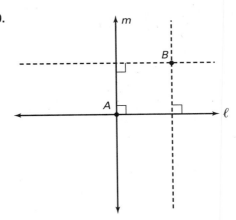

31. The shape appears to be a square.

32. The diagonals of a square are of equal length, and are perpendicular bisectors of each other.

PAGE 41, LOOK BACK

33. a line or a segment

34. a plane

35. a line or a segment

36. a point

37. $\overleftrightarrow{XY}$, $\overleftrightarrow{YX}$, or line ℓ

38. $\overline{MN}$ or $\overline{NM}$

39. $\angle MAP$, $\angle PAM$, or $\angle A$

40. $\overrightarrow{PS}$

41. Sample answer: Plane $\mathcal{F}$, Plane PQR, Plane QRP, or Plane RQP

42. point A

PAGE 42, LOOK BEYOND

43. Check student constructions.

44. No. Any triangle created with the segments will have the same size and shape.

1.5 PAGE 45, GUIDED SKILLS PRACTICE

4. The perpendicular bisectors meet in a single point.

5. The angle bisectors meet in a single point.

	Intersecting lines	Type of circle formed	Name of center
6.	perpendicular bisectors	circumscribed	circumcenter
7.	angle bisectors	inscribed	incenter

8.

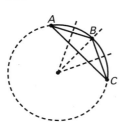

9.

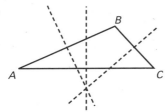

10.

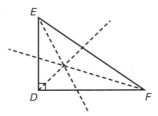

11.

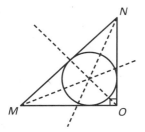

12.

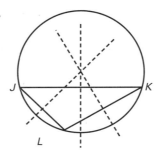

13.

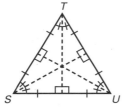

14.

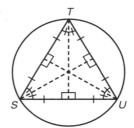

15 and 16.

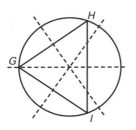

17.

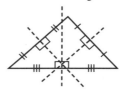

18.

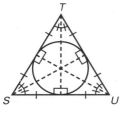

19. inside the triangle

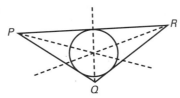

20. outside the triangle

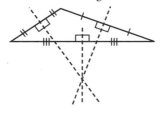

21. on the triangle

22.

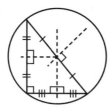

The longest side of the triangle divides the circle into two equal parts.

23. The center of the circumscribed circle of a right triangle is the midpoint of the longest side of the triangle.

24. The four triangles are exactly the same.

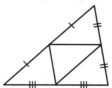

25. Sample answer:
The numbers in the top row are twice as large as the numbers in the bottom row. The distance from a vertex to the centroid is twice the distance from the centroid to the opposite side.

The distance from a vertex to the centroid is $\frac{2}{3}$, the distance from the vertex to the opposite side.

26. $AX = BX$; For any point, Y, on the perpendicular bisector of $\overline{AC}$, $AY = CY$. For any point, Z, on the perpendicular bisector of $\overline{BC}$, $BZ = CZ$. The intersection of the perpendicular bisectors is the same distance from each vertex of the triangle.

27. Y is the same distance from $\overline{DE}$ as it is from $\overline{EF}$; Any point on the angle bisector of $\angle D$ is the same distance from $\overline{DF}$ as it is from $\overline{DE}$. Any point on the angle bisector of $\angle F$ is the same distance from $\overline{EF}$ as it is from $\overline{DF}$. The intersection of the angle bisectors is the same distance from all 3 sides of the triangle.

28.

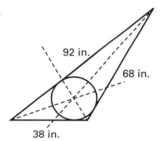

92 in.

68 in.

38 in.

The diameter of the largest duct can be estimated using a proportion.

$$\frac{\text{length of one side of triangle}}{\text{measure of scaled drawing of that side}} = \frac{\text{radius of inscribed circle}}{\text{measure of scaled drawing}}$$

Note: measurements may vary from those given below but the measurements will be proportional to those given. Thus, the final answer will be the same.

If the side labeled 92 in. measured 52 cm and the radius measured 6.5 cm, we could use the following proportion:

$$\frac{92 \text{ in.}}{52 \text{ cm}} = \frac{r}{6.5 \text{ cm}}$$
$$r = \frac{92(6.5)}{52} = 11.5$$

diameter $= 2r = 2(11.5) = 23$
The diameter of the largest duct is about 23 inches.

PAGE 48, LOOK BACK

---

29. collinear

30. coplanar

31. No. Through any two points there is exactly one line.

32. No. Through any three noncollinear points there is exactly one plane. If the points are collinear, they lie in an infinite number of planes.

33. Yes. If two non-parallel lines do not intersect, they are not in the same plane.

34. $4x - 3 + 3x + 2 = 48$
$7x = 49$
$x = 7$
$XY = 3x + 2 = 3(7) + 2 = 23$

35. $WX = 4x - 3 = 4(7) - 3 = 25$

36. $-10x + 3 = -x + 12$
$-9x + 3 = 12$
$-9x = 9$
$x = -1$
$m\angle BAC = -10x + 3 = -10(-1) + 3 = 13°$

37. $m\angle CAD = m\angle BAC = 13°$

PAGE 49, LOOK BEYOND

38. Fold a line through A perpendicular to $\overline{BC}$. Measure the length of the perpendicular.

39. All three altitudes meet in a single point inside the triangle.

40. If extended, all three altitudes meet in a single point. The point is outside the triangle.

1.6 PAGE 54, GUIDED SKILLS PRACTICE

8.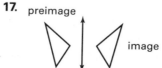

9.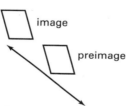

10.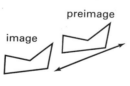

PAGES 55–57, PRACTICE AND APPLY

11.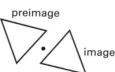

12.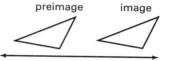

13.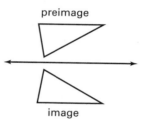

14.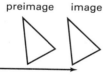

15.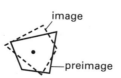

16.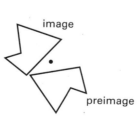

17.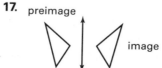

18.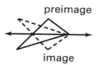

19.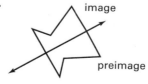

20.

21.

22. a. 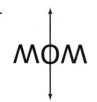   **b.** 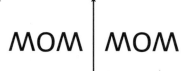   **c.**

23. The letters that stay the same in a reflection across a vertical line are A, H, I, M, O, T, U, V, W, X, and Y. The letters that stay the same in a reflection across a horizontal line are B, C, D, E, H, I, K, O, and X.

24. Sample answer: Each image point is the same distance from its preimage point. The segments are parallel and the same length.

25. The angles all measure 70°.

26. The line of reflection is the perpendicular bisector of $\overline{GG'}$, $\overline{HH'}$, and $\overline{II'}$.

27. a. Footprints made by the right and left foot, the pair on the left

   b. Footprints made by the same foot, the pair on the right

28. reflection; the preimage is the wood block on the left and the image is the print on the right.

29. Translate square *ABCD* from its preimage position one unit up, one unit down, and one unit to the left. Then translate the square twice to the right by one unit.

30. Rotate *ABCD* about each of its vertices by 90° counterclockwise. Then rotate *ABEF* 180° clockwise about the point *E*.

31. Reflect square *ABCD* from its preimage position across $\overline{CB}$, $\overline{CD}$, $\overline{DA}$, and $\overline{BA}$. Then reflect the square at *BEFA* across $\overline{EF}$.

32. The arrow in the direction of the translation is parallel to the line of reflection.

33. Transform the second image, then the third and so on.

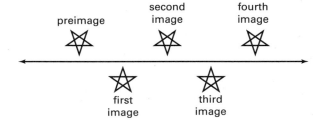

PAGE 57, LOOK BACK

34. a. $\overline{AB}$, $\overline{AC}$, $\overline{BC}$, $\angle A$, $\angle B$, $\angle C$

   b. *AB* = 2.6 cm, *AC* = 2.3 cm, *BC* = 1.7 cm

   c. m$\angle A$ = 40°; m$\angle B$ = 60°; m$\angle C$ = 80°

35. a. $\overline{DE}$, $\overline{DF}$, $\overline{EF}$, $\angle D$, $\angle E$, $\angle F$

   b. *DE* = 2.2 cm, *DF* = 3 cm, *EF* = 1.6 cm

c. $m\angle D = 33°$, $m\angle E = 103°$, $m\angle F = 44°$

36. a. $\overline{GH}$, $\overline{HI}$, $\overline{GI}$, $\angle G$, $\angle I$, $\angle H$

    b. $GH = 2.4$ cm, $HI = 1.9$ cm, $GI = 1.5$ cm

    c. $m\angle G = 50°$, $m\angle I = 90°$, $m\angle H = 40°$

37.

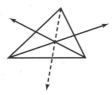

38.

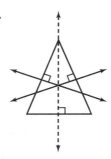

PAGES 57–58, LOOK BEYOND

39. The figure has been translated by $TS = 2.6$ cm.

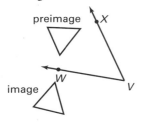

40.

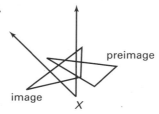

41. The figure has been rotated by $m\angle WVX = 55°$.

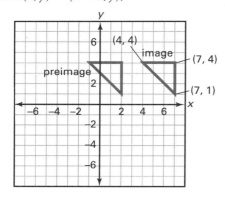

42.

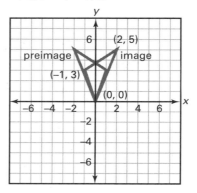

1.7   **PAGE 63, GUIDED SKILLS PRACTICE**

5. $H(x, y) = (x + 5, y)$;

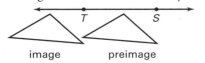

6. $N(x, y) = (-x, y)$;

7. $R(x, y) = (-x, -y)$;

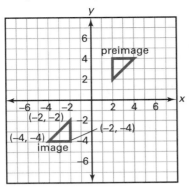

8. horizontal translation;

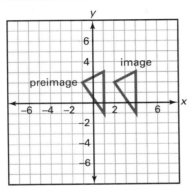

9. horizontal translation;

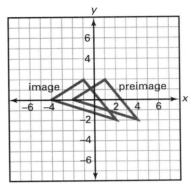

10. vertical translation;

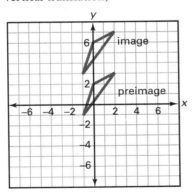

11. vertical translation;

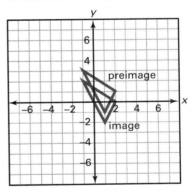

12. reflection across the x-axis;

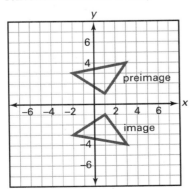

13. reflection across the x-axis;

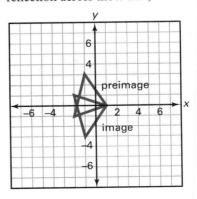

14. reflection across the *y*-axis;

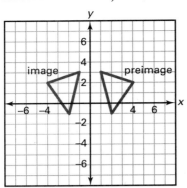

15. reflection across the *y*-axis;

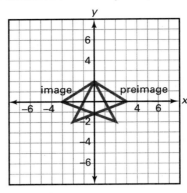

16. 180° rotation;

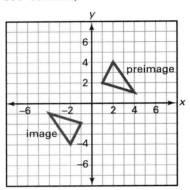

17. 180° rotation;

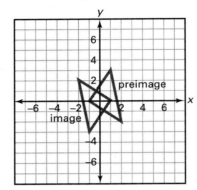

18.

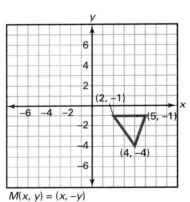

$M(x, y) = (x, -y)$

19.

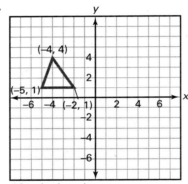

$N(x, y) = (-x, y)$

20.

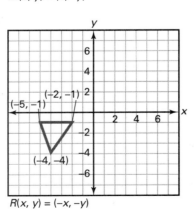

$R(x, y) = (-x, -y)$

21. The result is a 180° rotation about the origin as in Ex. 20.

22. The result is a 180° rotation about the origin as in Ex. 20 and 21.

23. translation 7 units to the right.

24. reflection across the *y*-axis

25. translation 6 units to the left and 7 units up

26. translation 4 units down

27. translation 7 units up

28. rotation of 180° about the origin

29. translation 7 units to the left

30. translation 2 units up

31.

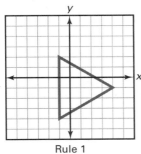
Rule 1

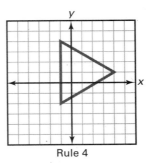
Rule 2

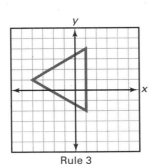
Rule 3

Rule 4

The triangle ends where it started.

32.

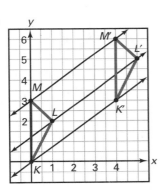

33. $\overleftrightarrow{KK'}: m = \frac{3-0}{4-0} = \frac{3}{4}$

$\overleftrightarrow{LL'}: m = \frac{4-1}{4-0} = \frac{3}{4}$

$\overleftrightarrow{mm'}: m = \frac{5-2}{5-1} = \frac{3}{4}$

34. The slope is the ratio of the amount of the vertical translation over the amount of the horizontal translation.

35. If K' is the result of the transformation of K under $T(x, y) = (x + h, y + k)$, then $\overleftrightarrow{KK'}$ has slope $\frac{k}{h}$.

36.

x	y
0	0
2	2
−1	−1

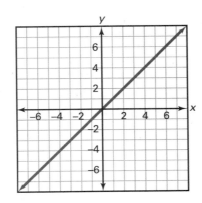

37. The line forms a 45° angle with the *y*-axis.

38–39.

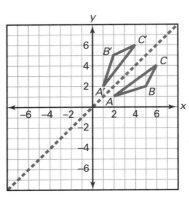

40. The two figures are reflections of each other across the line $y = x$.

41. $R(x, y) = (y, x)$

42. The line $y = x$ is the perpendicular bisector of each segment.

43. A point (x, y), where $x < y$, is above the line, and applying the transformation rule gives a point below the line. A point (x, y), where $x > y$, is below the line, and applying the transformation rule gives a point above the line. For a point (x, y), with $x = y$, the transformation rule has no effect.

44. Reflect $\triangle ABC$ across the *y*-axis and then translate the image 4 units up.
$T(x, y) = (-x, y + 4)$

45. It flips back and forth across the *y*-axis and moves up to form a pattern.

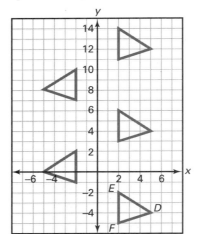

46. $G(x, y) = (-x, y + v)$

47. $G(x, y) = (x + h, -y)$

48. end points: $\left(\frac{144}{72}, \frac{72}{72}\right)$ and $\left(\frac{144}{72}, \frac{432}{72}\right)$ or $(2, 1)$ and $(2, 6)$

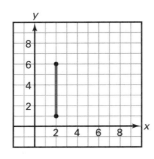

49. $(1, 1)$ and $(3, 4.5)$ in inches is equivalent to $(72, 72)$ and $(216, 324)$ in POSTSCRIPT units. The new POSTSCRIPT code is:

```
newpath
    72 72 moveto
    216 324 lineto
stroke
showpage
```

PAGE 67, LOOK BACK

50. 270

51. 315

52. 225

53. 202.5

54. Find the point where the perpendicular bisectors of the sides meet.

55. Find the point where the angle bisectors meet.

56. In a translation, all the points of a figure move a given distance in a given direction.

57. In a rotation, a figure turns around a given point called the turn center.

58. In a reflection, a figure is flipped over a line.

59. In a glide transformation, a figure is reflected across a line, while being translated parallel to that line.

PAGE 67, LOOK BEYOND

60. A 90° counterclockwise rotation about the origin;

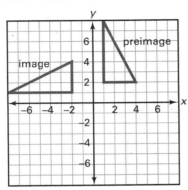

61. $Q(x, y) = (y, -x)$

CHAPTER REVIEW AND ASSESSMENT

1. $\overline{KJ}, \overline{JL}, \overline{LK}$

2. $\angle J, \angle K, \angle L$

3. $\overleftrightarrow{SQ}, \overleftrightarrow{RC}, \overleftrightarrow{EN}$

4. $\overrightarrow{MR}, \overrightarrow{MC}, \overrightarrow{MQ}, \overrightarrow{MO}, \overrightarrow{MN}, \overrightarrow{CR}, \overrightarrow{MS}, \overrightarrow{ME}, \overrightarrow{QS}, \overrightarrow{EN}, \overrightarrow{RC}, \overrightarrow{SQ}, \overrightarrow{NE}$

5. $CH = |-5 - (-1)| = |-5 + 1| = |-4| = 4;$
$CJ = |-5 - 3| = |-8| = 8;$
$CK = |-5 - 11| = |-16| = 16;$
$HJ = |-1 - 3| = |-4| = 4;$
$HK = |-1 - 11| = |-12| = 12;$
$JK = |3 - 11| = |-8| = 8$

6. $\overline{CJ} \cong \overline{JK}; \overline{CH} \cong \overline{HJ}$

7.

	25		13	
R		A		P

$RP = 25 + 13 = 38$

8.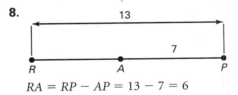

$RA = RP - AP = 13 - 7 = 6$

9. $m\angle PVT = 125°; m\angle QVT = 95°;$
$m\angle RVT = 65°; m\angle SVT = 30°;$
$m\angle PVS = 125 - 30 = 95°;$
$m\angle QVS = 95 - 30 = 65°;$
$m\angle RVS = 65 - 30 = 35°;$
$m\angle PVR = 125 - 65 = 60°;$
$m\angle QVR = 95 - 65 = 30°;$
$m\angle PVQ = 125 - 95 = 30°$

10. $\angle QVT \cong \angle PVS; \angle SVT \cong \angle PVQ \cong QVR;$
$\angle QVS \cong \angle RVT$

11. $m\angle LKN = 12 + 26 = 38°$

12. m∠LKM = 90 − 75 = 15°

13.
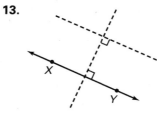

14.
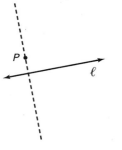

15.
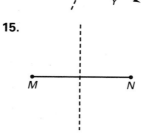

16.
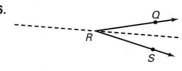

17. Sample answer:
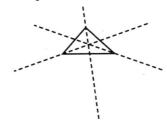

18. Sample answer:
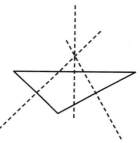

19.
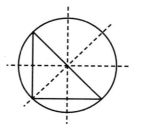

20.
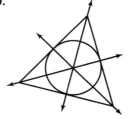

21. a. reflection

   b. translation

   c. rotation

22.

23.

24.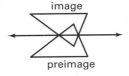

25.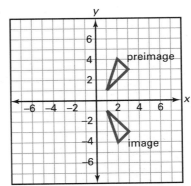

26. $T(x, y) = T(x - 3, y)$

27.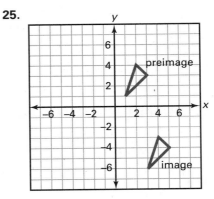

28. $R(x, y) = (-x, -y)$

29. $3(22) + 5 = 71$
Let x be the distance from Smithville to LaGrange.
Then $x + 22 = 71$
$\qquad\qquad x = 49$
It is 49 miles from Smithville to LaGrange and
71 miles from Bastrop to LaGrange.

30. To draw the complete circle, the student can choose any three points on the existing part of the circle, draw a triangle, and construct the circumscribd circle.

CHAPTER TEST

1. $\overleftrightarrow{MN}$

2. M, N, and P or M, N, and Q

3. $LR = 13 + 5(RT)$; $LT = 103$
$\qquad\qquad LR + RT = LT$
$\quad 13 + 5(RT) + RT = 103$
$\qquad\qquad\qquad RT = 15$
$\qquad\quad LR = 13 + 5(15) = 88$
From Limon to Rocky Ford is 88 miles.
From Rocky Ford to Timpas is 15 miles.

4. $BC + CF = BF$; $42 + CF = 60$; $CF = 12$

5. $BC + CF = BF$; $BC + 23 = 51$; $BC = 28$

6. $m\angle QPR = 32°$

7. $m\angle SPT = 180° - 32° - 32° = 116°$

8. $m\angle RPT = 116° + 32° = 148°$

9. Check students' work.

10. Check students' work.

11. Check students' work.

12. $m\angle XWY = 18°$

13. $m\angle XWZ = 18° + 18° = 36°$

14. Check students' work.

15. Check students' work.

16. Check students' work.

17. Check students' work.

18. Check students' work.

19. $S(x, y) = (x + 4, y)$

CHAPTER 1 CUMULATIVE ASSESSMENT

1. $NB = |-4 - (-1)| = |-3| = 3$
$BS = |-1 - 4| = |-5| = 5$
So $BS > NB$ and the answer is B.

2. $EH = EF + FH$
$FG = HG + FH$
Since $EF = HG$, $EH = FG$ and the answer is C.

3. Since $\overrightarrow{WZ}$ bisects $\angle XWY$, the answer is C.

4. $\overleftrightarrow{AB}$ or $\overleftrightarrow{BA}$

5. $\overrightarrow{CD}$

6. $\overline{EF}$ or $\overline{FE}$

7. $\angle HGI$, $\angle IGH$, or $\angle G$

8.

9.

10.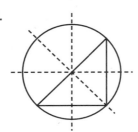

11.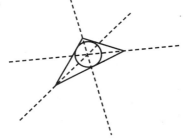

12. rotation

13. reflection

14. translation

15. translation

16. reflection

17. $AB = |-3 - (-1)| = |-2| = 2$

18. $AC = |-3 - 4| = |-7| = 7$

19. $m\angle AVC = m\angle AVD - m\angle CVD = 85 - 25 = 60°$

20. $\angle MON$ and $\angle MNO$ are supplementary to congruent angles, therefore they are congruent and $m\angle MON = 45°$.

Reasoning in Geometry

PAGE 83, GUIDED SKILLS PRACTICE

6. Two squares of each color need to be removed. Students may notice that certain conditions about the locations of the squares that are removed must also be met.

7. Sum of the first 10 odd numbers = 10^2 = 100.
   Sum of the first 100 odd numbers = 100^2 = 10,000

8.

Square	Area	Reason
B	36	Side of B = Side of A − Side of I $\qquad$ = 8 − 2 = 6 Area = 6^2 = 36
C	16	Side of C = Side of B − Side of I $\qquad$ = 6 − 2 = 4 Area = 4^2 = 16
D	16	Side of D = Side of Big Square − Side of B − Side of I $\qquad$ = Side of A + Side of B − Side of B − Side of I $\qquad$ = 8 + 6 − 6 − 2 = 4 Area = 4^2 = 16
E	36	Side of E = Side of C − Side of I + Side of D $\qquad$ = 4 − 2 + 4 = 6 Area = 6^2 = 36
F	16	Side of F = Side of Big Square − Side of D − Side of E $\qquad$ = Side of A + Side of B − Side of D − Side of E $\qquad$ = 8 + 6 − 4 − 6 = 4 Area = 4^2 = 16
G	4	Side of G = Side of Big Square − Side of A − Side of F $\qquad$ = Side of A + Side of B − Side of A − Side of F $\qquad$ = 8 + 6 − 8 − 4 = 2 Area = 2^2 = 4
H	4	Side of H = Side of E − Side of F Area = 2^2 = 4

Area of overall square is the sum of the areas of the individual squares:
64 + 36 + 16 + 16 + 36 + 16 + 4 + 4 + 4 = 196.

PAGES 83–86, PRACTICE AND APPLY

9. No. The squares must be removed in 2 × 2 groups of four (the same shape and size as the tiles).

10. No. The squares must be removed in 2×2 groups of four (the same shape and size as the tiles).

11. Sample answer: If the board is completely covered by the 16 tiles, the tiles will form a 4×4 grid. The 4 removed chessboard squares must all come from just one of the squares in the 4×4 grid.

12. Sample answer: The expression on the left side of the equation represents multiplying the length by the width of the overall figure and thus is the area of that figure. The expression on the right side of the equation represents the sum of the areas of the pieces that form the overall figure and thus is the area of that figure. Therefore, the expressions on the left and right must be equal.

13. 25^2: $2 \times 3 = 6$ $\Rightarrow$ $25^2 = 625$.
    75^2: $7 \times 8 = 56$ $\Rightarrow$ $75^2 = 5625$.
    105^2: $10 \times 11 = 110$ $\Rightarrow$ $105^2 = 11,025$.
    The shortcut works for these values.

14. Sample answer: The diagram shows a square with side length 35, so its area is 35^2. The pieces are then rearranged as shown:

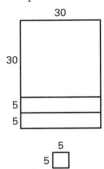

    The figure is now a rectangle of side lengths 30 and 40, with the small square "left over." The rectangle has area $30 \times 40 = 1200$. The "left over" square has area $5 \times 5 = 25$. Together, their area equals that of the original square, 35^2 or 1225.

15. Sample answer: Any number ending in 5 can be represented as a side of a square, similar to the diagram. Move one of the rectangles of length 5 so that it is adjacent (and parallel) to the other rectangle of length 5 (see the solution to Exercise 14). The length of one side of the figure will be a multiple of 10 and the width will be 10 more than the length. The area of the large rectangle plus the area of the small square, $5 \times 5 = 25$, will equal the area of the original square, or the square of the number.

16. Continuing to fill the table until 1 billion is reached is time and paper-consuming.

17. $1,000,000,000 = 1000^3$

18. Column B contains entries for the number $2^3, 5^3, 8^3, \ldots$; while column C contains entries for the numbers $3^3, 6^3, 9^3, \ldots$

19. Every entry in column C is the cube of a multiple of 3 (divisible by 3). No, columns A and B do not contain any cubes of multiples of 3.

20. The entry for 999^3 occurs in column C, because 999 is a multiple of 3.

21. Because 999^3 will occur in column C, the cube of the number immediately following 999 must occur in column A. Therefore, 1000^3 must occur in column A.

22. The next four terms are $\frac{1}{16 \cdot 2} = \frac{1}{32}$; $\frac{1}{32 \cdot 2} = \frac{1}{64}$ $\frac{1}{64 \cdot 2} = \frac{1}{128}$; $\frac{1}{128 \cdot 2} = \frac{1}{256}$

23. The numbers in the sequence are getting smaller. The numerator of the fraction remains 1, while the numbers in the denominators are getting larger, meaning that 1 is being divided into more and more parts. The value of the sequence fraction is therefore decreasing.

24.

Number of terms	Terms	Sum of terms
$\vdots$	$\vdots$	$\vdots$
5	$\frac{1}{2}, \frac{1}{4}, \frac{1}{8}, \frac{1}{16}, \frac{1}{32}$	$\frac{31}{32}$, or 0.96875
6	$\frac{1}{2}, \frac{1}{4}, \frac{1}{8}, \frac{1}{16}, \frac{1}{32}, \frac{1}{64}$	$\frac{63}{64}$, 0.984375
7	$\frac{1}{2}, \frac{1}{4}, \frac{1}{8}, \frac{1}{16}, \frac{1}{32}, \frac{1}{64}, \frac{1}{128}$	$\frac{127}{128}$, 0.9921875
8	$\frac{1}{2}, \frac{1}{4}, \frac{1}{8}, \frac{1}{16}, \frac{1}{32}, \frac{1}{64}, \frac{1}{128}, \frac{1}{256}$	$\frac{255}{256}$, 0.99609375

25. The sums seem to be getting closer and closer to 1. If the sum were continued infinitely, the sum would be 1.

26–27.

28. The area of the square is 1 unit^2. If this pattern of dividing the square is continued infinitely, the sum of the areas of all the rectangles is 1, which is the sum of the given sequence.

29. The number of dots is found by multiplying the number of rows (5) by the number of columns (5), so the number of dots represents the number 5^2.

30. The number of the number of dots added equals $5 + 6$, so the total number of dots is $5^2 + 5 + 6$.

31. A square array of dots with n rows of n dots represents the number n^2. When the square is increased to $n + 1$ rows of $n + 1$ dots $n + n + 1$ dots will have been added to the number of dots in the original square.

The diagram shows n^2 dots (as in Exercise 29) with the new rows divided so that one contains n dots and the other contains $n + 1$ dots. The total number of dots is: $(n + 1)^2 = n^2 + n + (n + 1)$.

32. $21^2 \Rightarrow n = 20$: $(20 + 1)^2 = 20^2 + 20 + (20 + 1)$
$$= 400 + 20 + 21 = 441$$

33. For $n = 1$: $1 = 1$ and $\frac{1(1 + 1)}{2} = \frac{1(2)}{2} = 1$.
For $n = 2$: $1 + 2 = 3$ and $\frac{2(2 + 1)}{2} = \frac{2(3)}{2} = 3$.
For $n = 3$: $1 + 2 + 3 = 6$ and $\frac{3(3 + 1)}{2} = \frac{3(4)}{2} = 6$.
For $n = 4$: $1 + 2 + 3 + 4 = 10$ and $\frac{4(4 + 1)}{2} = \frac{4(5)}{2} = 10$.
For $n = 5$: $1 + 2 + 3 + 4 + 5 = 15$ and $\frac{5(5 + 1)}{2} = \frac{5(6)}{2} = 15$.

34. 21

35. $1 + 2 + 3 + 4 + 5 + 6 = \frac{6(6 + 1)}{2} = \frac{6(7)}{2} = 21$
Yes, the results agree.

36. $6 \times 7 = 42$, as shown in text diagram.

37. The number of dots in the triangle is half the number of dots in the rectangle, or $\frac{n(n + 1)}{2}$.
Because the construction of the triangle and the rectangle can be repeated for any value of n, this proves the conjecture.

38. The conjecture gives the formula for the sum of the first n integers: $\frac{n(n+1)}{2}$.

Notice that if n is the row number, we get the following

For $n = 1$, $\frac{1(1+1)}{2} = 1$

For $n = 2$, $\frac{2(2+1)}{2} = 3$

For $n = 3$, $\frac{3(3+1)}{2} = 6$

For $n = 4$, $\frac{4(4+1)}{2} = 10$

These are the last numbers in each of the first four rows. Using this pattern, on the 45th row, the last number will be $\frac{45(45+1)}{2} = 1035$. Since there will be 45 numbers on the 45th row, the row begins with 991 and ends with 1035, so 1000 must occur on the 45th row.

39. The murder was committed by Colonel Mustard in the study with the candlestick. Sample answer: A disjunctive syllogism can often be used when at least one of two things must be true. If one of the two can be shown to be false, then the other must be true.

PAGE 86, LOOK BACK

40–41.

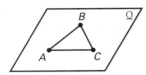

a triangle

42. $LN = LM + MN = (x + 3) + 4x$
Since $LN = 18$, $(x + 3) + 4x = 18$
$5x + 3 = 18$
$5x = 15$
$x = 3$

43. $LM = 7 \Rightarrow x + 3 = 7 \Rightarrow x = 4.$ $LN = (x + 3) + (4x) = (4 + 3) + (4 \cdot 4) = 7 + 16 = 23$

44. $m\angle KIT = (3x)° + (2x + 5)° = 80°.$ Solve for x:
$3x + 2x + 5 = 80$
$5x + 5 = 80$
$5x = 75$
$x = 15$

45. $m\angle KIY = (3x)° = (3(15))° = 45°$
$m\angle TIY = (2x + 5)° = (2(15) + 5)° = 35°$

PAGE 87, LOOK BEYOND

46. The first and last entry in each row is 1. All other entries are found by adding the numbers in the row above the entry, to the left and right.

47.
```
              1
            1   1
          1   2   1
        1   3   3   1
      1   4   6   4   1
    1   5  10  10   5   1
  1   6  15  20  15   6   1
1   7  21  35  35  21   7   1
```

48. 1st row: 1
2nd row: $1 + 1 = 2$
3rd row: $1 + 2 + 1 = 4$
4th row: $1 + 3 + 3 + 1 = 8$
5th row: $1 + 4 + 6 + 4 + 1 = 16$
The pattern of the sums appears to be powers of 2:
$2^0 = 1, 2^1 = 2, 2^2 = 4, 2^3 = 8, 2^4 = 16.$

49. Answers may vary. Students should check carefully to make sure that their pattern applies to *all* rows, columns, or diagonals.

2.2 **PAGE 94, GUIDED SKILLS PRACTICE**

6. If a worker is a United States Postal worker, then he is a federal employee. John is a United States Postal worker. Therefore, John is a federal employee.

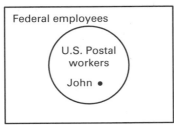

Federal employees

U.S. Postal workers

John •

7. Conditional:
If two lines are parallel, then the two lines do not intersect.
Converse: If two lines do not intersect, then the two lines are parallel.
The conditional is true. The converse is not true.

8. Form a logical chain:
If a number is divisible by 4, then the number is divisible by 2.
If a number is divisible by 2, then the number is even.
If a number is even, then the last digit is 0, 2, 4, 6, or 8.
Therefore, if a number is divisible by 4, then the last digit is 0, 2, 4, 6, or 8 by the If-Then Transitive Property.

PAGES 95–97, PRACTICE AND APPLY

9. If a person lives in Ohio, then the person lives in the United States.

10. Hypothesis: a person lives in Ohio
Conclusion: the person lives in the United States

11.

People who live in the USA

People who live in Ohio

12. Converse: If a person lives in the United States, then the person lives in Ohio.

13. If a plant is a tulip, then it is a flower.

14. If two angles form linear pairs, then the angles are supplementary angles.

15. If a person is a flutist, then the person is a musician.

16. If a thing is a boojum, then it is a snark.

17. Hypothesis: it is snowing in Chicago
Conclusion: it is snowing in Illinois
Converse: If it is snowing in Illinois, then it is snowing in Chicago.
The converse is false. Sample counterexample: It is snowing in Springfield Illinois, but not in Chicago, Illinois.

18. Hypothesis: two angles are complementary
Conclusion: the sum of the angle measures is 90°
Converse: If the sum of the measures of two angles is 90°, then the angles are complementary.
The converse is true.

19. Hypothesis: the measure of each angle in a triangle is less than 90°
Conclusion: the triangle is acute
Converse: If a triangle is acute, then the measure of each angle is less than 90°.
The converse is true.

20. Hypothesis: a figure is rotated
Conclusion: its size and shape stay the same
Converse: If a figure's size and shape stay the same, then the figure is rotated.
The converse is false. Sample counterexample: A figure which is translated has the same size and shape, but it is not rotated.

21. If ∠AXB and ∠BXD form a linear pair, then ∠AXB and ∠BXD are supplementary.

22. If ∠AXB and ∠BXD are supplementary, then m∠AXB + m∠BXD = 180°.

23. If m∠BXC + m∠CXD = 90°, then m∠AXB = 90°.

24. Conclusion: "Mikey" is a rodent.

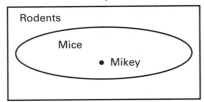

25. Conclusion: Socrates is a mortal.

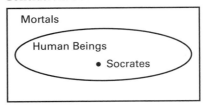

26. Conclusion: Jennifer will get a sunburn.

27. Conclusion: Ingrid lives in Scandinavia.

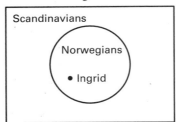

28. Conclusion: Figure ABCD is a rectangle.

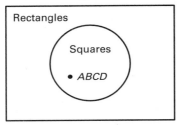

29. Conclusion: The line containing points S and T is in plane 𝒫.

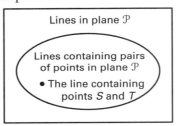

30. If it is winter, then the days are short.
If the days are short, then it is cold.
If it is cold, then the birds fly south.
Conclusion: If it is winter, then the birds fly south.

31. If Tim drives a car, then Tim drives too fast.
If Tim drives too fast, then the police catch Tim speeding.
If the police catch Tim speeding, then Tim gets a ticket.
Conditional: If Tim drives a car, then Tim gets a ticket.

32. If ruskers bleer, then homblers frain.
If homblers frain, then quompies plaun.
If quompies plaun, then romples gleer.
Conclusion: If ruskers bleer, then romples gleer.

33. If you clean your room, then you will go to the movie.
If you go to the movie, then you will spend all of your money.
If you spend all of your money, then you can't buy gas for your car.
If you can't buy gas for your car, then you will be stranded.
Conclusion: If you clean your room, then you will be stranded.

34. If the person is an independent farmer, then the person is disappearing.
The conclusion does not necessarily follow because the given statement refers to independent farmers as a group, not to individual independent farmers. In addition, as used in the given statement, "disappearing" means that the number of independent farmers is decreasing, not that they are no longer visible to the eye.

35. If a nail is lost, then the shoe is lost.
If a shoe is lost, then the horse is lost.
If a horse is lost, then the rider is lost.
If a rider is lost, then the battle is lost.
If a battle is lost, then the war is lost.
Conclusion: If a nail is lost, then the war is lost.

36.–37. Using the first conditional to lay the foundations, student arguments should include statements that support the claim that the object does or does not display form, beauty, and unusual perception on the part of its creator. Using the second conditional to lay the foundations, student arguments should include statements that support the claim that the object does or does not display creativity on the part of its creator. Check student arguments.

38. Conditional: If a person performs classical music, then the person dislikes jazz.
Converse: If a person dislikes jazz, then the person performs classical music.
The counterexample disproves the original conditional. Neither the conditional nor the converse is true.

PAGE 98, LOOK BACK

39. a plane

40. If lines form right angles at their place of intersection, they are perpendicular.
Fold the paper in half lengthwise, then unfold it and then fold it in half widthwise. The lengthwise crease is the perpendicular bisector of the widthwise crease, and vice versa.

41. An angle bisector is a line that divides an angle into two congruent angles. Fold over one side of the angle until it meets the edge of the opposite side of the angle. The line formed is the bisector of the angle.

42. circumcenter, the center of the circumscribed circle of the triangle

43. incenter, the center of the inscribed circle of the triangle

44. obtuse; acute; right

45. Three types of rigid transformations are reflection, rotation, and translation.

46.

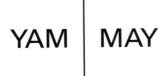

PAGE 98, LOOK BEYOND

47. Pattern: Add 3. The next number is $14 + 3 = 17$.

48. Pattern: Add 7, then 9, then 11, and so on.
The next number is $60 + 15 = 75$.

49. Pattern: Multiply by 3.
The next number is $54 \times 3 = 162$.

50. Pattern: Multiply by $\frac{1}{3}$ (or divide by 3).
The next number is $\frac{1}{9} \times \frac{1}{3} = \frac{1}{27}$.

51. Pattern: Each number (after the first 'ones' in the sequence) is the sum of the two numbers immediately before it. The next five numbers are 13, 21, 34, 55, and 89.

52. Every third number in the sequence is even. Pattern: odd, odd, even, odd, odd, even, … This pattern occurs because the sum of an even and odd is odd. To get an even number, two odds (or two evens) must be added.

53. Every fourth number in the sequence is divisible by 3. Every fifth number is divisible by 5.

2.3 **PAGE 102, GUIDED SKILLS PRACTICE**

6. Sample answer: A glosh is a figure which has 6 sides, 2 of which are parallel, and the remaining 4 sides are equal in length. Shapes *a*, *c*, *d*, and *f* fit this definition.

7. $\angle 1$ and $\angle 3$; $\angle 1$ and $\angle 2$; $\angle 2$ and $\angle 4$; $\angle 3$ and $\angle 4$.

PAGES 103–105, PRACTICE AND APPLY

8. a. Conditional: If a person is a teenager, then the person is 13 years old or older.

b. Converse: If a person is 13 years old or older, then the person is a teenager.

c. Biconditional: A person is a teenager if and only if the person is 13 years old or older.

d. The statement is not a definition because the converse is not true. A person who is 20 years old or older is over 13, but is not a teenager.

9. a. Conditional: If a person is a teenager, then the person is from 13 to 19 years old.

b. Converse: If a person is from 13 to 19 years old, then the person is a teenager.

c. Biconditional: A person is a teenager if and only if the person is from 13 to 19 years old.

d. The statement is a definition because the conditional and the converse are both true.

10. a. Conditional: If an integer is zero, then it is between -1 and 1.

b. Converse: If an integer is between -1 and 1, then it is zero.

c. Biconditional: An integer is zero if and only if it is between -1 and 1.

d. The statement is a definition because the conditional and the converse are both true.

11. a. Conditional: If a number is even, then it is divisible by 2.

b. Converse: If a number is divisible by 2, then it is even.

c. Biconditional: A number is even if and only if it is divisible by 2.

d. The statement is a definition because the conditional and the converse are both true.

12. a. Conditional: If something is an angle, then it is formed by two rays.

b. Converse: If something is formed by two rays, then it is an angle.

c. Biconditional: Something is an angle if and only if it is formed by two rays.

d. Not a definition. Converse is not true. Angles must have a common endpoint.

13. a. Conditional: If an angle is a right angle, then it has a measure if 90°.

b. Converse: If an angle has a measure of 90°, then it is a right angle.

c. Biconditional: An angle is a right angle if and only if it has a measure of 90°.

d. The statement is a definition because the conditional and the converse are both true.

14. a. Conditional: If a rock is granite, then it is hard and crystalline.

b. Converse: If a rock is hard and crystalline, then it is granite.

c. Biconditional: A rock is granite if and only if it is hard and crystalline.

d. The statement is not a definition because the converse is not true. A rock such as quartz is hard and crystalline, but is different from granite.

15. a. Conditional: If a substance is hydrogen, then it is the lightest of all known substances.

b. Converse: If a substance is lightest of all known substances, then it is hydrogen.

c. Biconditional: A substance is hydrogen if and only if it is the lightest of all known substances.

d. The statement is a definition because the conditional and the converse are both true.

16. a. Conditional: If an animal is an otter, then it is a small furry mammal with webbed feet that are used for swimming.

b. Converse: If an animal is a small furry mammal with webbed feet that are used for swimming, then it is an otter.

c. Biconditional: An animal is an otter if and only if it is a small furry mammal with webbed feet that are used for swimming.

d. The statement is not a definition because the converse is not true. A beaver is a small furry mammal with webbed feet that are used for swimming, but is not an otter.

17. $\angle WVX$ and $\angle XVY$; $\angle XVY$ and $\angle YVZ$; $\angle WVY$ and $\angle YVZ$; $\angle WVX$ and $\angle XVZ$

18. They do not share a common vertex or a common side.

19. They do not share a common side.

20. They do not share a common vertex.

21. They do not share a common vertex.

22. The angles overlap.

23. Shapes b and d are flitches because each is a triangle with a "tail" consisting of five small lines in the pattern shown. These are drawn at one vertex. Because a has seven small lines and c is a circle, they are not flitches.

24. Only a is a zobble because is a spiral-like shape with a "tail" consisting of two identical markings. b is not a spiral, c has three identical "tail" markings, and d has only one "tail" marking.

25. Shapes b and c are parallelograms because they are quadrilaterals with both pairs of opposite sides parallel. a is a triangle, not a quadrilateral, while d has only one pair of opposite sides parallel.

26. A figure is a polygon if and only if it is a closed plane figure consisting of line segments which intersect exactly two other line segments, once at each endpoint.

27. A closed plane figure is a regular polygon if and only if all of its sides and angles are congruent.

28. Student definitions should give enough examples of what are and what are not considered to be their chosen objects, and should illustrate those properties that make their chosen objects unique.

29. *Ciconiiformes:*
Great Blue Heron
Roseate Spoonbill
Cuculiformes:
Greater Roadrunner
Yellow-Billed Cuckoo
Apodiformes:
White-Collared Swift
Ruby-Throated Hummingbird

PAGE 106, LOOK BACK

30. $-2x + 1 = x + 10$
$-3x + 1 = 10$
$-3x = 9$
$x = -3$
$\Rightarrow AB = -2(-3) + 1 = 7$

31.–32.

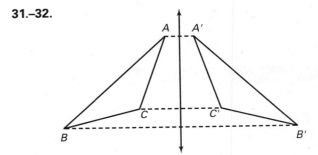

33. They are parallel.

34. The line of reflection is the perpendicular bisector of the line segments.

35.–36.

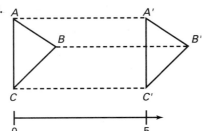

37. They are parallel.

38. They are parallel.

PAGE 106, LOOK BEYOND

39.

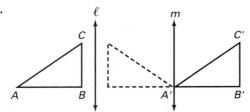

40. The image is △*ABC* moved right. A translation would produce the same image as the two reflections.

41.

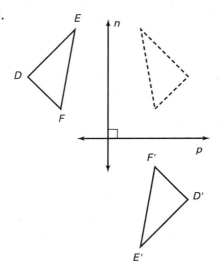

42. The image △*DEF* is rotated. A rotation of 180° around the intersection of lines *n* and *p* produces the same image.

2.4 **PAGES 111–112, GUIDED SKILLS PRACTICE**

5. *XZ* = 19 by part 1 of the Overlapping Segments Theorem.

6. 30

7. △*CDE* ≅ △*C″D″E″*

8. The Transitive Property of Congruence

PAGES 112–115, PRACTICE AND APPLY

9. Subtraction

10. Addition; Division

11. Transitive

12. Symmetric; Transitive

13. m∠*MLN* + m∠*NLP*

14. m∠*NLP* + m∠*PLQ*

15. Angle Addition Postulate

16. m∠*MLN* + m∠*NLP* = m∠*PLQ* + m∠*NLP*

17. m∠MLP = m∠NLQ

18. m∠MLP − m∠NLP = m∠NLQ − m∠NLP

19. m∠MLN = m∠PLQ

20. Answers will vary depending on letters chosen.

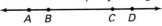

Given: $AC = BD$
Prove: $AB = CD$

Statements	Reasons
$AC = BD$	Given
$AC = AB + BC$	Segment Addition Postulate
$BD = CD + BC$	Segment Addition Postulate
$AB + BC = CD + BC$	Substitution Property
$AB = CD$	Subtraction Property

21. $XZ = 15$ (by the Overlapping Segments Theorem)

22. Since $WX = YZ$, then $XZ = WY$ by the Overlapping Segments Theorem. So $XZ = 9$. Since $XZ = XY + YZ$ by the Segment Addition Postulate, $9 = 2a + 1 + a^2$. Solving for a:
$9 = 2a + 1 + a^2$
$0 = a^2 + 2a - 8$
$0 = (a - 2)(a + 4)$
$a - 2 = 0 \quad \text{or} \quad a + 4 = 0$
$\quad a = 2 \quad \text{or} \quad\quad a = -4$
a cannot equal -4 since the length of a segment must be positive, $a = 2$.
$WZ = WY + YZ = 9 + 4 = 13$

23. m∠HDF = 85° (by the Overlapping Angles Theorem)

24. Since m∠EDG = m∠FDH (by the Overlapping Angles Theorem) and m∠FDH = m∠FDG + m∠GDH (by the Angle Addition Postulate), m∠EDG = m∠FDG + m∠GDH (by the Transitive Property). Using subtraction, $9x - 8 = 6x + 8 + 2x - 4$. Solving for x gives:
$9x - 8 = 8x + 4$
$\quad x = 12$
m∠HDF = m∠EDG = $9x - 8 = 9(12) - 8 = 100°$
m∠HDG = m∠GDH = $2x - 4 = 2(12) - 4 = 20°$

25. GH

26. m∠GHI

27. Transitive Property

28. Transitive Property

29. Yes; Transitive Property

30. Given

31. Transitive Property

32. Substitution Property

33. m∠BAC = m∠DEC

34. m∠1 + m∠4 + m∠5 = 180°

35. Substitution Property

36. m∠3

37. m∠1 + m∠2 + m∠3

38. Angle Addition Postulate

39. 90°

40. Given

41. m∠CDE + m∠CDB = m∠EDB

42. Transitive (or Substitution) Property

43. Transitive (or Substitution) Property

44. Given

45. Substitution Property

46. m∠ABC = m∠CDE

47. Transitive; the relation is not an equivalence relation. Counterexample to the Reflexive Property: 5 ≮ 5. Counterexample to the Symmetric Property: 2 < 3 but 3 ≮ 2.

48. Reflexive and Transitive; the relation is not an equivalence relation. Counterexample to the Symmetric Property: 1 ≤ 2 but 2 ≰ 1.

49. Reflexive and Transitive; the relation is not an equivalence relation. Counterexample to the Symmetric Property: 10 is divisible by 5, but 5 is not divisible by 10.

50. Symmetric; the relation is not an equivalence relation. Counterexample to the Reflexive Property: the letter b is not a reflection if itself (it is the reflection of the letter d). Counterexample to the Transitive Property: the letter b is the reflection of letter d, and the letter d is the reflection of the letter b, but letter b is not the reflection of the letter b.

51. Reflexive, Symmetric, and Transitive; it is an equivalence relation. Note: Any figure is a rotation of itself, by an amount of 360°.

52. The relation is not an equivalence relation. Counterexample to the Reflexive Property: Cassie is not a sister of herself. Counterexample to the Symmetric Property: Cassie is a sister of John, but John is not a sister of Cassie. Counterexample to the Transitive Property: Cassie is a sister of Diana, and Diana is a sister of Cassie, but Cassie is not a sister of Cassie.

53. Reflexive, Symmetric, and Transitive; it is an equivalence relation.

PAGE 115, LOOK BACK

54. Sample answer: point *S*, point *E*, and point *I*

55. Postulate 1.1.7. Through any three noncollinear points there is exactly one plane.

56. Sample answer: point *F* and point *Z*

57. Postulate 1.1.6. Through any two points there is exactly one line.

58. line ℓ

59. Postulate 1.1.5. The intersection of two planes is a line.

60. Sample answer: By Postulate 1.1.7 three points determine a plane, so any three points will rest securely on a flat surface. Four given points, however, are not necessarily coplanar.

PAGE 116, LOOK BEYOND

61. Lines of longitude and the equator

62. Sample answer: Lines are defined this way because the shortest path between any two points on a sphere is along a circle that divides the sphere (Earth) into two equal halves.

63. Sample answer: On Earth, the north and south poles are two points through which many lines of longitude pass.

64. Sample answer: The route would appear to curve on a traditional map: the pilot would fly north from Washington D.C., then fly south to get to London.

7.

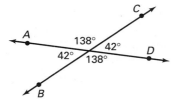

8.

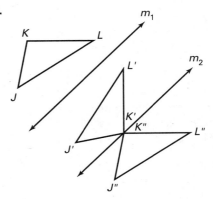

9.

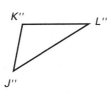

The first set of reflections translates the figure right and downward.
The second set of reflections translates the figure left and upward.

PAGES 121–124, PRACTICE AND APPLY

10. ∠DES

11. ∠LEP

12. ∠FEI

13. m∠ABC = m∠EBD = 160°

14. ∠ABC is supplementary to ∠EBA, so
m∠ABC = 180° − m∠EBA = 180° − 112° = 68°.

15. ∠ABC is supplementary to ∠ABE,
∠ABE is a right angle ⇒
m∠ABC = 180° − 90° = 90°.

16. ∠ABC is supplementary to ∠ABE and
m∠ABC = m∠ABE, so
2m∠ABC = 180° or m∠ABC = 90°.

17. m∠ABC = m∠EBD,
so $3x + 15 = 4x \Rightarrow x = 15$
m∠ABC = (3(15) + 15)° = 60°

18. m∠ABE = m∠DBC,
so $7x − 19 = 4x + 2 \Rightarrow x = 7$
m∠ABC = 180° − m∠DBC = 180° − (4 · 7 + 2)° = 150°

19. $\angle ABC$ is supplementary to $\angle CBD$, so
$-12x + 13 + (-21x + 2) = 180$
$-33x = 165 \Rightarrow x = -5$
$m\angle ABC = (-12(-5) + 13)° = 73°$

20. $2(3x + 5) + 10x - 6 = 180$
$6x + 10 + 10x - 6 = 180$
$16x + 4 = 180$
$16x = 176$
$x = 11$

$m\angle ABC = m\angle ABP + m\angle PBC$
$= m\angle RBQ + m\angle QBC$
$= [10(11) - 6]° + [3(11) + 5]° = 142°$

21. Inductive reasoning. The argument is not a proof because it has not been shown to be true for all possible times he eats strawberries.

22. Deductive reasoning. The argument is a proof. Because the hypothesis occurred, the conclusion is proved.

23. Deductive reasoning. The argument is a proof because the diagram shows that $\angle 1$ and $\angle 2$ form a linear pair.

24. Inductive reasoning. The argument uses many years' data but it cannot be proven that the data will show an increase in the future.

25. Transitive or Substitution Property

26. Given (Angle Congruence Postulate)

27. Subtraction Property

28. You are given that $\angle 1$ and $\angle 2$ and $\angle 3$ and $\angle 4$ are pairs of supplementary angles. Then $m\angle 1 + m\angle 2 = 180°$ and $m\angle 3 + m\angle 4 = 180°$. So $m\angle 1 + m\angle 2 = m\angle 3 + m\angle 4$ by substitution. Since $\angle 1 \cong \angle 3$ is given, $m\angle 1 = m\angle 3$. Using substitution again, $m\angle 1 + m\angle 2 = m\angle 1 + m\angle 4$. Now, substracting $m\angle 1$ from both sides, $m\angle 2 = m\angle 4$. Therefore $\angle 2 \cong \angle 4$.

29. Double the distance between reflecting lines is $2(5 \text{ cm}) = 10 \text{ cm}$.

30. Double the distance between reflecting lines is $2(10 \text{ cm}) = 20 \text{ cm}$.

31. Double the distance between reflecting lines is $2(x \text{ cm}) = 2x \text{ cm}$.

32. The parallel lines should be $\frac{1}{2}(10 \text{ cm}) = 5$ cm apart.

33. The parallel lines should be drawn perpendicular to the direction of the arrow.

34. Yes. Any pair of lines which are 5 centimeters apart and perpendicular to the direction of the arrow will produce the desired translation. Check students' diagrams.

35.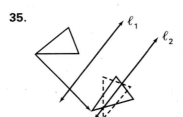

36.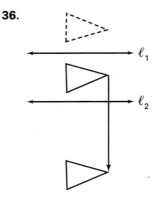

37.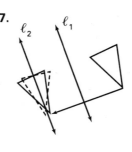

38.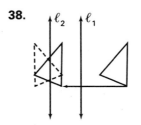

39.

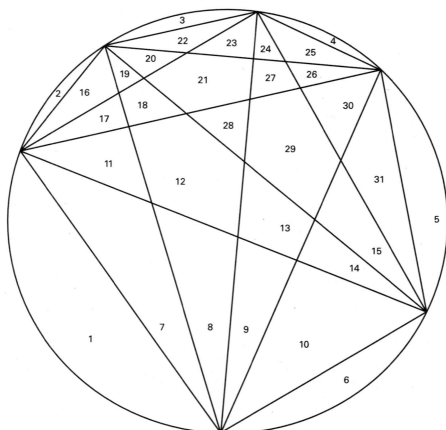

31 regions are possible.

40.

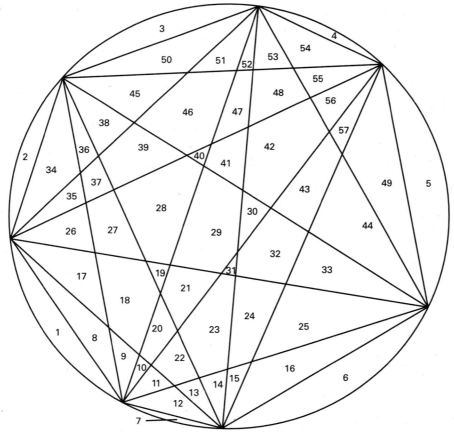

57 regions are possible.

41.

Number of regions	1		2		4		8		16		31		57	99
1st differences		1		2		4		8		15		26		42
2nd differences			1		2		4		7		11		16	
3rd differences				1		2		3		4		5		

42. I conjecture that 8 points would have 99 regions, 9 points would have 163 regions, and 10 points would have 256 regions.

43. No, an induction proof cannot be a proof for all cases

PAGE 125, LOOK BACK

44. Sample answer: point A, point B, point C, and point D

45. Sample answer: $\overline{AB}$, $\overline{BC}$, $\overline{CD}$, and $\overline{AD}$

46. Sample answer: plane ABG, plane DCH, plane ADE, and plane EGH

47. If it is Saturday, then I sleep until 8:00 A.M.
If I sleep until 8:00 A.M., then I am well rested during the afternoon.
If I am well rested in the afternoon, then I am in a good mood.

48. If it is Saturday, then I am in a good mood.

49.

Culver HS Students
- Football Team Members
  - • Brady

50.

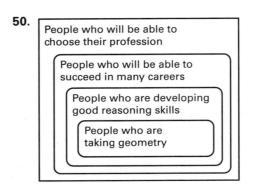

People who will be able to choose their profession
  - People who will be able to succeed in many careers
    - People who are developing good reasoning skills
      - People who are taking geometry

you will be able to choose your profession

PAGE 125, LOOK BEYOND

51.

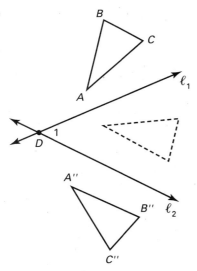

52. A rotation

53. D is the center of rotation.

54. They are twice the measure of $\angle 1$.

55. Conjecture: The reflection of a figure across two intersecting lines is equivalent to a rotation about the point of intersection of the two lines, through an angle twice the measure of the angle between the lines.

CHAPTER REVIEW AND ASSESSMENT

1. 7; 10

2.

Lines	1	2	3	4
Sections	4	7	10	13

First differences   3   3   3

3. Each additional horizontal line will cut the curve into 3 additional pieces. The curve is divided by n horizontal lines into $1 + 3n$ sections.

4. Proof: The first line divides the figure into 4 pieces, two above and two below. The next line creates regions above, below, and between the lines. There are two pieces above and two below as before, with three pieces between the lines. Each additional line will create a new region between two lines, with three pieces of the curve in it. Therefore, each additional line creates 3 additional pieces of the curve. Since the curve starts with 1 section the number of sections for n lines is $1 + 3n$.

5. The evening star is a planet.

6. Converse: If a "star" is really a planet, then it doesn't flicker. It is a true conditional because a planet does not flicker.

7. c, b, a

8. If Darren tells Charles a secret, then Andrew tells Tina the secret.

9. Objects a and b do not have a hole in the center so they are not tori. Objects c and d are tori.

10. A torus is a doughnut-shaped object.

11. No. The converse is not true because a figure with four sides is not always a square.

12. Yes. Both the conditional and its converse are true.

13. Linear Pair Property

14. Linear Pair Property

15. Vertical Angles Theorem

16. Two-column proof:

Statements	Reason
$m\angle 1 + m\angle 2 = 180°$	Linear Pair Property
$m\angle 1 = 90°$	Given
$90° + m\angle 2 = 180°$	Substitution Property
$m\angle 2 = 90°$	Subtraction Property
$m\angle 1 + m\angle 4 = 180°$	Linear Pair Property
$90° + m\angle 4 = 180°$	Substitution Property
$m\angle 4 = 90°$	Subtraction Property
$m\angle 1 = m\angle 3$	Vertical Angles Theorem
$m\angle 3 = 90°$	Transitive Property

17. Let $a = 3$, $b = 4$, $c = 5$. Then $a^2 + b^2 = c^2$ gives: $3^2 + 4^2 = 5^2$, or $9 + 16 = 25$, which is true so 3, 4, and 5 form a Pythagorean triple. Let $a = 5$, $b = 12$, $c = 13$. Then $a^2 + b^2 = c^2$ gives: $5^2 + 12^2 = 13^2$ or $25 + 144 = 169$, which is true, so 5, 12, and 13 form a Pythagorean triple.

18. Let $a = 3$, $b = 4$, $c = 12$ and $d = 13$. Then $a^2 + b^2 + c^2 = d^2$ gives: $3^2 + 4^2 + 12^2 = 13^2$, or $9 + 16 + 144 = 169$, which is true, so 3, 4, 12, and 13 form a Pythagorean quadruple.

19. Conjecture: If a, b, and c are a Pythagorean triple such that $a^2 + b^2 = c^2$ and c, d, and e are another Pythagorean triple such that $c^2 + d^2 = e^2$, then $a^2 + b^2 + d^2 = e^2$.

20. Paragraph proof: $a^2 + b^2 = c^2$ and $c^2 + d^2 = e^2$ because a, b, c and c, d, e each form Pythagorean triples. Using the Substitution Property, substitute $a^2 + b^2$ for c^2 in the equation $c^2 + d^2 = e^2$: $a^2 + b^2 + d^2 = e^2$.

21. $3^2 - 4 \times 1 \times 4 = 9 - 16 = -7$. The program would display "NO REAL SOLUTIONS".

22. Yes. It is a mammal because it has hair and produces milk. Since the platypus is a mammal and it lays eggs, it is a monotreme.

CHAPTER TEST

1. The next four terms are: $\frac{1}{25}$, $\frac{1}{36}$, $\frac{1}{49}$, and $\frac{1}{64}$.

2. The terms are described by $\frac{1}{n^2}$, where n is the number of the term.

3. Hypothesis: A quadrilateral is a square.
Conclusion: The quadrilateral is a rectangle.

4. Converse: If a quadrilateral is a rectangle, then it is a square.
The converse is false. Counterexample: A 4-by-5 rectangle is not a square.

5. c: If it is windy, then Brett flies his kite.
a: If Brett flies his kite, then it is cool.
b: If it is cool, then it is autumn.

6. If it is windy, then it is autumn.

7. a: If an angle is obtuse, then it has a measure of 100°.
b: If an angle has a measure of 100°, then it is obtuse.
c: An angle is obtuse if and only if it has a measure of 100°.
d: The statement is not a definition because the biconditional is false.
    A 120° angle is obtuse, but it does not have a measure of 100°.

8. $m\angle ADB + m\angle BDC = m\angle ADC$
$(8x - 31) + (3x + 9) = 110$
$11x - 22 = 110$
$11x = 132$
$x = 12$
$m\angle BDC = 3x + 9$
$m\angle BDC = 3(12) + 9$
$m\angle BDC = 45°$

9. Given

10. Transitive Property

11. Given

12. Substitution Property

13. Subtraction Property of Equality

14. Given

15. Definition of supplementary or Linear Pair Property

16. Substitution Property

17. Subtraction Property of Equality

18. $4x + 18 = 7x - 24$
$42 = 3x$
$14 = x$
$m\angle ABC = 4(14) + 18 = 74°$

19. $(3x - 13) + (6x - 14) = 180$
$9x - 27 = 180$
$9x = 207$
$x = 23$
$m\angle ABC = 3(23) - 13 = 56°$

1. C. ∠*ABC* is supplemental to a right angle, so it is a right angle.

2. D. There is not enough information. The sequence may be 1, 4, 16... or 1, 4, 7, 10, ...

3. B. Make a drawing and count the regions.

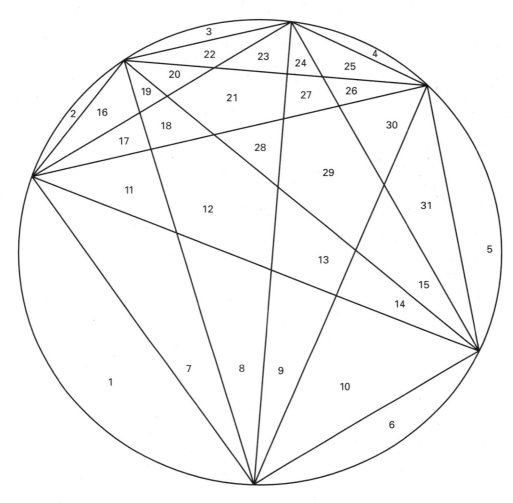

4. C. By the Vertical Angles Theorem, we know that vertical angles are equal.

5. 16; 25

6. The number of regions formed by *n* pairs of perpendicular lines is $(n + 1)^2$.

7. *Paragraph proof:*
The vertical line of the first pair of perpendicular lines will divide the square into 2 vertical strips and the horizontal line will divide each of these strips in half for a total of $2(2) = 4$ regions. For each additional perpendicular line, the vertical line will add a vertical strip and the horizontal line will divide each vertical strip into one more region. For n perpendicular lines, we have $n + 1$ vertical strips which are divided into $n + 1$ pieces for a total of $(n + 1)(n + 1)$ or $(n + 1)^2$ regions.

8. $(4, -1), (2, -2),$ and $(3, 0)$

9. $(1, 0), (-1, 1),$ and $(0, -1)$

10. If a figure is a square, then it is a rectangle.

11. Sample answer: A splort is any figure with a 90° rotational symmetry.

12. Sample answer:
point A, point B, point C, point D, and point E

13. Sample answer: $\overleftrightarrow{AX}, \overleftrightarrow{CF},$ and $\overleftrightarrow{BE}$

14. Sample answer: $\overrightarrow{XB}, \overrightarrow{XD}, \overrightarrow{EB},$ and $\overrightarrow{BE}$

15. Sample answer:
$\angle AXB, \angle CXB, \angle DXB, \angle FXB, \angle AXC, \angle CXD,$
$\angle DXF,$ and $\angle FXA$

16. a. Definition; the conditional (If a point is a midpoint, then it divides the segment into two equal parts) and its converse (If a point divides a segment into two equal parts, then it is a midpoint) are true.

    b. Not a definition; the converse (If a triangle has an angle of less than 90°, then it is an acute triangle) is false.

    c. Not a definition; the conditional (If something is determined by two points, then it is a line) is false.

    d. Not a definition; the converse (If a triangle is isosceles, then it is equilateral) is false.

17. $m\angle 1 = 180° - 130° = 50°.$

18. $m\angle 3 = 130°$ by the Vertical Angles Theorem.

19. $m\angle QPR + m\angle RPS + m\angle SPT = m\angle QPT$
Using substitution: $(2x - 5)° + (4x + 10)° + (3x)° = 140°$
$$2x - 5 + 4x + 10 + 3x = 140$$
$$9x + 5 = 140$$
$$9x = 135$$
$$x = 15$$

20. $m\angle RPS = (4x + 10)° = (4(15) + 10)° = 70°$

3.1   PAGE 143, GUIDED SKILLS PRACTICE

6. Yes. The axis of symmetry bisects both the side and the angle through which it passes.

7. A square has 4 sides so the measure of a central angle $\theta = \frac{360°}{4} = 90°$.

A pentagon has 5 sides so the measure of a central angle $\theta = \frac{360°}{5} = 72°$.

A hexagon has 6 sides so the measure of a central angle $\theta = \frac{360°}{6} = 60°$.

8. A windmill with 4 blades forms a square with central angle 90°, and 4 rotations through a central angle will place the square in its original position. Thus there is 4-fold rotational symmetry. Similarly, 6 blades form a hexagon with central angle 60°, through which 6 rotations will place the hexagon in its original position. Thus there is 6-fold rotational symmetry.

PAGES 143–146, PRACTICE AND APPLY

9. 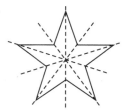   **10.** 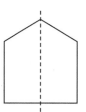   **11.** 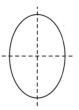   **12.**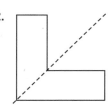

13. A circle has an infinite number of axes of symmetry. Any line through the center of the circle is an axis of symmetry; an infinite number of such lines can be drawn.

14. Both sides should be identical.   **15.** Both sides should be identical.   **16.** Both sides should be identical.

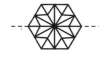

17. Exercise 16 has 6-fold rotational symmetry; rotating the figure 60°, 120°, 180°, 240°, and 360° results in an identical figure.

18. 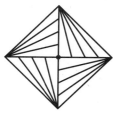 **19.** 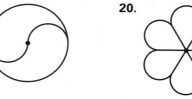 **20.**  **21.**

22. Exercise 20 has 6 axes of reflection in its completed figure.

23. Sample answer: an isosceles triangle that is not equilateral.

24. Sample answer: a rectangle that is not a square.

25. Sample answer: an equilateral triangle.

26. Sample answer: a regular pentagon.

27. Sample answer: a regular octagon.

28. The axis must pass through the midpoint of $\overline{AB}$. The axis of symmetry is the perpendicular bisector of $\overline{AB}$.

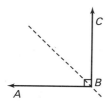

29. B is the image of A, A is the image of B, $\overline{BA}$ is the image of $\overline{AB}$.

30. The axis of symmetry goes through the vertex of $\angle ABC$. The axis of symmetry is the angle bisector of $\angle ABC$.

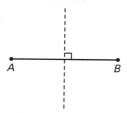

31. $\overrightarrow{BC}$ is the image of $\overrightarrow{BA}$, $\overrightarrow{BA}$ is the image of $\overrightarrow{BC}$, $\angle CBA$ is the image of $\angle ABC$.

32. Draw the bisector of the angle.

33. Reflect $\overline{BX}$ across the horizontal axis of symmetry $\overleftrightarrow{AC}$ or rotate $\overline{BX}$ through 180° about X.

34. Reflection of $\overline{AD}$ across the horizontal axis of symmetry $\overleftrightarrow{AC}$ or rotation of $\overline{AD}$ through m$\angle BAD$ about A.

35. Rotation of $\overline{AD}$ through 180° about X.

36. Reflection of $\angle BCX$ across the vertical axis of symmetry $\overleftrightarrow{BD}$.

37. Reflection of $\angle BCX$ across the horizontal axis of symmetry $\overleftrightarrow{AC}$.

38. Rotation of $\angle BCX$ through 180° about X.

39. Reflection of $\angle AXD$ across the vertical axis of symmetry $\overleftrightarrow{BD}$.

40.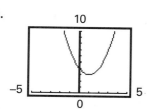

Axis of symmetry: $x = 1$

41.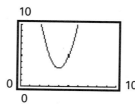

Axis of symmetry: $x = 4$

42.

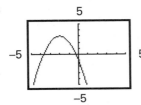

Axis of symmetry: $x = -2$

43.

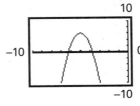

Axis of symmetry: $x = -5$

44.

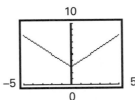

Axis of symmetry: $x = 0$

45.

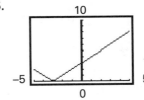

Axis of symmetry: $x = -3$

46. Since a chiliagon has 1000 sides, the measure of a central angle $\theta = \frac{360°}{1000} = 0.36°$. Since a myriagon has 10,000 sides, the measure of a central angle is $\theta = \frac{360°}{10,000} = 0.036°$.

47. Sample answer:

48. Sample answer:

49. The table extension preserves the angle measures but increases the length of 2 sides of the table. The extended table is not a regular polygon. It is not equilateral. It is equiangular.

50. The measure of a central angle of a 16-gon is $\theta = \frac{360°}{16} = 22.5°$.

51. The Egyptian bowl has no reflection symmetry axes. There are nontrivial rotational symmetries for rotations of 72°, 144°, 216°, and 288°.

52. The Egyptian bowl has no reflection symmetry axes. There are nontrivial rotational symmetries of 120° and 240°.

53. The Egyptian bowl has 2 axes of symmetry—vertically and horizontally through the center. There is a nontrivial rotational symmetry for a rotation of 180°.

54. The Egyptian bowl has 4 axes of symmetry—vertically and horizontally through the center—and 2 diagonally through the center of the diamond shaped figures. There are nontrivial rotational symmetries for rotations of 90°, 180°, and 270°.

55. The sign is a regular octagon with 8 axes of symmetry—4 lines through the midpoints of opposite sides and 4 lines through opposite vertices. It has nontrivial rotational symmetries for rotations of 45°, 90°, 135°, 180°, 225°, 270° and 315°.

56. The sign is a regular quadrilateral or square. It has 4 axes of symmetry—2 lines through midpoints of opposite sides and 2 lines through opposite vertices. It has nontrivial rotational symmetries for rotations of 90°, 180°, and 270°.

57. The sign is a regular triangle or equilateral triangle. It has 3 axes of symmetry—each axis passes through a vertex and the midpoint of the opposite side. It has nontrivial rotational symmetries for rotations of 120° and 240°.

58. The sign is an equiangular quadrilateral or rectangle. It is not regular. It has 2 axes of symmetry—2 lines through midpoints of opposite sides. It has a nontrivial rotational symmetry for a rotation of 180°.

59. The two lines intersect at point O.

61. The two planes intersect at $\overleftrightarrow{AB}$.

63. $m\angle ABD = 25° + 60° = 85°$

60. Sample answer: point B, point O, and point M

62. $m\angle CBD = 80° - 30° = 50°$

64. $m\angle ABC + m\angle CBD = m\angle ABC$
$m\angle ABC + m\angle ABC = 88°$
$2m\angle ABC = 88°$
$m\angle ABC = 44°$

PAGE 146, LOOK BEYOND

65. France, Japan, Qatar, Micronesia, Macedonia, Israel, Guyana, and Iceland

67. Japan, United Kingdom, Micronesia, Macedonia, and Israel

66. Japan, Syria, St. Lucia, Micronesia, Canada, Macedonia, Israel, Russia, and Ethiopia

68. United States

3.2 **PAGE 151, GUIDED SKILLS PRACTICE**

5. a. $AB = DC = 3$ in. because opposite sides of a parallelogram are congruent.

 b. $m\angle ABC = m\angle CDA = 50°$ because opposite angles of a parallelogram are congruent.

 c. $AE = EC = 1$ in. because the diagonals of a parallelogram bisect each other.

 d. $m\angle BCD = 180° - m\angle CDA = 180° - 150° = 130°$ because adjacent angles of a parallelogram are supplementary.

6. a. $FG = IH = 12$ cm because opposite sides of a rhombus are congruent.

 b. $m\angle FGH = m\angle FIH = 30°$ because opposite angles of a rhombus are congruent.

 c. $m\angle GHI = 180° - m\angle HIF = 180° - 30° = 150°$ because adjacent angles of a rhombus are supplementary.

 d. $m\angle FJG = 90°$ because the diagonals of a rhombus are perpendicular.

7. $LN = KM = 2 \cdot KO = 2 \cdot 2 = 4$ ft because the diagonals of a rectangle are congruent and bisect each other.

8. $m\angle PTQ = 90°$ and $m\angle QTR = 90°$ because the diagonals of a square are perpendicular.

PAGES 152–153, PRACTICE AND APPLY

9. $YZ = WX = 10$ because opposite sides of a parallelogram are congruent.

11. $WV = \frac{1}{2}(WY) = \frac{1}{2}(13) = 6.5$ because the diagonals of a parallelogram bisect each other.

13. $m\angle WXY = m\angle WZY = 130°$ because opposite angles of a parallelogram are congruent.

15. $m\angle XYZ = 180° - m\angle WZY = 180° - 130° = 50°$ because adjacent angles of a parallelogram are supplementary.

10. $XY = WZ = 4$ because opposite sides of a parallelogram are congruent.

12. $VY = \frac{1}{2}(WY) = \frac{1}{2}(13) = 6.5$ because parallelogram diagonals bisect each other.

14. $m\angle XWZ = 180° - m\angle WZY = 180° - 130° = 50°$ because adjacent angles of a parallelogram are supplementary.

16. $GH = FG = 21$ because all sides of a rhombus are congruent.

17. $HI = FG = 21$ because all sides of a rhombus are congruent.

18. $FJ = \frac{1}{2}(FH) = \frac{1}{2}(15) = 7.5$ because the diagonals of a rhombus bisect each other.

19. $JH = \frac{1}{2}(FH) = \frac{1}{2}(15) = 7.5$ because the diagonals of a rhombus bisect each other.

20. $m\angle FIH = m\angle FGH = 70°$ because opposite angles of a rhombus are congruent.

21. $m\angle GFI = 180° - m\angle FGH = 180° - 70° = 110°$ because adjacent angles of a rhombus are supplementary.

22. $m\angle FJG = 90°$ because the diagonals of a rhombus are perpendicular.

23. $m\angle HJI = 90°$ because the diagonals of a rhombus are perpendicular.

24. $CD = AB = 6$ because opposite sides of a rectangle are congruent.

25. $BC = AD = 8$ because opposite sides of a rectangle are congruent.

26. $BD = AC = 10$ because the diagonals of a rectangle are congruent.

27. $AE = \frac{1}{2}(AC) = \frac{1}{2}(10) = 5$ because the diagonals of a rectangle bisect each other.

28. $BE = \frac{1}{2}(BD) = \frac{1}{2}(AC) = \frac{1}{2}(10) = 5$ because the diagonals of a rectangle bisect each other and are congruent.

29. $LM = KL = 50$ because all sides of a square are congruent.

30. $LN = KM \approx 70.7$ because the diagonals of a square are congruent.

31. $m\angle KOL = 90°$ because the diagonals of a square are perpendicular.

32. $m\angle LOM = 90°$ because the diagonals of a square are perpendicular.

33. Adjacent angles of a parallelogram are supplementary. Therefore
$$(2x)° + x° = 180°$$
$$3x° = 180°$$
$$x = 60$$
$m\angle P = (2x)° = (2 \cdot 60)° = 120°$. $m\angle R = m\angle P = 120°$ because opposite angles of a parallelogram are congruent. $m\angle Q = x° = 60°$. $m\angle S = m\angle Q = 60°$ because opposite angles of a parallelogram are congruent.

34. Diagonals of a rectangle are congruent. Therefore,
$$x - 2 = \sqrt{x}$$
$$(x - 2)^2 = (\sqrt{x})^2$$
$$x^2 - 4x + 4 = x$$
$$x^2 - 5x + 4 = 0$$
$$(x - 4)(x - 1) = 0$$
Therefore $x = 4$ or $x = 1$. Since $x - 2$ is the length of a diagonal, $x = 1$ does not make sense because this would give a negative diagonal length. Thus, $x = 4$, so $WY = x - 2 = 4 - 2 = 2$, and $XZ = WY = 2$.

35. False. Every rectangle is a parallelogram so there are some parallelograms which are rectangles.

36. True. Squares are parallelograms, so a figure that is not a parallelogram cannot be a square.

37. True. A parallelogram has two pairs of parallel sides but a trapezoid has only one pair of parallel sides.

38. True. A trapezoid has one and only one pair of parallel sides, but a rectangle has 2 pairs of parallel sides.

39. True. Squares have 4 sides of equal length therefore they are rhombuses.

40. False. A square is both a rectangle and a rhombus.

41. False. A square is a rhombus as well as a rectangle.

42. First write the statement as a conditional: If a quadrilateral is a parallelogram, then opposite sides are congruent. Now, the converse must be: If opposite sides of a quadrilateral are congruent, then it is a parallelogram. The converse is true.

43. If a figure is a rhombus, then its opposite sides are congruent, so it must be a parallelogram. The statement is true.

44. The beam connecting points B and D is 73 in. because diagonals of a rectangle are congruent.

45. If the creature is a whale, then it is a mammal. Converse: If the creature is a mammal, then it is a whale. This is not a definition because not all mammals are whales.

46. If a figure is a square, then it is a four-sided polygon. Converse: If a figure is a four-sided polygon, then it is a square. This is not a definition because not all four-sided polygons are squares.

47. If a figure is a square, then it is a rectangle. Converse: If a figure is a rectangle, then it is a square. This is not a definition because not all rectangles are squares.

48. 90°, 180°, and 270° because each of these rotations gives an image which coincides with its preimage.

49.

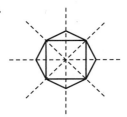

50. First, write the statement as a conditional: If a quadrilateral is a rhombus, then its diagonals are perpendicular. Now the converse must be: If the diagonals of a quadrilateral are perpendicular, then it is a rhombus. The converse is not true.

51.

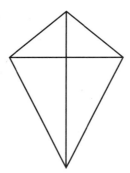

Any kite has perpendicular diagonals, but is not a rhombus, thus showing that the converse in Exercise 50 is not true.

52. First, write the statement as a conditional: If a quadrilateral is a rectangle, then its diagonals are congruent. Now the converse must be: If the diagonals of a quadrilateral are congruent, then it is a rectangle. The converse is not true.

53.

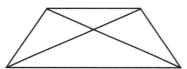

Any isosceles trapezoid has congruent diagonals, but is not a rectangle, thus showing that the converse in Exercise 52 is not true.

3.3 **PAGE 159, GUIDED SKILLS PRACTICE**

5. ∠3 and ∠6; ∠4 and ∠5

6. ∠1 and ∠8; ∠2 and ∠7

7. ∠3 and ∠5; ∠4 and ∠6

8. Sample answer: ∠1 and ∠5; ∠2 and ∠6

9. $\angle 4 \cong \angle 1$, because $\angle 1$ and $\angle 4$ are vertical angles. $\angle 5 \cong \angle 1$ because $\angle 1$ and $\angle 5$ are corresponding angles. $\angle 8 \cong \angle 1$, because $\angle 5$ and $\angle 8$ are vertical angles and $\angle 5$ and $\angle 1$ are corresponding angles.

10. $\angle 3 \cong \angle 2$, because $\angle 2$ and $\angle 3$ are vertical angles. $\angle 6 \cong \angle 2$ because $\angle 2$ and $\angle 6$ are corresponding angles. $\angle 7 \cong \angle 2$ because $\angle 6$ and $\angle 7$ are vertical angles and $\angle 6$ and $\angle 2$ are corresponding angles.

11. No. Every angle either is vertical to, corresponds to, or is vertical to a corresponding angle of $\angle 1$ and $\angle 2$. Hence every other angle is congruent to $\angle 1$ or $\angle 2$.

12. $\angle 1$ is vertical to $\angle 4$, $\angle 1$ corresponds to $\angle 5$, and $\angle 8$ is vertical to $\angle 5$ which corresponds to $\angle 1$. So $\angle 4$, $\angle 5$, and $\angle 8$ all measure 130°. $\angle 2$ is linear to $\angle 1$, hence supplementary. So $m\angle 2 = 50°$. $\angle 2$ is vertical to $\angle 3$, $\angle 2$ corresponds to $\angle 6$, and $\angle 7$ is vertical to $\angle 6$ which corresponds to $\angle 2$. So $\angle 3$, $\angle 6$, and $\angle 7$ all measure 50°.

13. Use line s as a transversal of lines p and q.
$\angle 1 \cong \angle 6$ by the Vertical Angles Theorem.
$\angle 1 \cong \angle 9$ by the Corresponding Angles Postulate.
$\angle 9 \cong \angle 14$ by the Vertical Angles Theorem, so $\angle 1 \cong \angle 14$ by the Transitive Property of Congruence.

14. $\angle 2 \cong \angle 5$ by the Vertical Angles Theorem.

15. Use line r as a transversal of lines p and q. $\angle 3 \cong \angle 4$ by the Vertical Angle Theorem. $\angle 3 \cong \angle 8$ by the Corresponding Angles Postulate. $\angle 8 \cong \angle 11$ by the Vertical Angle Theorem, so $\angle 3 \cong \angle 11$ by the Transitive Property of Congruence.

16. Use line r as a transversal of lines p and q. By the Vertical Angles Theorem, the angle formed by $\angle 1$ and $\angle 2$ is congruent to the angle formed by $\angle 5$ and $\angle 6$. By the Corresponding Angles Postulate, the angle formed by $\angle 1$ and $\angle 2$ is congruent to $\angle 7$. By the Vertical Angles Theorem, $\angle 7 \cong \angle 12$, so by the Transitive Property of Congruence, $\angle 12$ is congruent to the angle formed by $\angle 1$ and $\angle 2$.

17. Use line s as a transversal of lines p and q. By the Vertical Angle Theorem, the angle formed by $\angle 2$ and $\angle 3$ is congruent to the angle formed by $\angle 4$ and $\angle 5$; and by the Corresponding Angles Postulate, is congruent to $\angle 10$. By the Vertical Angles Theorem, $\angle 10 \cong \angle 13$, so by the Transitive Property of Congruence, $\angle 13$ is congruent to the angle formed by $\angle 2$ and $\angle 3$.

18. Line p is not a transversal because it intersects lines r and s in 1 point.

19. Line q is is transversal, intersecting lines r and s.

20. Line r is a transversal, intersecting lines p and q.

21. Line s is a transversal, intersecting lines p and q.

22. $m\angle ADE = 50°$ because it is a corresponding angle to $\angle DCB$.

23. $m\angle AED = m\angle ADE = 50°$

24. $m\angle DEB = 180° - m\angle AED = 180° - 50° = 130°$

25. $m\angle BDE = 25°$ because it is an alternate interior angle to $\angle DBC$.

26. $m\angle CDB = 180° - m\angle DCB - m\angle DBC$
$= 180° - 50° - 25° = 105°$

27. $m\angle ABD = 180° - m\angle DEB - m\angle BDE$
$= 180° - 130° - 25° = 25°$

28. Because $\angle 2$ and $\angle 6$ are corresponding angles, they are congruent angles. Thus
$$x = 3x - 60$$
$$-2x = -60$$
$$x = 30$$
So $m\angle 2 = x° = 30°$

29. $m\angle 1 = 180° - m\angle 2 = 180° - 30° = 150°$ because $\angle 1$ and $\angle 2$ are supplementary.

30. $m\angle 3 = m\angle 2 = 30°$ because $\angle 3$ and $\angle 2$ are vertical angles.

31. $m\angle 5 = m\angle 1 = 150°$ because $\angle 5$ and $\angle 1$ are corresponding angles.

32. $m\angle 7 = m\angle 3 = 30°$ because $\angle 7$ and $\angle 3$ are corresponding angles.

33. $m\angle 8 = m\angle 5 = 150°$ because $\angle 8$ and $\angle 5$ are vertical angles.

34. Given: Line $\ell \parallel$ line m. Line p is a transversal.
Prove: $\angle 1 \cong \angle 2$

Statement	Reasons
Line $\ell \parallel$ line m Line p is a transversal.	Given
$\angle 1 \cong \angle 3$	Corresponding Angles Postulate
$\angle 3 \cong \angle 2$	Vertical Angles Theorem
$\angle 1 \cong \angle 2$	Transitive Property of Congruence

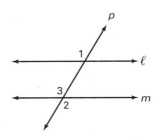

35. Given: Line $\ell \parallel$ line m. Line p is a transversal.
Prove: $m\angle 2 + m\angle 1 = 180°$

Statement	Reasons
Line $\ell \parallel$ line m Line p is a transversal.	Given
$m\angle 2 + m\angle 3 = 180°$	Linear Pair Property
$m\angle 3 = m\angle 1$	Corresponding Angles Postulate
$m\angle 2 + m\angle 1 = 180°$	Substitution Property

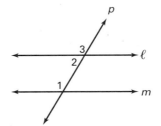

36. Given: Line $\ell \parallel$ line m. Line p is a transversal. $m\angle 1 = 90°$
Prove: $m\angle 2 = m\angle 3 = m\angle 4 = m\angle 5 = m\angle 6 = m\angle 7 = 90°$

Statement	Reasons
Line $\ell \parallel$ line m Line p is a transversal. $m\angle 1 = 90°$	Given
$m\angle 1 + m\angle 2 = 180°$	Linear Pair Property
$90 + m\angle 2 = 180°$	Substitution Property
$m\angle 2 = 90°$	Subtraction Property
$\angle 1 \cong \angle 4$ and $\angle 2 \cong \angle 3$ ($m\angle 1 = m\angle 4$ and $m\angle 2 = m\angle 3$)	Vertical Angles Theorem
$\angle 1 \cong \angle 8$ and $\angle 2 \cong \angle 7$ ($m\angle 1 = m\angle 8$ and $m\angle 2 = m\angle 7$)	Alternate Exterior Angles Theorem
$\angle 1 \cong \angle 5$ and $\angle 2 \cong \angle 6$ ($m\angle 1 = m\angle 5$ and $m\angle 2 = m\angle 6$)	Corresponding Angles Postulate
$m\angle 1 = m\angle 4 = m\angle 8 = m\angle 5 = 90°$ $m\angle 2 = m\angle 3 = m\angle 7 = m\angle 6 = 90°$	Substitution Property

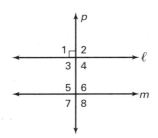

37. It is a transversal to them because it intersects them at two different points.

38. They are congruent because they are alternate interior angles.

39. It is a transversal to them because it intersects them at two different points.

40. $\angle 1$ and $\angle 2$ are congruent because they are corresponding angles of two vertical beans crossed by $\overline{PT}$.

41. $\overleftrightarrow{AB}$ is a transversal to the lines that are formed by extending the parallel mirrors.

42. No. $\angle 2$ and $\angle 3$ are alternate interior angles. The vertical angles of $\angle 2$ and $\angle 3$ that are formed by extending $\overleftrightarrow{AB}$ are alternate exterior angles.

43. $\angle 2$ and $\angle 3$ are congruent by the Alternate Interior Angles Theorem. By the given property of reflection, $\angle 1 \cong \angle 2$ and $\angle 3 \cong \angle 4$, Thus $\angle 1 \cong \angle 4$ by the Transitive Property of Congruence.

44. $\angle 1 \cong \angle 2$, so $m\angle 2 = 45°$. $\angle 2 \cong \angle 3$, so $m\angle 3 = 45°$. $\angle 3 \cong \angle 4$ so $m\angle 4 = 45°$.

45. ray

46. perpendicular

47. parallel

48. vertical angles

49. polygon

50. trapezoid

51. rhombus

PAGE 161, LOOK BEYOND

52. Given: trapezoid $ABCD$ with $\overline{AB} \parallel \overline{CD}$

Prove: m$\angle DAB$ + m$\angle ABC$ + m$\angle BCD$ + m$\angle CDA$ = 360°

1. Trapezoid $ABCD$, $\overline{AB} \parallel \overline{CD}$

2. m$\angle DAB$ + m$\angle CDA$ = 180°
   m$\angle ABC$ + m$\angle BCD$ = 180°

3. m$\angle DAB$ + m$\angle CDA$ + m$\angle ABC$ + m$\angle BCD$ = 360°

1. Given

2. Same-Side Interior Angles Theorem

3. Addition Property

Statement	Reasons
Trapezoid $ABCD$, $\overline{AB} \parallel \overline{CD}$	Given
m$\angle DAB$ + m$\angle CDA$ = 180° m$\angle ABC$ + m$\angle BCD$ = 180°	Same-Side Interior Angles Theorem
m$\angle DAB$ + m$\angle CDA$ + m$\angle ABC$ + m$\angle BCD$ = 360°	Addition Property

53. If $ABCD$ is a trapezoid, rectangle or parallelogram, then two sides are parallel, say $\overline{AB} \parallel \overline{CD}$. Thus the other edges form transversals to the parallel edges. By the Same-Side Interior Angles Theorem, m$\angle DAB$ + m$\angle CDA$ = 180° and m$\angle ABC$ + m$\angle BCD$ = 180°. Thus, the Addition Property of Equality implies that m$\angle DAB$ + m$\angle CDA$ + m$\angle ABC$ + m$\angle BCD$ = 360°.

3.4 **PAGE 164, GUIDED SKILLS PRACTICE**

5. Original statement: If two lines cut by a transversal are parallel, then alternate interior angles are congruent.

Converse: If alternate interior angles are congruent, then the two lines cut by the transversal are parallel.

6. The indicated angles are corresponding and congruent, so by the Converse of the Corresponding Angles Postulate, lines m and n are parallel.

7. Extend $\overline{AB}$ into a ray $\overrightarrow{BA}$. Measure $\angle ABC$. Using a protractor, draw a line through A such that the new angle corresponds to $\angle ABC$ and has the same measure as $\angle ABC$.

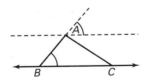

PAGES 164–167, PRACTICE AND APPLY

8. Alternate Exterior Angles Theorem

9. Alternate Exterior Angles Theorem

10. Alternate Interior Angles Theorem

11. Alternate Interior Angles Theorem

12. Corresponding Angles Postulate

13. Corresponding Angles Postulate

14. Same-Side Interior Angles Theorem

15. Same-Side Interior Angles Theorem

16. Given: m∠DFH = 55°; m∠GHF = 125°
　　　∠DFH and ∠GHE are corresponding angles.

Statement	Reasons
m∠DFH = 55°; m∠GHF = 125° ∠DFH and ∠GHE are corresponding angles.	Given
m∠FHG + m∠GHE = 180°	Linear Pair Property
125° + m∠GHE = 180°	Substitution Property
m∠GHE = 55°	Subtraction Property
∠DFH ≅ ∠GHE	Angle Congruence Postulate
$\overline{DF} \parallel \overline{GH}$	Converse of the Corresponding Angles Postulate

17. Given: m ∥ n; m∠RSU = m∠RTU = 70°
Prove: RSUT is a parallelogram.

Statement	Reasons
m ∥ n; m∠RSU = m∠RTU = 70°	Given
∠RSU ≅ ∠RTU	Angle Congruence Postulate
∠RTU ≅ ∠1 where ∠1 is the alternate interior angle of ∠RTU with transversal $\overline{RT}$.	Alternate Interior Angles Theorem
∠RSU ≅ ∠1	Transitive Property of Congruence
$\overline{RT} \parallel \overline{SU}$	Converse of the Corresponding Angles Postulate
RSUT is a parallelogram.	Definition of parallelograms

18. Given

19. Definition of supplementary angles

20. Linear Pair Property

21. Transitive or Substitution Property

22. Subtraction Property

23. Converse of the Corresponding Angles Postulate

24. Given: ∠1 ≅ ∠2
Prove: $\ell_1 \parallel \ell_2$

Statement	Reasons
∠1 ≅ ∠2	Given
m∠1 = m∠2	Angle Congruence Postulate
m∠1 + m∠3 = 180°	Linear Pair Property
m∠2 + m∠3 = 180°	Substitution Property
∠2 and ∠3 are supplementary.	Definition of supplementary angles
$\ell_1 \parallel \ell_2$	Converse of the Same-Side Interior Angles Theorem

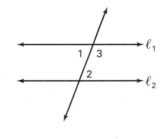

25. Given: ∠1 ≅ ∠2
Prove: $\ell_1 \parallel \ell_2$

Statement	Reasons
∠1 ≅ ∠2	Given
∠2 ≅ ∠4	Vertical Angles Theorem
∠1 ≅ ∠4	Transitive Property of Congruence
$\ell_1 \parallel \ell_2$	Converse of the Corresponding Angles Postulate

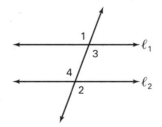

26. transversal　　　**27.** both are right angles　　　**28.** line m　　　**29.** line p

30. Converse of the Corresponding Angles Postulate　　　**31.** ∠3

32. Corresponding Angles Postulate　　　**33.** ∠2

34. Corresponding Angles Postulate　　　**35.** ∠2

36. Transitive Property of Congruence　　　**37.** Converse of the Corresponding Angles Postulate

38. Using the T square as the transversal, each line drawn with the triangle will have congruent corresponding angles to every other line drawn with the triangle. Thus, by the Converse of the Corresponding Angles Postulate, the lines will be parallel.

39. Paint lines that have congruent corresponding angles to the first line. By the Converse of the Corresponding Angles Postulate, each of these lines will be parallel to the first.

40. The angle formed by the wall and the roof corresponds to the angle formed by the plumb line and the roof, and they are congruent. Thus the Converse of the Corresponding Angles Postulate implies the wall and the plumb lines are parallel. Since the plumb line is vertical, so is the wall.

PAGE 167, LOOK BACK

41. If a figure is a rectangle, then it is a parallelogram.

42. Hypothesis: the figure is a rectangle
Conclusion: the figure is a parallelogram

43.

Parallelograms

Rectangles

44. Converse: If a figure is a parallelogram, then it is a rectangle.

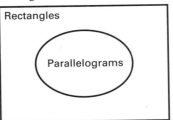

Rectangles

Parallelograms

45. It is false. For example, a rhombus is a parallelogram, but it is not necessarily a rectangle.

46. A pilot flying in the opposite direction is 180° from the heading of this pilot so we add 180.
$155 + 180 = 335$

47. A pilot flying perpendicular is either 90° clockwise from the heading of this pilot or 90° counterclockwise from the heading of this pilot so we add 90° to get one heading and subtract 90° to get the other. $235 + 90 = 325$; $235 - 90 = 145$

PAGE 167, LOOK BEYOND

48. Sample answer: By showing that both tracks are perpendicular to the railroad ties. This would guarantee that the tracks are parallel to each other.

49. Since objects at a distance appear smaller, so too will the distance between the railroad tracks. "Meet at infinity" refers to the illusion of the lines meeting far away, although the lines do not meet at any finite distance from any particular vantage point. In a way, it does *not* make sense because "infinity" is not a place at which lines could meet. On the other hand, it *does* make if "at infinity" is taken to mean "not any finite distance from here."

50. They are parallel because they are opposite sides of a cube.

51. They would intersect at a single point.

3.5   PAGE 173, GUIDED SKILLS PRACTICE

5. ∠1 and ∠4; ∠3 and ∠5

6. $m\angle 1 + m\angle 2 + m\angle 3 = 180°$ because the angles fit together to form a straight line.

7. Because the sum of angles in a triangle is 180°,
$65° + 50° + m\angle 2 = 180°$
$m\angle 2 = 180° - 65° - 50° = 65°.$

8. $m\angle 3 = 180° - 85° - 45° = 50°$

9. $m\angle B = 180° - 45° - 90° = 45°$

10. $m\angle K = 180° - 60° - 60° = 60°$

11. $m\angle Y = 180° - 90° - 90° = 0°$
No such triangle exists.

12. $m\angle H = 180° - 105° - 80° = -5°$
No such triangle exists.

13. Using $\overline{BA}$ as a transversal, $m\angle B = m\angle ADE = 60°$ because they are corresponding angles of parallel lines.

14. The sum of the angle measures in $\triangle ABC = 180°$.
$m\angle A = 180° - m\angle B - m\angle ACB$
$\qquad = 180° - 60° - 50° = 70°$

15. Using $\overline{AC}$ as a transversal, $m\angle AED = m\angle C = 50°$ because they are corresponding angles of parallel lines.

16. $m\angle EDB + m\angle ADE = 180°$ because they are a linear pair. Thus, $m\angle EDB = 180° - 60° = 120°$.

17. $m\angle AED + m\angle DEC = 180°$ because they are a linear pair. Thus, $m\angle DEC = 180° - 50° = 130°$.

18. $m\angle FEC = m\angle AED = 50°$ because $\angle AED$ and $\angle FEC$ are vertical angles.

19. Using $\overline{AC}$ as a transversal, $m\angle ECF = m\angle A = 70°$ because they are alternate interior angles of parallel lines.

20. The sum of angle measures in $\triangle EFC = 180°$.
$m\angle F = 180° - m\angle FEC - m\angle ECF$
$\qquad = 180° - 50° - 70° = 60°$

21. $m\angle 1 = 180° - 90° - 60° = 30°$

22. All three angles are the same measure.
$3m\angle 2 = 180°$
$\ m\angle 2 = 60°$

23. The angle linear to the 140° angle must have measure 40° since linear pairs are supplementary.
$m\angle 3 = 180° - 40° - 40° = 100°$

24. $30x + 40x + 10x^2 = 180$
$10x^2 + 70x - 180 = 0$
$\quad x^2 + 7x - 18 = 0$
$\quad (x + 9)(x - 2) = 0$
$\ x + 9 = 0 \qquad x - 2 = 0$
$\qquad x = -9 \ \text{or} \qquad x = 2$
Since angle measures are positive and $m\angle A = 30x$, x cannot equal -9. Thus $x = 2$,
$m\angle A = [30(2)]° = 60°$,
$m\angle B = [40(2)]° = 80°$, and
$m\angle C = [10(2)]° = 40°$.

25. $(4x^2 - 10) + (8x + 5) + (x^2 + 2x + 10) = 180$
$5x^2 + 10x + 5 = 180$
$5x^2 + 10x - 175 = 0$
$x^2 + 2x - 35 = 0$
$(x + 7)(x - 5) = 0$
$x + 7 = 0 \qquad x - 5 = 0$
$x = -7 \quad \text{or} \qquad x = 5$
Since angle measures are positive and
$m\angle E = 8x + 5$, x cannot equal -7. Thus $x = 5$,
$m\angle D = [4(5)^2 - 10]° = 90°$,
$m\angle E = [8(5) + 5]° = 45°$, and
$m\angle F = [(5)^2 + 2(5) + 10]° = 45°$.

26. $30° + 70° = 100°$

27. $m\angle 3 = 180° - (m\angle 1 + m\angle 2)$
$\qquad = 180° - 100° = 80°$

28. $m\angle 4 = 180° - m\angle 3 = 180° - 80° = 100°$

29. $30° + 80° = 110°$

30. $m\angle 3 = 180° - (m\angle 1 + m\angle 2)$
$\qquad = 180° - 110° = 70°$

31. $m\angle 4 = 180° - m\angle 3 = 180° - 70° = 110°$

32. $40° + 80° = 120°$

33. $m\angle 3 = 180° - (m\angle 1 + m\angle 2)$
$\qquad = 180° - 120° = 60°$

34. $m\angle 4 = 180° - m\angle 3 = 180° - 60° = 120°$

35. $40° + 90° = 130°$

36. $m\angle 3 = 180° - (m\angle 1 + m\angle 2)$
$\qquad = 180° - 130° = 50°$

37. $m\angle 4 = 180° - m\angle 3 = 180° - 50° = 130°$

38. There are two possible exterior angles at each vertex. These angles can be seen by extending the sides of the triangle. Their measures are the same because both exterior angles are supplementary to the vertex angle.

39. the sum of the measures of the remote interior angles

40. Software will show many examples in which $m\angle 1 + m\angle 2 = m\angle 4$.

41. Linear Pair Property

42. Triangle Sum Theorem

43. Transitive or Substitution Property

44. Subtraction Property

45.

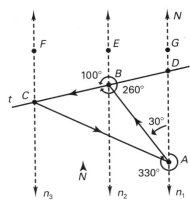

Because the $\overrightarrow{AB}$ flight path was heading 330, m$\angle DAB = 360° - 330° = 30°$. Because the $\overrightarrow{BC}$ flight path was heading 260, m$\angle EBC = 360° - 260° = 100°$. Because $n_1 \parallel n_2$, corresponding angles on transversal t are congruent. Thus, m$\angle GDB =$ m$\angle EBC = 100°$ and m$\angle BDA = 180° - 100° = 80°$.

$$\begin{aligned} \text{m}\angle DBA &= 180° - \text{m}\angle BDA - \text{m}\angle DAB \\ &= 180° - 80° - 30° \\ &= 70° \end{aligned}$$

Since m$\angle BCA +$ m$\angle BAC =$ m$\angle DBA$ and $\angle BCA \cong \angle BAC$,

$$\begin{aligned} \text{m}\angle BCA + \text{m}\angle BCA &= 70° \\ 2\text{m}\angle BCA &= 70° \\ \text{m}\angle BCA &= 35° \end{aligned}$$

Since $n_1 \parallel n_3$, alternate interior angles on transversal t are congruent. Thus m$\angle FCB =$ m$\angle BDA = 80°$.

To return to town A, the pilot's direction clockwise from north must be m$\angle FCB +$ m$\angle BCA = 80° + 35° = 115°$. Thus, the pilot's heading must be 115.

PAGES 175–176, LOOK BACK

46. intersection

47. 2

48. 3 noncollinear

49. linear pair

50.

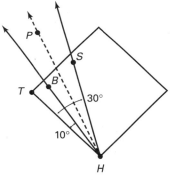

Angle that path of ball bisects:

$$\begin{aligned} \text{m}\angle THS - \text{m}\angle THB &= 30° - 10° \\ &= 20° \end{aligned}$$

Angle that path of ball makes with third-base line:

$$10° + \frac{20°}{2} = 10° + 10° = 20°$$

51. Yes, the vertical lines are great circles and the horizontal line at the equator is a great circle.

52. Answers will vary. It be helpful to have students trace a rhumb line for a specific direction, such as northwest. They should notice that the path must curve as it approaches the poles.

3.6 PAGE 180, GUIDED SKILLS PRACTICE

4. An octagon has 8 sides, so $n = 8$ in the formula $180°(n - 2)$: $180°(8 - 2) = 1080°$.

5. A 13-gon has 13 sides so $n = 13$ in the formula $180°(n - 2)$: $180°(13 - 2) = 1980°$.

6. The sum of the measures of the exterior angles of any polygon is 360°.

7. The sum of the measures of the exterior angles of any polygon is 360°.

8. a. $x = 180 - 54 - 48 = 78$

    b. $y = 180 - 78 = 102$

    c. The sum of the interior angles of a pentagon $= 540°$, and the angle x is vertical to the fifth angle.
$z = 540 - 115 - 120 - 100 - 78 = 127$

9. The sum of the interior angle measures is $360°$.
$360° - 90° - 75° - 85° = 110°$

10. The sum of the exterior angle measures is $360°$.
$360° - 132° - 120° = 108°$

11. The sum of the interior angle measures is $540°$.
$540° - 130° - 100° - 110° - 90° = 110°$

12. The sum of the base angle measures is $180° - 90° = 90°$. Since the base angles are congruent, they measure $\frac{90°}{2} = 45°$

13. The sum of the interior angle measures is $540°$.
$540° - 2(110°) - 2(100°) = 120°$

14. The angle measure is congruent to an angle that forms a linear pair with an angle measuring $75°$.
$180° - 75° = 105°$

15. Because rectangles have all right angles, interior angles measure $90°$ and exterior angles measure $90°$.

16. Equilateral triangles are also equiangular, and the 3 congruent interior angles must have measures that add to $180°$. So interior angles measure $60°$ and exterior angles measure $180° - 60° = 120°$.

17. A regular dodecagon has 12 sides and 12 congruent exterior angles that must add to $360°$. So each exterior angle has measure $\frac{360°}{12} = 30°$. Hence each interior angle has measure $180° - 30° = 150°$.

18. An equiangular pentagon has 5 sides and 5 congruent exterior angles that must add to $360°$. So each exterior angle has measure $\frac{360°}{5°} = 72°$. Hence each interior angle has measure $180° - 72° = 108°$.

19. Let n represent the number of sides of the regular polygon.
$$135° = \frac{180°(n-2)}{n}$$
$$135°n = 180°n - 360°$$
$$45°n = 360°$$
$$n = 8$$

20. Let n represent the number of sides of the regular polygon.
$$150° = \frac{180°(n-2)}{n}$$
$$150°n = 180°n - 360°$$
$$30°n = 360°$$
$$n = 12$$

21. Let n represent the number of sides of the regular polygon.
$$165° = \frac{180°(n-2)}{n}$$
$$165°n = 180°n - 360°$$
$$15°n = 360°$$
$$n = 24$$

22. Let n represent the number of sides of the regular polygon.
$$60°n = 360°$$
$$n = 6$$

23. Let n represent the number of sides of the regular polygon.
$$36°n = 360°$$
$$n = 10$$

24. Let n represent the number of sides of the regular polygon.
$$24°n = 360°$$
$$n = 15$$

For Exercises 25–28, the sum of the measures of the interior angles of the figure is 360°.
Thus, $(x)° + (2x)° + (3x)° + (4x)° = 360°$
$$(10x)° = 360°$$
$$x = 36$$

25. $m\angle A = x° = 36°$

26. $m\angle B = (2x)° = (2 \cdot 36)° = 72°$

27. $m\angle C = (3x)° = (3 \cdot 36)° = 108°$

28. $m\angle D = (4x)° = (4 \cdot 36)° = 144°$

For Exercises 29–32, the sum of the measures of the interior angles of the figure is 360°. Thus, $(8x - 10)° + (x^2 + 10)° + (2x + 30)° + (x^2 + 2x + 10)° = 360°$

$$(2x^2 + 12x + 40)° = 360°$$
$$2x^2 + 12x - 320 = 0$$
$$x^2 + 6x - 160 = 0$$
$$(x + 16)(x - 10) = 0$$
$$x + 16 = 0 \qquad x - 10 = 0$$
$$x = -16 \quad \text{or} \quad x = 10$$

Since angle measures are positive and m$\angle E = 8x - 10$, x cannot equal -16. Thus, $x = 10$.

29. m$\angle E = (8x - 10)° = [8(10) - 10]° = 70°$

30. m$\angle F = (x^2 + 10)° = [(10)^2 + 10]° = 110°$

31. m$\angle G = (2x + 30)° = [2(10) + 30]° = 50°$

32. m$\angle H = (x^2 + 2x + 10)°$
$= [(10)^2 + 2(10) + 10]°$
$= 130°$

For Exercises 33–37, the sum of the measures of the interior angles of the figure is 540°. Thus, $(4x + 32)° + (5x - 14)° + (4x - 37)° + (5x + 15)° + (2x + 4)° = 540°$

$$(20x)° = 540°$$
$$x = 27$$

33. m$\angle I = (4x + 32)° = [4(27) + 32]° = 140°$

34. m$\angle J = (5x - 14)° = [5(27) - 14]° = 121°$

35. m$\angle K = (4x - 37)° = [4(27) - 37]° = 71°$

36. m$\angle L = (5x + 15)° = [5(27) + 15]° = 150°$

37. m$\angle M = (2x + 4)° = [2(27) + 4]° = 58°$

38. 3. Sample answer: an equilateral triangle has three 60°-angles. No; if all three angles measure 90° or more, then the sum of the measures of the interior angles would be more than 180°.

39. 3. Sample answer: a quadrilateral with angle measures 60°, 70°, 70°, 160° has 3 acute angles, but if a quadrilateral had 4 acute angles, then the sum of their measures would be less than 360°. Yes; for example, a rectangle has 4 right angles.

40. 3. Sample answer: a pentagon with angle measurements 80°, 80°, 80°, 150°, 150° has 3 acute angles, but if apentagon had 4 acute angles then the sum of the measures of all 5 angles would be less than 360° + 180° = 540°. Yes; for example, a regular pentagon has 108° interior angles.

41. The angle labeled 1 + 4 is an exterior angle with remote interior angles labeled 1 and 4. The angle labeled 3 + 5 is an exterior angle for the triangle with remote interior angles labeled 3 and 5. The angle labeled 1 + 4 + 3 + 5 is an exterior angle with remote interior angles labeled 1 + 4 and 3 + 5. Angles 2 and 1 + 4 + 3 + 5 form a straight edge, so 2 + 1 + 4 + 3 + 5 = 180°. Thus, the sum of the vertex angles of the star is 180°.

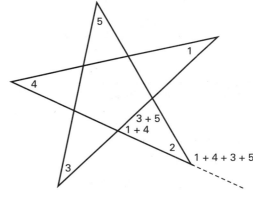

42. The upper portion of the gem is an isosceles trapezoid with base angles of 35° and upper angles of $x°$, due to the vertical axis of symmetry. Since the sum of the interior angle measures in a quadrilateral is 360°,

$$x° + x° + 35° + 35° = 360°$$
$$(2x)° + 70° = 360°$$
$$(2x)° = 290°$$
$$x = 145$$

The lower portion of the gem is an isosceles triangle with base angles of 41°, due to the vertical axis of symmetry. Since the sum of the interior angle measures in a triangle is 180°,

$$41° + 41° + y° = 180°$$
$$y° = 98°$$
$$y = 98$$

43.

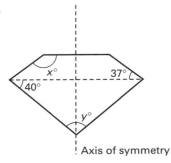

Axis of symmetry

The upper portion of the gem is an isosceles trapezoid with base angles of 37° and upper angles of $x°$, due to the vertical axis of symmetry. Since the sum of interior angle measures in a quadrilateral is 360°,

$$x° + x° + 37° + 37° = 360°$$
$$(2x)° + 74° = 360°$$
$$(2x)° = 286°$$
$$x = 143$$

The lower portion of the gem is an isosceles triangle with base angles of 40°, due to the vertical axis of symmetry. Since the sum of interior angle measures in a triangle is 180°,

$$40° + 40° + y° = 180°$$
$$y° = 100°$$
$$y = 100$$

PAGE 182, LOOK BACK

44. The distance from a point to a line is determined by measuring the perpendicular segment from the point to the line.

45. ∠1 and ∠2, ∠2 and ∠3, ∠3 and ∠4, and ∠4 and ∠1; these angles are called linear pair angles.

46. $m\angle 1 + m\angle 4 = 180°$
$75° + m\angle 4 = 180°$
$m\angle 4 = 180° - 75°$
$m\angle 4 = 105°$

47. ∠1 and ∠3, ∠2 and ∠4; these angles are called vertical angles.

PAGE 182, LOOK BEYOND

48. Squares have all right angles, so the measure of each angle is 90°. There are four 90° angles at this center point, adding to 360°.

49. This is true because the sum of the degrees of all the angles that fit around a point must be 360°. The interior angles of a regular n-gon are all congruent, so if a certain number of n-gons fit around a point, the interior angle measure of each n-gon must divide evenly into 360°.

50. $n = 3 \Rightarrow$ interior angle measure = 60° $\frac{360°}{60°} = 6 \Rightarrow$ Regular triangles fit.

$n = 4 \Rightarrow$ interior angle measure = 90° $\frac{360°}{90°} = 4 \Rightarrow$ Squares fit.

$n = 5 \Rightarrow$ interior angle measure = 108° $\frac{360°}{108°} = \frac{10}{3} \Rightarrow$ Pentagons will *not* fit.

$n = 6 \Rightarrow$ interior angle measure = 120° $\frac{360°}{120°} = 3 \Rightarrow$ Regular hexagons fit.

No other n-gons are possible because n-gons with $n > 6$ have interior angle measures $> 120°$ and not even 3 such n-gons will fit without overlapping.

51. Sample answer: Hexagons fit together with no gaps or overlaps.

5. The length of the triangle midsegment is one-half of the length of the triangle base.

$\frac{1}{2}(12) = 6$

6. The length of the midsegment of the triangular lake is one-half the width, w, of the lake at its base.

$\frac{1}{2}w = 7$

$w = 2 \cdot 7 = 14$

The width of the lake is 14 miles.

7. The length of the midsegment of a trapezoid is one-half the sum of its bases.

$\frac{1}{2}(7 + 15) = \frac{1}{2}(22) = 11$

8. width $= \frac{180 + 48}{2} = \frac{228}{2} = 114$

The width is 114 feet.

9. Length of triangle midsegment $= \frac{\text{base}}{2}$;

length of trapezoid midsegment $= \frac{\text{base}_1 + \text{base}_2}{2}$;

the formula for the length of the trapezoid midsegment can be used to find the length of a triangle midsegment by letting the short base $= 0$.

PAGES 187–189, PRACTICE AND APPLY

10. $\frac{1}{2}AB = 20$

$AB = 40$

11. $IJ = \frac{1}{2}(50) = 25$

12. $PQ = \frac{1}{2}(40 + 50) = \frac{1}{2}(90) = 45$

13. $90 = \frac{1}{2}(RS + 100)$

$90 = \frac{1}{2}RS + 50$

$40 = \frac{1}{2}RS$

$80 = RS$

14. $\frac{1}{2}(5x - 5) = 2x + 1$

$5x - 5 = 2(2x + 1)$

$5x - 5 = 4x + 2$

$x = 7$

$DE = 2(7) + 1 = 15$

15. $8 = \frac{1}{2}((x^2 + x - 2) + (x^2 + 3x - 12))$

$8 = \frac{1}{2}(2x^2 + 4x - 14)$

$8 = x^2 + 2x - 7$

$0 = x^2 + 2x - 15$

$0 = (x + 5)(x - 3)$

$x + 5 = 0 \qquad x - 3 = 0$

$x = -5 \quad \text{or} \quad x = 3$

If $x = -5$, $IH = (-5)^2 + 3(-5) - 12 = -2$

Since IH cannot be negative, x cannot be -5.

Thus $x = 3$, $IH = (3)^2 + 3(3) - 12 = 6$

$FG = (3)^2 + (3) - 2 = 10$

16. FG is the midsegment between bases $\overline{BC}$ and vertex A, where the upper base may be taken as 0.

$DE = \frac{1}{2}(20 + 0) = 10$

$FG = \frac{1}{2}(40 + 0) = 20$

$HI = \frac{1}{2}(40 + 20) = 30$

17. DE, FG, and HI are $\frac{1}{4}$, $\frac{1}{2}$, and $\frac{3}{4}$, respectively, of BC.

18. If parallel segments divide two sides of a triangle into four congruent segments, the length of the shortest segment is $\frac{1}{4}$ of the length of the base, the length of the middle segment is $\frac{2}{4}$ or $\frac{1}{2}$ of the length of the base and the length of the longest segment is $\frac{3}{4}$ of the length of the base.

19. If parallel segments divide two sides of a triangle into three congruent segments, the length of the shortest segment is $\frac{1}{3}$ of the length of the base and the length of the other segment is $\frac{2}{3}$ of the length of the base. If parallel segments divide two sides of a triangle into eight congruent segments, the length of the shortest segment is $\frac{1}{8}$ of the length of the base and the length of each of the remaining segments is $\frac{m}{8}$ of the length of the base where $m = 2, 3, 4, 5, 6, 7$. If parallel segments divide two sides of a triangle into n congruent segments, the length of each segment is $\frac{m}{n}$ of the length of the base where m is the segment number from 1 to $n - 1$.

20. No, because both bases of a trapezoid must be taken into account. For example, for $n = 4$, the midsegment divides the trapezoid into two smaller trapezoids, which have midsegments of length $\frac{3b_1 + b_2}{4}$ and $\frac{b_1 + 3b_2}{4}$.

21. By the argument of Exercise 19, each successive segment decreases by $\frac{1}{n} = \frac{1}{3}$ if the base length KL is taken as one unit. Therefore $MN = \frac{1}{3}KL$ and $PQ = \frac{2}{3}KL$.

22. The lengths of the midsegments are $\frac{1}{2}(16) = 8$, $\frac{1}{2}(18) = 9$, and $\frac{1}{2}(26) = 13$. Perimeter of the outer triangle is $16 + 18 + 26 = 60$. The sum of the sides of the inner triangle is $8 + 9 + 13 = 30$. The perimeter of the inner triangle is one-half the perimeter of the outer triangle.

23.

Parallelogram; Two sides of the figure are triangle midsegments making opposite sides parallel, therefore a parallelogram is formed inside the triangle.

24.

Rhombus, Two sides of the figure are triangle midsegments making opposite sides parallel, therefore a parallelogram is formed. Because the triangle is isosceles, the adjacent sides that are part of the triangle are congruent and half the length of the congruent sides of the triangle. The remaining adjacent sides are midsegments of the congruent triangle sides, and thus half of their length. So the lower sides are congruent to each other and the upper sides. Thus, the figure is a rhombus.

25.

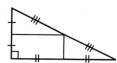

Rectangle; The two sides of the figure are triangle midsegments making opposite sides parallel, therefore a parallelogram is formed. Consecutive interior angles are supplementary, so all four angles measure 90°. Thus, the figure is a rectangle.

26.

Square; Two sides of the figure are triangle midsegments, so a parallelogram is formed. Because the triangle is isosceles, the sides of the quadrilateral that are part of the triangle are congruent. The sides in the interior of the triangle are midsegments of the congruent triangle sides and thus half their length. Thus, they are congruent to each other and to the other sides of the quadrilateral. Consecutive interior angles are supplementary, so all four angles measure 90°. Thus, a square is formed inside the triangle.

27. Midsegment lengths are $\frac{1}{2}, \frac{1}{4}, \frac{1}{8}, \frac{1}{16}, \cdots$

Looking at the bottom edge of the box, we can figure out the lengths of the segments along the bottom by the lengths of the midsegments. The first segment is $1 - \frac{1}{2} = \frac{1}{2}$. The 2nd segment is $\frac{1}{2} - \frac{1}{4} = \frac{1}{4}$. The 3rd segment is $\frac{1}{4} - \frac{1}{8} = \frac{1}{8}$, and so on. Thus $\frac{1}{2} + \frac{1}{4} + \frac{1}{8} + \ldots$ must add up to the length of bottom edge of the square, or 1.

28. If FC is the midsegment of the trapezoid,

$FC = \frac{1}{2}(AB + ED) = \frac{1}{2}(25 + 70) = 47.5$ ft. Since $FC = 45$ ft, it is not the midsegment.

29. distance from ladder to wall $= \frac{1}{2}(5) = 2.5$ ft

30. distance from ladder to wall $= \frac{1}{4}(5) = 1.25$ ft

31. $p \cdot 5 = 2$

$p = \frac{2}{5} = .40$

The painter can reach the wall from the top 40% of the ladder.

32. length of center string $= \frac{1}{2}(17 + 38) = 27.5$ in.

PAGE 189, LOOK BACK

33.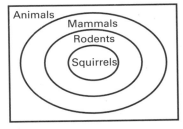

34. False; rhombuses do not have four right angles in general, so they need not be squares.

35. True; all rectangles have two sets of parallel sides, thus all rectangles are parallelograms.

36. True; squares have four right angles and opposite sides congruent, thus all squares are rectangles.

37. False; parallelograms need not have all four sides congruent.

38. $x° = 180° - 60° - 30° = 90°$
$x = 90$

39. The angle adjacent to the 120° angle is
$180° - 120° = 60°$.
$x° = 180° - 60° - 50° = 70°$
$x = 70$

40. The angle adjacent to x measures
$180° - 86° - 45° = 49°$.
$x° = 180° - 49° = 131°$
$x = 131$

PAGE 189, LOOK BEYOND

41. If the midsegments of a triangle are used to divide a triangle into 4 triangles, the four small triangles are congruent.

42. Draw a triangle, with midsegments, such as the one in the diagram. Since the inner triangle is made up of midsegments of the larger triangle, each of its sides is half the length of the parallel side of the larger triangle. Rotate the inner triangle to see that the four small triangles match up. Thus, the ratio of the area of each of the four triangles to the area of the original triangle is 1:4.

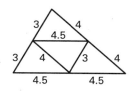

43. The 9 shaded triangles are equilateral with side length $\frac{1}{4}$, hence each has perimeter $\frac{3}{4}$. Thus the side lengths of all the shaded triangles $= 9 \cdot \frac{3}{4} = \frac{27}{4} = 6.75$.

44. Less than; because the area of the outer triangle includes the areas of the unshaded triangles.

45. The shaded areas of the new figure would be less than the area of the shaded regions in the original figure because center triangular areas are being removed from the area. The sum of all the side lengths would be more because triangular borders are being added.

3.8 | PAGE 194, GUIDED SKILLS PRACTICE

6.

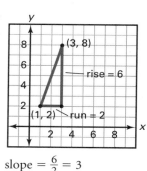

slope $= \frac{6}{2} = 3$

7. $\frac{4-0}{4-0} = \frac{4}{4} = 1$

8. $\frac{5-3}{4-(-1)} = \frac{2}{5}$

9. $\frac{-6-1}{4-2} = -\frac{7}{2}$

10.

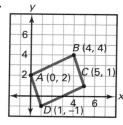

This is a parallelogram because the slopes of opposite sides are equal, so the opposite sides are parallel.

$m_{\overline{AD}} = \frac{-1-2}{1-0} = -3$ and $m_{\overline{BC}} = \frac{1-4}{5-4} = -3$

$m_{\overline{AB}} = \frac{4-2}{4-0} = \frac{2}{4} = \frac{1}{2}$ and $m_{\overline{CD}} = \frac{1-(-1)}{5-1} = \frac{2}{4} = \frac{1}{2}$

11.

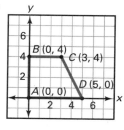

midpoint of $\overline{AB}$: $\left(\frac{0+0}{2}, \frac{0+4}{2}\right) = (0, 2)$;

midpoint of $\overline{CD} = \left(\frac{3+5}{2}, \frac{4+0}{2}\right) = (4, 2)$; midsegment length = 4;

The bases and the midsegment are all horizontal, so they are parallel.

The base lengths are 3 and 5.

Using the bases, the midsegment length $= \frac{1}{2}$(base 1 + base 2) $= \frac{1}{2}(3 + 5)$ $= \frac{1}{2}(8) = 4$, confirming the trapezoid midsegment conjecture.

12. $m = \dfrac{2 - 0}{4 - 0} = \dfrac{2}{4} = \dfrac{1}{2}$

$MP = \left(\dfrac{0 + 4}{2}, \dfrac{0 + 2}{2}\right) = (2, 1)$

13. $m = \dfrac{-1 - 1}{1 - (-1)} = \dfrac{-2}{2} = -1$

$MP = \left(\dfrac{-1 + 1}{2}, \dfrac{1 + (-1)}{2}\right) = (0, 0)$

14. $m = \dfrac{3 - (-1)}{3 - (-3)} = \dfrac{4}{6} = \dfrac{2}{3}$

$MP = \left(\dfrac{-3 + 3}{2}, \dfrac{-1 + 3}{2}\right) = (0, 1)$

15. $m = \dfrac{-3 - 2}{1 - (-5)} = \dfrac{-5}{6}$

$MP = \left(\dfrac{-5 + 1}{2}, \dfrac{2 + (-3)}{2}\right) = \left(-2, \dfrac{-1}{2}\right)$

16. $m_1 = \dfrac{3 - 1}{2 - (-1)} = \dfrac{2}{3};\ m_2 = \dfrac{4 - 2}{5 - 2} = \dfrac{2}{3}$

Because the slopes are equal, the segments are parallel.

17. $m_1 = \dfrac{-2 - 1}{1 - (-2)} = \dfrac{-3}{3} = -1;$

$m_2 = \dfrac{3 - (-1)}{3 - (-1)} = \dfrac{4}{4} = 1$

$m_1 \cdot m_2 = -1 \cdot 1 = -1$

Because the product of the slopes is -1, the segments are perpendicular.

18. $m_1 = \dfrac{2 - 2}{3 - (-2)} = \dfrac{0}{5} = 0$ (horizontal);

$m_2 = \dfrac{4 - (-1)}{2 - 2} = \dfrac{5}{0} =$ undefined (vertical)

Because all horizontal lines are perpendicular to all vertical lines, the segments are perpendicular.

19. $m_1 = \dfrac{-2 - 2}{1 - (-1)} = \dfrac{-4}{2} = -2;$

$m_2 = \dfrac{1 - (-2)}{2 - 1} = \dfrac{1}{1} = 1$

$m_1 \cdot m_2 = -2 \cdot 1 = -2$

Because the slopes are neither the same nor opposite reciprocals, the segments are neither parallel nor perpendicular.

20. $m_{\overline{SA}} = \dfrac{0 - 6}{7 - (-2)} = \dfrac{-6}{9} = \dfrac{-2}{3}$

21. $m_{\overline{TB}} = \dfrac{10 - 1}{4 - (-2)} = \dfrac{9}{6} = \dfrac{3}{2}$

22. $m_{\overline{SA}} \cdot m_{\overline{TB}} = \dfrac{-2}{3} \cdot \dfrac{3}{2} = -1$

$\overline{SA}$ is perpendicular to $\overline{TB}$ because their slopes are negative reciprocals. Thus, $\angle 1$ is a right angle.

23. $m_1 = \dfrac{3 - 4}{4 - (-1)} = \dfrac{-1}{5},$

$m_2 = \dfrac{1 - 3}{1 - 4} = \dfrac{-2}{-3} = \dfrac{2}{3},$

$m_3 = \dfrac{1 - 4}{1 - (-1)} = \dfrac{-3}{2}$

$m_2 \cdot m_3 = \dfrac{2}{3}\left(\dfrac{-3}{2}\right) = -1$

Yes, it is a right triangle.

24. $m_1 = \dfrac{0 - 3}{2 - 1} = -3,$

$m_2 = \dfrac{2 - 0}{-3 - 2} = \dfrac{2}{-5},$

$m_3 = \dfrac{2 - 3}{-3 - 1} = \dfrac{-1}{-4} = \dfrac{1}{4}$

Because no pair of slopes have a product of -1, it is not a right triangle.

25. $m_1 = \dfrac{-1 - 3}{3 - (-2)} = \dfrac{-4}{5},$

$m_2 = \dfrac{-1 - (-1)}{-2 - 3} = \dfrac{0}{-5} = 0$ (horizontal)

$m_3 = \dfrac{-1 - 3}{-2 - (-2)} = \dfrac{-4}{0} =$ undefined (vertical)

Because one side is horizontal and another is vertical, it is a right triangle.

26. $m_1 = \dfrac{1 - 0}{0 - 1} = \dfrac{1}{-1} = -1,$

$m_2 = \dfrac{0 - 1}{-1 - 0} = \dfrac{-1}{-1} = 1,$

$m_3 = \dfrac{0 - 0}{-1 - 1} = \dfrac{0}{-2} = 0$ (horizontal)

$m_1 \cdot m_2 = -1 \cdot 1 = -1$

Yes, it is a right triangle.

27. $m_1 = \dfrac{3 - 2}{3 - 1} = \dfrac{1}{2},$

$m_2 = \dfrac{0 - 3}{4 - 3} = -3,$

$m_3 = \dfrac{0 - 2}{4 - 1} = \dfrac{-2}{3}$

Because no pair of slopes have a product of -1, it is not a right triangle.

28. $m_1 = \frac{-6-5}{2-9} = \frac{-11}{-7} = \frac{11}{7},$

$m_2 = \frac{-1-(-6)}{-1-2} = \frac{5}{-3},$

$m_3 = \frac{-1-5}{-1-9} = \frac{-6}{-10} = \frac{3}{5}$

$m_2 \cdot m_3 = \frac{5}{-3} \cdot \frac{3}{5} = -1$

Yes, it is a right triangle.

29.

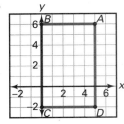

Rectangle; The slopes of $\overline{CD}$ and $\overline{AB}$ are 0 and the slopes of $\overline{DA}$ and $\overline{BC}$ are undefined. Therefore the opposite sides of the polygon are parallel. Vertical lines have undefined slopes and horizontal lines have zero slopes. Thus adjacent sides are perpendicular and $ABCD$ is a rectangle.

30.

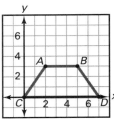

Trapezoid; The slope of $\overline{CA}$ is $\frac{3}{2}$, the slope of $\overline{BD}$ is $\frac{-3}{2}$, and the slopes of $\overline{AB}$ and $\overline{CD}$ are 0. Therefore $\overline{AD}$ and $\overline{CD}$ are parallel and the other pair is not. Because it has one set of parallel sides, $ABDC$ is a trapezoid.

31.

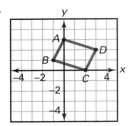

Parallelogram; Sides $\overline{AB}$ and $\overline{CD}$ have slope 2, hence are parallel. Sides $\overline{AD}$ and $\overline{BC}$ have slope $\frac{-1}{3}$, hence are parallel. Since it has two pairs of parallel lines, $ABCD$ is a parallelogram.

32.

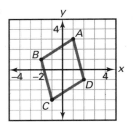

Parallelogram; Sides $\overline{AB}$ and $\overline{CD}$ have slope $\frac{2}{3}$, hence are parallel. Sides $\overline{AD}$ and $\overline{BC}$ have slope -4, hence are parallel. Since it has two pairs of parallel sides, $ABCD$ is a parallelogram.

33.

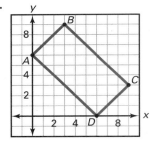

Rectangle; Sides $\overline{AB}$ and $\overline{CD}$ have slope 1, hence are parallel. Sides $\overline{BC}$ and $\overline{AD}$ have slope -1, hence are parallel. There are two sets of parallel sides, and the adjacent sides have slopes whose product is -1, therefore the adjacent sides are perpendicular. Thus, $ABCD$ is a rectangle.

34.

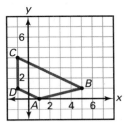

Trapezoid; Sides $\overline{AD}$ and $\overline{BC}$ have slope $\frac{-1}{2}$, hence are parallel. $\overline{AB}$ has slope $\frac{1}{4}$ and $\overline{CD}$ has undefined slope; hence these sides are not parallel. Since there is exactly one pair of parallel sides, $ABCD$ is a trapezoid.

35.

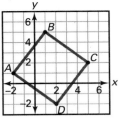

The slope of diagonal $\overline{AC} = \frac{(1-2)}{(-2-5)} = \frac{1}{7}$, the slope of diagonal $\overline{BD} = \frac{(5-(-2))}{(1-2)} = -7$.

$m_{\overline{AC}} \cdot m_{\overline{BD}} = \frac{1}{7}(-7) = -1$

Since the product of the two slopes is -1, the diagonals are perpendicular.

36.

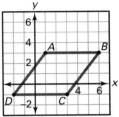

The slope of the diagonal $\overline{AC} = \frac{-1-3}{3-1} = \frac{-4}{2}$ $= -2$ and the slope of the diagonal

$\overline{BD} = \frac{-1-3}{-2-6} = \frac{-4}{-8} = \frac{1}{2}.$

$m_{\overline{AC}} \cdot m_{\overline{BD}} = -2 \cdot \frac{1}{2} = -1$

Since the product of the slopes is -1, the diagonals are perpendicular.

37.

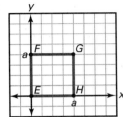

The slope of the diagonal $\overline{EG} = \frac{a-0}{a-0} = \frac{a}{a} = 1$, and the slope of the diagonal $\overline{FH} = \frac{0-a}{a-0} = \frac{-a}{a}$ $= -1.$

$m_{\overline{EG}} \cdot m_{\overline{FH}} = 1(-1) = -1$

Since the product of the slopes is -1, the diagonals are perpendicular.

38. Sample answers: $(4, 3)$ or $(-4, -3)$.

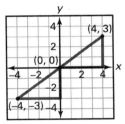

39. Sample answers: $(7, 1)$ or $(4, -5)$.

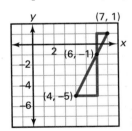

40. Sample answers: $(-4, 9)$ or $(2, -1)$.

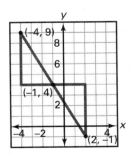

41. Sample answers: $(3, 4)$ or $(5, 2)$.

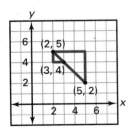

42. $\overline{AB}$: $m = \dfrac{0-8}{2-0} = \dfrac{-8}{2} = -4$, $MP = \left(\dfrac{0+2}{2}, \dfrac{8+0}{2}\right) = (1,4)$

$\overline{BC}$: $m = \dfrac{4-0}{3-2} = 4$, $MP = \left(\dfrac{2+3}{2}, \dfrac{0+4}{2}\right) = (2.5, 2)$

$\overline{AC}$: $m = \dfrac{4-8}{3-0} = \dfrac{-4}{3}$, $MP = \left(\dfrac{0+3}{2}, \dfrac{8+4}{2}\right) = (1.5, 6)$

Let $E = (1,4)$, $F = (2.5, 2)$, $G = (1.5, 6)$ denote the midsegment endpoints.

Slope of $\overline{EF} = \dfrac{2-4}{2.5-1} = \dfrac{-2}{1.5} = \dfrac{-4}{3} =$ slope of $\overline{AC}$

Slope of $\overline{EG} = \dfrac{6-4}{1.5-1} = \dfrac{2}{.5} = \dfrac{4}{1} = 4 =$ slope of $\overline{BC}$

Slope of $\overline{FG} = \dfrac{6-2}{1.5-2.5} = \dfrac{4}{-1} = -4 =$ slope of $\overline{AB}$

Thus each midsegment is parallel to a side of the triangle.

43. $\overline{KL}$ and $\overline{MN}$ are vertical segments; hence, they are parallel and the bases of the trapezoid.

Midpoint of $\overline{KM} = \left(\dfrac{0+4}{2}, \dfrac{0+0}{2}\right) = (2,0)$;

midpoint of $\overline{LN} = \left(\dfrac{0+4}{2}, \dfrac{7+9}{2}\right) = (2,8)$.

Thus the midsegment is vertical, so it is parallel to the bases.

44. The slopes of $\overline{AB}$ and $\overline{CD}$ must be equal. Hence

$$\dfrac{(x+1)-(x-1)}{x-(-1)} = \dfrac{-1-1}{(x-2)-3}$$

$$\dfrac{2}{x+1} = \dfrac{-2}{x-5}$$

$$2(x-5) = -2(x+1)$$

$$2x - 10 = -2x - 2$$

$$4x = 8$$

$$x = 2$$

In exercises 45–49 the figures will vary. Sample graphs are shown:

45.

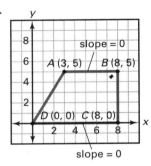

46.

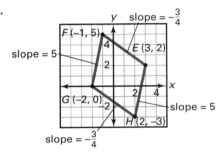

47.

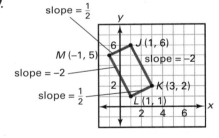

48.

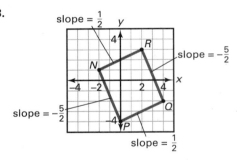

49. Denote the midpoints by $A = (5, 5)$, $B = (10, 5)$, $C = (7, 1)$, $D = (2, 1)$

$\overline{AC}$ is parallel to $\overline{TU}$ and $\overline{VS}$, and has slope $\frac{1-5}{7-5} = \frac{-4}{2} = -2$. So $\overline{TU}$ and $\overline{VS}$ have

slopes -2. Since $STUV$ is a rectangle, the other two sides $\overline{ST}$ and $\overline{UV}$ are

perpendicular to the first two and hence must have opposite reciprocal slope $\frac{1}{2}$. To

find the corners, draw right triangles of slope $\frac{1}{2}$ from midpoints A and C and of

slope -2 from midpoints B and D.

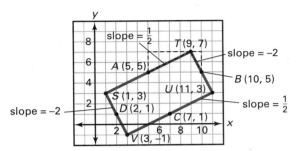

50. Let r be the run.

$$\frac{30}{r} = \frac{1}{12}$$
$$30 \cdot 12 = r \cdot 1$$
$$r = 360$$

The minimum run is 360 inches or 30 feet.

$$\frac{30}{r} = \frac{1}{20}$$
$$30 \cdot 20 = r \cdot 1$$
$$r = 600$$

The maximum run is 600 inches or 50 feet.

51. The roof rises $23.0 - 10.5 = 12.5$ ft over a run of

$\frac{25}{2} = 12.5$ ft. The slope or pitch is then $\frac{12.5}{12.5} = 1.0$.

The house violates the building codes.

Sample answer: Adjust the roof so that $\frac{\text{rise}}{12.5} = .7$

$\text{rise} = .7(12.5) = 8.75$

Adjust the peak of the roof to $(12.5, 19.25)$.

PAGE 196, LOOK BACK

52. A postulate is accepted without proof. A theorem must be proven using definitions, postulates, or previous theorems.

54. $\angle 2$ and the given $145°$ angle form a linear pair, hence are supplementary.
Thus, $m\angle 2 = 180° - 145° = 35°$.

56. The interior angles of a triangle add to $180°$.
Thus, $m\angle 1 = 180° - 20° - 35° = 125°$.

53. $\angle 3$ and the given $20°$ angle are alternate interior angles of two parallel lines, hence are congruent.
Thus, $m\angle 3 = 20°$.

55. $\angle 4$ and $\angle 2$ are alternate interior angles of two parallel lines, hence are congruent.
Thus, $m\angle 4 = 35°$.

57. Let x be the unknown angle. The sum of the measures of the interior angles of a pentagon is $540°$.
$$90° + 90° + x° + x° + 144° = 540°$$
$$2x + 324 = 540$$
$$2x = 216$$
$$x = 108°$$

58. Since 1 palm = 4 fingers and 1 cubit = 7 palms, 1 cubit = 28 fingers. Multiply cubits by 28 and palms by 4 to find equivalent amounts in fingers.

x	y
0	98
$28 = 1 \cdot 28$	$95 = 3 \cdot 28 + 2 \cdot 4 + 3$
$56 = 2 \cdot 28$	$84 = 3 \cdot 28$
$84 = 3 \cdot 28$	$68 = 2 \cdot 28 + 3 \cdot 4$
$112 = 4 \cdot 28$	$41 = 1 \cdot 28 + 3 \cdot 4 + 1$
$140 = 5 \cdot 28$	0

59.

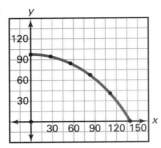

CHAPTER REVIEW AND ASSESSMENT

1.

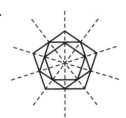

2. This figure has 5-fold rotational symmetry. The image will coincide with the original figure after rotations of 72°, 144°, 216°, and 288°.

3.

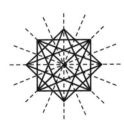

4. This figure has 8-fold rotational symmetry. The image will coincide with the original figure after rotations of 45°, 90°, 135°, 180°, 225°, 270°, and 315°.

5. Opposite angles of a parallelogram are congruent. Thus $\angle ADC = 48°$.

6. Diagonals of a parallelogram bisect each other. Thus $AC = 2(AX) = 2(7) = 14$.

7. Diagonals of a rhombus are perpendicular. Thus $\angle EYF = 90°$.

8. Diagonals of a rectangle are congruent. Thus $KM = JL = 13$.

9. Sample answer: $\angle 2$ and $\angle 4$

10. Sample answer: $\angle 3$ and $\angle 6$

11. Because $\angle 1$ and $\angle 6$ are vertical angles, $\angle 1$ and $\angle 3$ are corresponding angles, $\angle 1$ and $\angle 8$ are alternate exterior angles, $\angle 1$ is congruent to $\angle 6$, $\angle 3$, and $\angle 8$.

12. Because $\angle 1$ and $\angle 2$ form a linear pair, $m\angle 2 = 180° - m\angle 1 = 180° - 130° = 50°$. Because $\angle 2 \cong \angle 5 \cong \angle 4 \cong \angle 7$, $m\angle 2 = m\angle 5 = m\angle 4 = m\angle 7 = 50°$ and $m\angle 1 = m\angle 6 = m\angle 3 = m\angle 8 = 130°$.

13. Yes; because there are two 115° corresponding angles, by the Converse of the Corresponding Angles Postulate they are parallel.

14. Yes; because there are two 115° alternate exterior angles, by the Converse of the Alternate Exterior Angles Theorem they are parallel.

15. No; because if p and q were parallel, then same-side exterior angles would be supplementary, but $115° + 70° = 185° \neq 180°$.

16. Given: $ABCD$ is a rectangle
Prove: $\overline{AB} \parallel \overline{CD}$ and $\overline{BC} \parallel \overline{AD}$

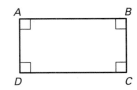

Statement	Reasons
$\overline{AB}$ and $\overline{CD}$ are perpendicular to $\overline{BC}$	Definition of rectangle
$\overline{AB} \parallel \overline{CD}$	Converse of Same-Side Interior Angles Theorem
$\overline{BC}$ and $\overline{AD}$ are perpendicular to $\overline{AB}$	Definition of rectangle
$\overline{BC}$ and $\overline{AD}$ are perpendicular to $\overline{AB}$	Converse of Same-Side Interior Angles Theorem

17. Vertical angles are congruent.
Thus $m\angle STR = m\angle PTQ = 125°$.
$m\angle RST = m\angle RSQ = 30°$
By the Triangle Sum Theorem,
$m\angle TRS = 180° - m\angle STR - m\angle RST$
$= 180° - 125° - 30° = 25°$.

18. $m\angle PST = m\angle PSQ = 83°$
Because $\angle PTS$ and $\angle PTQ$ are a linear pair,
$m\angle PTS = 180° - m\angle PTQ = 180° - 125° = 55°$.
By the Triangle Sum Theorem,
$m\angle SPT = 180° - m\angle PTS - m\angle PST$
$= 180° - 55° - 83° = 42°$.

19. Given: $m\angle PQR = 57°$ and
$m\angle QRP = m\angle PRQ = 90°$.
By the Triangle Sum Theorem,
$m\angle QPR = 180° - m\angle PQR - m\angle QRP$
$= 180° - 57° - 90° = 33°$.

20. Given: $m\angle PTQ = 125°$ and
$m\angle QPT = m\angle QRP = 33°$ (see Exercise 19).
By the Triangle Sum Theorem,
$m\angle PQT = 180° - m\angle QPT - m\angle PTQ$
$= 180° - 33° - 125° = 22°$.

21. Two exterior angles can be found by their corresponding interior angles with which they are supplementary. Interior angle $130°$ has a $50°$ exterior angle and interior angle $60°$ has a $120°$ exterior angle. Thus 5 of the 6 exterior angles have measures $80°$, $45°$, $50°$, $120°$, $30°$. All exterior angle measures must sum to $360°$, so the remaining exterior angle measures
$360° - 80° - 45° - 50° - 120° - 30° = 35°$.

22. The interior angle measures of a regular pentagon sum to $540°$, so each must be $108°$.

23. The exterior angle measures must sum to $360°$, so each of those is $\left(\frac{360}{17}\right)°$. Hence each interior angle is $180° - \left(\frac{360}{17}\right)° = \left(\frac{2700}{17}\right)° \approx 158.8°$.

24. The exterior angle measures must sum to $360°$, so each of those is $\left(\frac{360}{12}\right)° = 30°$.

25. midsegment length $= \frac{1}{2}(\text{base}) = \frac{1}{2}(12) = 6$

26. $45 = \frac{1}{2}(\text{base}) \Rightarrow \text{base} = 90$

27. $RS = \frac{1}{2}(KL + NM) = \frac{1}{2}(30 + 50) = 40$

28. $TU = \frac{1}{2}(RS + NM) = \frac{1}{2}(40 + 50) = 45$

29. Yes; $\overline{FG}$ has slope $\frac{0-1}{2-1} = \frac{-1}{1} = -1$,
$\overline{FH}$ has slope $\frac{-1-1}{-1-1} = \frac{-2}{-2} = 1$,
$\overline{GH}$ has slope $\frac{-1-0}{-1-2} = \frac{-1}{-3} = \frac{1}{3}$. Since the slopes of $\overline{FG}$ and $\overline{FH}$ multiply to -1, they are perpendicular.

30. Name the ordered pairs $A(0, 2)$, $B(2, 3)$, $C(5, 2)$, $D(3, 1)$.
slope of $\overline{AB} = \frac{1}{2}$, slope of $\overline{BC} = \frac{-1}{3}$,
slope of $\overline{CD} = \frac{1}{2}$, slope of $\overline{AD} = \frac{-1}{3}$.
Opposite sides have same slope, so they are parallel.
b. parallelogram.

31. Name the ordered pairs $A(2, 1)$, $B(1, 3)$, $C(5, 5)$, $D(6, 3)$. Slope of $\overline{AB} = -2$, slope of $\overline{BC} = \frac{1}{2}$, slope of $\overline{CD} = -2$, slope of $\overline{AD} = \frac{1}{2}$. Opposite sides have same slope, so are they parallel. Also adjacent sides have opposite reciprocal slopes, so they are perpendicular.
c. rectangle

32. Name the ordered pairs $A(0, 1)$, $B(3, 3)$, $C(7, 3)$, $D(1, -1)$. Slope of $\overline{AB} = \frac{2}{3}$, slope of $\overline{BC} = 0$ (horizontal), slope of $\overline{CD} = \frac{2}{3}$, slope of $\overline{AD} = -2$. Only $\overline{AB}$ and $\overline{CD}$ have same slope, so they are the only pair of parallel sides.
a. trapezoid

33. The quilt has 2 axes of symmetry—2 lines through the center along the diagonals. It has nontrivial rotational symmetry for a rotation of 180°.

34. If the frame is a rectangle, then its diagonals will be congruent.

35. First section: slope $= \frac{10}{35} \approx 28.6\%$ grade.
Second section: slope $\frac{10}{25} = 40\%$ grade.

CHAPTER TEST

1.

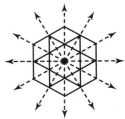

2. The figure has 6-fold rotational symmetry. The image will coincide with the original figure after rotations of 60°, 120°, 180°, 240°, 300°, and 360°.

3. The leaf changes the square to a nonsquare rectangle, which is not regular, not equilateral, but is equiangular.

4. $BC = 38$ because a rhombus has four congruent sides.

5. $ED = \frac{1}{2}(43) = 21.5$ because the diagonals of a parallelogram bisect each other.

6. $m\angle DEC = 90°$ because the diagonals of a rhombus are perpendicular.

7. $m\angle DAB = 75°$ because opposite angles of a parallelogram are congruent.

8. $\angle ADC = 180° - 75° = 105°$ because consecutive angles of a parallelogram are supplementary.

9. $\angle AEB = 90°$ because the diagonals of a rhombus are perpendicular.

10. 25 inches because diagonals of a rectangle are congruent.

11. $\angle 1$ and $\angle 8$ or $\angle 2$ and $\angle 7$

12. $\angle 3$ (vertical angles), $\angle 6$ (corresponding angles), and $\angle 7$ (vertical angles and corresponding angles)

13. $m\angle 1 = 60°$, $m\angle 4 = 60°$, $m\angle 5 = 60°$, $m\angle 8 = 60°$, $m\angle 2 = 180° - 60° = 120°$, $m\angle 3 = 180° - 60° = 120°$, $m\angle 6 = 180° - 60° = 120°$, and $m\angle 7 = 180° - 60° = 120°$

14. transversal

15. Both are right angles.

16. m

17. n

18. Converse of Corresponding Angles Postulate

19. $m\angle R = 50°$ (Corresponding Angles Postulate)

20. $m\angle T = 180° - 50° - 75° = 60°$ (Triangle Sum Theorem)

21. $m\angle SUT = 70°$ (Corresponding Angles Postulate)

22. $m\angle QPU = 50°$ (Alternate Interior Angles Theorem)

23. $m\angle PQU = 60°$ (Alternate Interior Angles Theorem with $\angle T$)

24. $m\angle PUQ = 180° - 50° - 60° = 70°$ (Triangle Sum Theorem)

25. The measure of an interior angle of a square is $180° - \frac{360°}{4} = 90°$. The measure of an exterior angle of a square is $\frac{360°}{4} = 90°$.

26. The measure of an interior angle of a regular octagon is $180° - \frac{360°}{8} = 135°$. The measure of an exterior angle of a regular octagon is $\frac{360°}{8} = 45°$.

For Items 27–29, $\overline{OP}$ is the midsegment between bases $\overline{JK}$ and vertex L, where the length of $L = 0$.

27. $OP = \frac{1}{2}(68 + 0) = 34$

28. $QR = \frac{1}{2}(34 + 0) = 17$

29. $MN = \frac{1}{2}(68 + 34) = 51$

30. Check students' drawings.
Figure ABC is a right triangle. The slope of $\overline{AC}$ is $\frac{2-5}{-1+3} = -\frac{3}{2}$ and the slope of $\overline{BC}$ is $\frac{2-4}{-1-2} = \frac{2}{3}$. Therefore, $\overline{AC}$ is perpendicular to $\overline{BC}$ because the product of their slopes is -1.

CHAPTERS 1–3 CUMULATIVE ASSESSMENT

1. A
$m\angle 1 = \frac{180°(5-2)}{5} = 108°$; $m\angle 2 = \frac{360°}{5} = 72°$

2. D
$\angle 1$ and $\angle 2$ are alternate interior angles, but we need to know whether the two apparently horizontal lines are, in fact, parallel.

3. D
The Triangle Sum Theorem states that $m\angle A + m\angle B + m\angle C = 180°$, but $\angle A$ could be acute, right or obtuse.

4. point; line

5. lies in the same plane

6. perpendicular line segment

7. circumscribed circle

8. $T(x, y) = (x - 3, y - 3)$ because this is a vertical and horizontal translation.

9. $T(x, y) = (-x, y)$ because this is a reflection across the y-axis.

10. $T(x, y) = (x, -y)$ because this is a reflection across the x-axis.

11. $T(x, y) = (-x, -y)$ because this is 180° rotation about the origin.

12. $\angle HIG$; because $\overline{HJ} \parallel \overline{FG}$ (since $\overline{HJ}$ is the midsegment of trapezoid $DEFG$) and $\angle HIG$ is an alternate interior angle to $\angle EGF$.
$\angle EIJ$; because $\angle EIJ \cong \angle HIG$ (since they are vertical angles) and $\angle HIG \cong \angle EGF$.
$\angle DEG$; because $\overline{DE} \parallel \overline{FG}$ (given) and $\angle DEG$ is an alternate interior angle to $\angle EGF$.

13. $IJ:FG = 1:2$ because $IJ = \frac{1}{2}(FG)$ since $\overline{IJ}$ is the midsegment of $\triangle GEF$.

14. Because it is the midsegment of trapezoid $DEFG$,
$HJ = \frac{1}{2}(\text{base 1} + \text{base 2}) = \frac{1}{2}(3 + 8) = \frac{11}{2} = 5.5$
Because IJ is the midsegment of $\triangle GEF$,
$HI = HJ - IJ$
$= 5.5 - \frac{1}{2}(FG)$
$= 5.5 - \frac{1}{2}(8)$
$= 5.5 - 4 = 1.5$

15. $\angle GFE = m\angle EJH = 42°$ because the two angles are alternate interior angles.
$m\angle HJF = 180° - m\angle EJH = 180° - 42° = 138°$ because the angles form a linear pair.
$m\angle GIJ = 180° - m\angle EIJ = 180° - 34° = 146°$ because the two angles form a linear pair.
$m\angle EIJ = m\angle EGF = 34°$.
$m\angle GEJ = 180° - m\angle EIJ - m\angle EJI$
$\quad\quad\quad = 180° - 34° - 42° = 104°$
by the Triangle Sum Theorem.

16. $m\angle HIE = m\angle GIJ = 146°$ because they are vertical angles.
$m\angle GHJ = 180° - m\angle EGD - m\angle HIG$ by the Triangle Sum Theorem. $m\angle HIG = m\angle EGF = 34°$ because they are corresponding angles. Thus, $m\angle GHJ = 180° - 30° - 34° = 116°$.
$m\angle GDE = m\angle GHJ = 116°$ because the two angles are corresponding angles.
$\angle DHI = 180° - m\angle GDE = 180° - 116° = 64°$ because the two angles are same-side interior angles.

17. $90° - 25° = 65°$

18. $180° - 75° = 105°$

19. $x + x = 90°$
$2x = 90°$
$x = 45°$

20. $x + x = 180°$
$2x = 180°$
$x = 90°$

CHAPTER 4

Triangle Congruence

7. QPTSR   QRSTP
PTSRQ   RSTPQ
TSRQP   STPQR
SRQPT   TPQRS
RQPTS   PQRST

8. Sample answer:
$PQRST \cong VZYXW$

9. It is given that $\angle A \cong \angle D$, $\angle ABC \cong \angle DBC$, $\angle ACB \cong \angle DCB$, $\overline{AB} \cong \overline{DB}$, and $\overline{AC} \cong \overline{DC}$. By the Reflexive Property of Congruence, $\overline{BC} \cong \overline{BC}$, so $\triangle ABC \cong \triangle DBC$ by the Polygon Congruence Postulate.

PAGES 214–216, PRACTICE AND APPLY

10. Sample answer:
QUTSR, RSTUQ, UTSRQ

11. $\angle QRS$ $\angle TUQ$
$\angle RST$ $\angle UQR$
$\angle STU$

12. $\angle QRS \cong \angle VWX$ $\angle TUQ \cong \angle YZV$
$\angle RST \cong \angle WXY$ $\angle UQR \cong \angle ZVW$
$\angle STU \cong \angle XYZ$

13. a. $\overline{XY}$
    b. $\overline{RS}$
    c. $\overline{VZ}$

14. No; the segments have different lengths.

15. Yes; the segments have the same length.

16. No; the sides have different lengths.

17. Yes; all corresponding sides and angles are congruent.

18. Yes; all sides have the same length and all angles are right angles.

19. Yes; the measure of the two angles is equal.

20. a. $\angle E$
    b. $\angle D$
    c. $\angle F$

21. a. $\overline{ED}$
    b. $\overline{DF}$
    c. $\overline{EF}$

22. a. $\angle NMP$
    b. $\angle MNO$
    c. $\angle LQP$
    d. $\angle MPQ$

23. a. $\overline{OP}$
    b. $\overline{MP}$
    c. $\overline{LM}$
    d. $\overline{MP}$

24. a. $\angle F$
    b. $\angle J$
    c. $\angle H$
    d. $\angle B$
    e. $\angle D$

25. a. $\overline{FG}$
    b. $\overline{DE}$
    c. $\overline{CD}$
    d. $\overline{GH}$
    e. $\overline{BC}$

26. $LQ = LR + RQ = 5 + 3 = 8$

27. $RP = RQ + QP = 3 + 5 = 8$

28. The triangles are congruent. It is given that each of the angles are congruent and also two pairs of sides are congruent. From Exercise 26 and 27, the remaining sides $\overline{LQ}$ and $\overline{PR}$ have the property $LQ = PR$; thus, $\overline{LQ} \cong \overline{PR}$.

29. $m\angle DAB = 180 - (x + x) = (180 - 2x)°$
$m\angle DCB = 180 - (x + x) = (180 - 2x)°$

30.

Statements	Reasons
$\angle ADB \cong \angle ABD \cong \angle CDB \cong \angle CBD$ ($m\angle ADB = m\angle ABD = m\angle CDB = m\angle CBD$)	Given
$m\angle ADB + m\angle ABD + m\angle DAB = 180°$ $m\angle CDB + m\angle CBD + m\angle DCB = 180°$	Triangle Sum Theorem
$m\angle ADB + m\angle ABD + m\angle DAB =$ $m\angle ADB + m\angle ABD + m\angle DCB$	Substitution Property
$m\angle DAB = m\angle DCB$	Subtraction Property
$\overline{DB} \cong \overline{DB}$	Reflexive Propety of Congruence
$\overline{AB} \cong \overline{BC} \cong \overline{CD} \cong \overline{DA}$	Definition of rhombus
$\triangle ABD \cong \triangle CBD$	Polygon Congruence Postulate

31. $\triangle XYZ \cong \triangle JKL$

32. Transitive Property of Congruence

33. $\triangle ABC$ is isosceles; $\overline{AB}$ and $\overline{AC}$ are corresponding sides, so $\overline{AB} \cong \overline{AC}$.

34. $\triangle ABC$ is equilateral; $\overline{AB}$ and $\overline{BC}$ are corresponding sides, and $\overline{BC}$ and $\overline{CA}$ are corresponding sides, so $\overline{AB} \cong \overline{BC} \cong \overline{CA}$.

35. a. $\triangle RNO \cong \triangle AKC$
$FRNT \cong BAKX$
$FRAB \cong TNKX$
$FBXT \cong RAKN$
$RACO \cong NKCO$
$FRONT \cong BACKX$

b. Sample answer:
$\angle ORN \cong \angle CAK$
$\angle ONR \cong \angle CKA$
$\angle RON \cong \angle ACK$
$\angle RFT \cong \angle ABX$
$\angle RFB \cong \angle NTX$

c. Sample answer:
$\overline{FR} \cong \overline{TN}$
$\overline{BA} \cong \overline{XK}$
$\overline{RA} \cong \overline{NK}$
$\overline{FB} \cong \overline{TX}$
$\overline{FT} \cong \overline{BX}$

36. 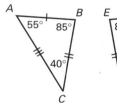 The quilt contains congruent isosceles right triangles (2 sizes, large and small), parallelograms, and squares.

37. Sample answer:
$\triangle ABC \cong \triangle DEF$

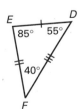

38. Yes; Transitive Property of Congruence

PAGE 216, LOOK BACK

39. definition of a parallelogram

40. definition of a parallelogram

41. Alternate Interior Angles Theorem

42. Alternate Interior Angles Theorem

43. Reflexive Property of Congruence

44. True; every square is a quadrilateral with four sides of equal length.

45. False; parallelograms have 2 pairs of parallel sides but trapezoids only have 1 pair of parallel sides.

46. False; a parallelogram is not required to have sides of equal length.

47. True; every rectangle is a quadrilateral with two pairs of parallel sides.

PAGE 216, LOOK BEYOND

48.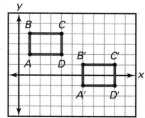

The image is congruent to the preimage because the corresponding sides and angles are congruent.

49.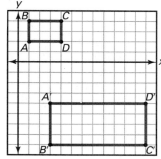

The image is not congruent to the preimage because the corresponding sides are not congruent.

4.2 PAGE 221, GUIDED SKILLS PRACTICE

5. SAS

6. SSS

7. ASA

PAGES 221–223, PRACTICE AND APPLY

8. Can't be proven congruent because only 3 angles are given to be congruent.

9. △FDR ≅ △WXY by ASA

10. △LMN ≅ △PON by SAS, since ∠MNL ≅ ∠ONP because vertical angles are congruent.

11. △FGH ≅ △FDH by ASA because $\overline{FH} \cong \overline{HF}$ by the Reflexive Property of Congruence.

12. △PRT ≅ △PSQ by ASA, since ∠RTP ≅ ∠PQS because they are alternate interior angles and $\overline{TR} \parallel \overline{SQ}$ and ∠TPR ≅ ∠QPS because vertical angles are congruent.

13. Can't be proven congruent because only two sides and a nonincluded angle are given to be congruent.

14. Yes; SSS

15. Yes; SAS

16. No, many triangles can be constructed.

17. Yes; ASA

18. Yes; ASA because m∠O can be determined from m∠M and m∠N.

19. No, more than one triangle can be constructed.

20.

Any segment that connects this endpoint to a point on the circle is congruent to the second segment, because the compass setting used to create the circle was the length of the second segment.

21.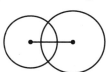

Any segment that connects this endpoint to a point on the circle is congruent to the third segment, because the compass setting used to create the circle was the length of the third segment.

22.

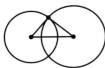

The figure is a triangle with side lengths equal to the original three line segments.

23. Converse of the Same-Side Interior Angles Theorem

24. Alternate Interior Angles Theorem

25. Converse of the Same-Side Interior Angles Theorem

26. Alternate Interior Angles Theorem

27. Reflexive Property of Congruence

28. $\triangle CBD$

29. ASA

30. By definition of a regular polygon, all the sides of the polygon are congruent. If a segment is drawn from the center to each vertex, the polygon is divided into triangles. By definition of the center of a regular polygon, the segments from the center to the vertices are all congruent, so the triangles are all congruent by SSS. Thus, the central angles are congruent because CPCTC.

31. Given: Rhombus $ABCD$

Prove: Diagonal $\overline{DB}$ divides rhombus $ABCD$ into two congruent triangles.

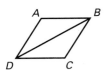

Statements	Reasons
$\overline{AB} \cong \overline{CD}$ $\overline{AD} \cong \overline{BC}$	Definition of a rhombus
$\overline{BD} \cong \overline{DB}$	Reflexive Property of Congruence
$\triangle ABD \cong \triangle CDB$	SSS

32. Sometimes; two congruent triangles which share a common side will either form a quadrilateral or a triangle. If they form a quadrilateral, they will form a parallelogram or a kite.

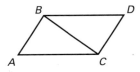 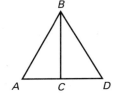 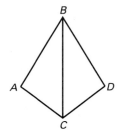

33. The diagonal board turns the rectangle into two congruent, rigid triangles.

34. Check if the triangles are congruent by measuring the 3 sides of each triangle. By SSS, the triangles are congruent if the lengths of the corresponding sides are the same.

35. a. Yes. Because vertical angles are congruent, $\angle FOA \cong \angle COD$. It is given that $\overline{FO} \cong \overline{CO}$ and $\overline{OA} \cong \overline{OD}$. By SAS, $\triangle FOA \cong \triangle COD$.

b. Yes; Because it is a regular hexagon, $\overline{AB} \cong \overline{DC}$. It is given that $\overline{BO} \cong \overline{CO}$ and $\overline{OA} \cong \overline{OD}$. By SSS, $\triangle BOA \cong \triangle COD$.

c. $\triangle FOE \cong \triangle COB$ by either SAS or SSS.

36. Yes, they are congruent by SAS.

37. Yes, they are congruent by SSS.

38. No, AAA does not prove congruence.

39. Yes, they are congruent by ASA.

40. Given

41. Given

42. Corresponding Angles Postulate

43. Transitive Property of Congruence

44. Converse of the Corresponding Angles Postulate

45. $\angle X$ **46.** $\overline{XY}$ **47.** Possible

PAGE 224, LOOK BEYOND

48. The quadrilateral is not rigid. Here are two possible shapes.

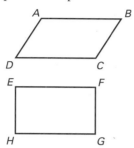

49. When the congruent segments are opposite each other, it appears that a parallelogram is formed. When the congruent segments are adjacent, it appears that a kite is formed.

4.3   **PAGE 230, GUIDED SKILLS PRACTICE**

6. No. AAA is not a valid test for congruence.

7. Yes. AAS is a valid test for congruence.

8. No. SSA is not a valid test for congruence.

9.

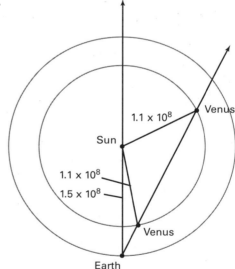

Set up a proportion with measured distances on scaled drawing and actual distances to estimate the distances.

$$\frac{1 \text{ cm}}{3 \text{ cm}} = \frac{x}{1.5 \times 10^8 \text{ km}}$$

$x = 0.5 \times 10^8 \text{ km}$

$$\frac{4.5 \text{ cm}}{3 \text{ cm}} = \frac{x}{1.5 \times 10^8 \text{ km}}$$

$x = 2.25 \times 10^8 \text{ km} \approx 2 \times 10^8 \text{ km}$

The distance from one position is 0.5×10^8 km; the distance from the other position is 2×10^8 km.

10. $\triangle ABC \cong \triangle YXZ$; ASA

11. $\triangle DEF \cong \triangle FGD$; SSS

12. $\triangle HJK \cong \triangle MNL$; AAS

13. $\triangle PQR \cong \triangle TSU$; SAS

14. Can't be proven congruent

15. Can't be proven congruent

16. $\triangle ABE \cong \triangle DCF$; AAS

17. $\triangle WXY \cong \triangle WZY$; ASA

18. Can't be proven congruent

19. Can't be proven congruent

20.

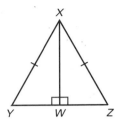

21. $\triangle XWY$ has hypotenuse $\overline{XY}$;
$\triangle XWZ$ has hypotenuse $\overline{XZ}$.

22. The hypotenuses are congruent because it is given that $\overline{XY} \cong \overline{XZ}$.

23. Reflexive Property of Congruence

24. $\triangle WXY \cong \triangle WXZ$ by HL

25.

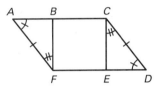

Statements	Reasons
$\angle A \cong \angle D$ $\overline{AF} \cong \overline{DC}$ $\angle BFA \cong \angle ECD$	Given
$\triangle AFB \cong \angle DCE$	ASA

26.

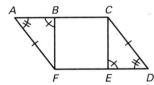

Statements	Reasons
$\angle 1 \cong \angle 4$ $\overline{AF} \cong \overline{DC}$ $\angle A \cong \angle D$	Given
$\triangle AFB \cong \angle DCE$	AAS

27. AAS

28. SAS

29. Transitive Property of Congruence

30. Yes; ASA

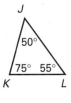

31. No

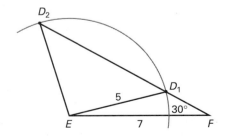

32. No, AAA does not determine a unique triangle.

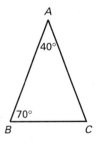

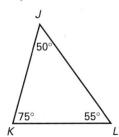

33. Yes; HL

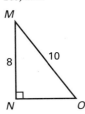

34. Yes; AAS

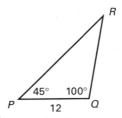

35.

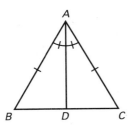

a. △ABD ≅ △ACD by SAS.

b. ∠ADB ≅ ∠ADC because CPCTC. Because they form a linear pair and are congruent, ∠ADB and ∠ADC are right angles, so m∠ADB = m∠ADC = 90°.

36. Given: ∠ABC is equilateral.

$\overline{AD}$ bisects ∠A.

$\overline{BE}$ bisects ∠B.

$\overline{CF}$ bisects ∠C.

Prove: △AFX ≅ △BFX ≅ △BDX ≅ △CDX ≅ △CEX ≅ △AEX

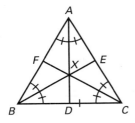

Because △ABC is equilateral, ∠A ≅ ∠B ≅ ∠C. Because ∠XAE, ∠XAF, ∠XBF, ∠XBD, ∠XCD, and ∠XCE are formed by bisecting the congruent vertices of △ABC, they are congruent. From Exercise 35, we know that each angle bisector forms 2 congruent triangles. By CPCTC, we know $\overline{BD} ≅ \overline{DC}$, $\overline{CE} ≅ \overline{EA}$, and $\overline{AF} ≅ \overline{FB}$. Thus, each side of the equilateral triangle has been bisected so $\overline{BD} ≅ \overline{DC} ≅ \overline{CE} ≅ \overline{EA} ≅ \overline{AF} ≅ \overline{FB}$. From Exercise 35, we also know that the bisector forms two right angles with the side opposite the vertex. Thus, ∠XEA ≅ ∠XEC ≅ ∠XDC ≅ ∠XDB ≅ ∠XFB ≅ XFA. By ASA, △AFX ≅ △BFX ≅ △BDX ≅ △CDX ≅ △CEX ≅ △AEX.

37. Given: $\angle B \cong \angle E$
$\qquad \angle A \cong \angle D$
$\qquad \overline{BC} \cong \overline{EF}$

Prove: $\triangle ABC \cong \triangle DEF$ (AAS is valid.)

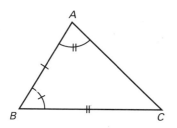 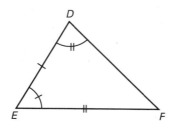

Statements	Reasons
$\angle B \cong \angle E$ (m$\angle B$ = m$\angle E$)	Given
$\angle A \cong \angle D$ (m$\angle A$ = m$\angle D$)	
$\overline{BC} \cong \overline{EF}$	
m$\angle F = 180° -$ m$\angle E -$ m$\angle D$	Triangle Sum Theorem
m$\angle C = 180° -$ m$\angle B -$ m$\angle A$	
m$\angle C = 180° -$ m$\angle E -$ m$\angle D$	Substitution Property
m$\angle C =$ m$\angle F$ ($\angle C \cong \angle F$)	Substitution Property
$\angle C \cong \angle F$	Angle Congruence Postulate
$\triangle ABC \cong \triangle DEF$	ASA

38. a.

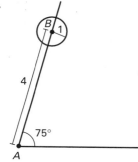

b.

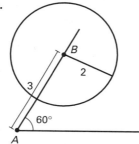

c.

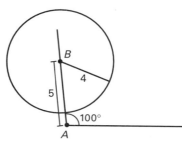

None.   None   None

Reason: Let A be the origin and assume C lies on the positive x-axis. Place point B (AB) units from A on the ray which forms $\angle A$. Make a circle of radius (BC) centered at B. Notice that this circle never touches the x-axis, hence the triangle cannot be completed.

39. a.

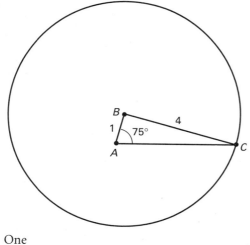

b.

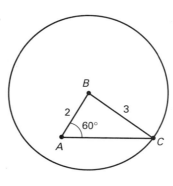

One   One

c.

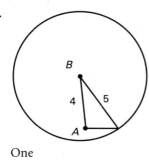

One

d. longer than; tests will vary. Students should construct several more examples like the ones shown in a, b, and c.

40.

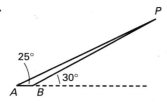

$\triangle ABP$ is unique by ASA.

41. The actual height, which may be computed using trigonometry, is about 461 feet. Student answers will vary due to inaccuracies in the scale drawings. Sample answer:

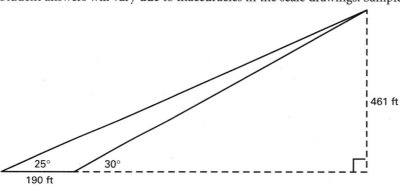

From the scaled drawing, the height of the pyramid measures 4.4 cm.

$$\frac{x \text{ ft}}{4.4 \text{ cm}} = \frac{100 \text{ ft}}{1 \text{ cm}}$$
$$\frac{x}{4.4} = \frac{100}{1}$$
$$x = 440$$

The height of the pyramid is about 440 ft.

42.

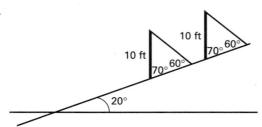

The vertical poles make a 70° angle with the embankment. Assuming the wires are mounted to the poles at the same location, the wires must be the same length by AAS.

43. a.

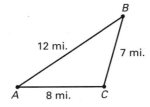

Yes; SSS

b.

Yes; ASA

c.

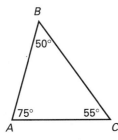

No. AAA is not a valid congruence postulate.

PAGE 234, LOOK BACK

44. Sample answer:

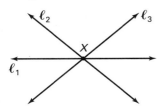

45. Sample answer:

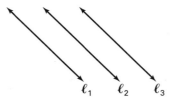

46. Sample answer:

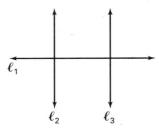

47. $\angle 1 \cong \angle 3 \cong \angle 5 \cong \angle 7$
$\angle 2 \cong \angle 4 \cong \angle 6 \cong \angle 8$

48. Sample answer:
$m\angle 1 + m\angle 4 = 180°$; Same-side exterior angles

49. $2x + 3x + 4x = 180°$
$9x = 180°$
$x = 20°$
The angle mesaures are
$2x = 2 \cdot 20° = 40°$,
$3x = 3 \cdot 20° = 60°$, and
$4x = 4 \cdot 20° = 80°$.

50. If two exterior angles are congruent, then two interior angles are congruent.

51. No; the side lengths would have to be the same.

52. Yes; yes. Each hexagon has three common sides with other hexagons on the soccer ball. Thus, the hexagons must all have the same side length, s. Since each pentagon has five common sides with hexagons on the soccer ball, the pentagons must also have side length s. Thus, all the hexagons are congruent to each other and all the pentagons are congruent to each other.

4.4 **PAGE 239, GUIDED SKILLS PRACTICE**

5. $\triangle ABC \cong \triangle AED$

6. CPCTC

7. Overlapping Segments Theorem

8. HL Congruence Theorem

9. $\angle A$ is supplementary to an angle measuring 50° so $m\angle A = 180° - 50° = 130°$.
By the Triangle Sum Theorem,
$$m\angle B = 180° - m\angle A - m\angle C$$
$$= 180° - 130° - 25°$$
$$= 25°.$$
By the Converse of the Isosceles Triangle Theorem and the definition of an isosceles triangle, $\triangle ABC$ is isosceles with $\overline{AB} \cong \overline{AC}$. Thus, the distance across the river is 100 feet.

PAGES 240–242, PRACTICE AND APPLY

10. $m\angle Z = m\angle Y = 70°$

11. $KL = KM = 23$

12. $QR = RP = 7$

13. Because $\triangle EFG$ is equilateral, $m\angle F = 60°$.

14. $m\angle CBD = 180° - 65° - 90° = 25°$

15. By SAS $\triangle HFJ \cong \triangle GFJ$ and by CPCTC $\overline{GJ} \cong \overline{HJ}$ thus, $GH = GJ + HJ = GJ + GJ = 12 + 12 = 24$.

16.
$$PR = PQ$$
$$x + 1 = \sqrt{x + 3}$$
$$(x + 1)^2 = x + 3$$
$$x^2 + 2x + 1 = x + 3$$
$$x^2 + x - 2 = 0$$
$$(x + 2)(x - 1) = 0$$
$$x + 2 = 0 \quad x - 1 = 0$$
$$x = -2 \quad \text{or} \quad x = 1$$
Because lengths are not negative, $x = 1$.
$$PR = x + 1 = 1 + 1 = 2$$

17.
$$m\angle L = m\angle N$$
$$x = \sqrt{39x + 40}$$
$$x^2 = (\sqrt{39x + 40})^2$$
$$x^2 = 39x + 40$$
$$x^2 - 39x - 40 = 0$$
$$(x + 1)(x - 40) = 0$$
$$x + 1 = 0 \quad x - 40 = 0$$
$$x = -1 \quad x = 40$$
Because angle measures are not negative, $x = 40$.
$$\angle L = x° = 40°$$

18.

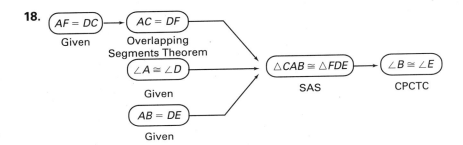

19.

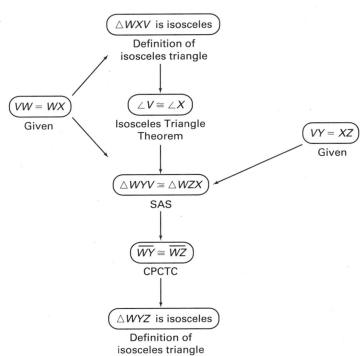

20.

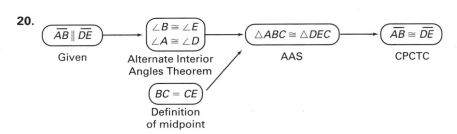

21.

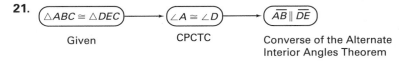

22.

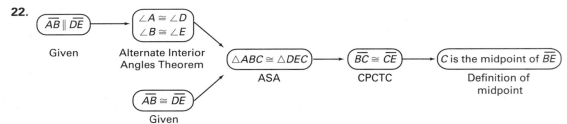

23. Reflexive Property of Congruence

25. AAS

27. Reflexive Property of Congruence

29. CPCTC

31. bisects the side opposite the angle

24. Transitive or Substitution Property

26. CPCTC

28. SAS

30. bisector

32. Let ABC be an equilateral triangle. By the Isosceles Triangle Theorem $\angle B \cong \angle C$ because $\overline{AB} \cong \overline{AC}$ and $\angle A \cong \angle C$ because $\overline{BC} \cong \overline{BA}$. Hence $m\angle A = m\angle C = m\angle B$. By the Triangle Sum Theorem, $m\angle A + m\angle B + m\angle C = 180°$. Using substitution, $m\angle A + m\angle A + m\angle A = 180°$. Thus, $m\angle A = 60°$ so $m\angle B = 60°$ and $m\angle C = 60°$.

33. The bisector of the vertex angle of an isosceles triangle is the perpendicular bisector of the base.

Proof:

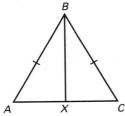

Let $\triangle ABC$ be an isosceles triangle with $\angle A \cong \angle C$ and $\overline{BX}$ be the bisector of $\angle B$. By the proof in Exercises 27–29, $\overline{BX}$ bisects $\overline{AC}$. By the Reflexive Property of Congruence, $\overline{BX} \cong \overline{BX}$, so $\triangle ABX \cong \triangle CBX$ by SAS or SSS. Thus by CPCTC, $\angle AXB \cong \angle CXB$. Since $\angle AXB$ and $\angle CXB$ form a linear pair, $m\angle AXB + m\angle CXB = 180°$. Since $\angle AXB \cong \angle CXB$, $m\angle AXB = m\angle CXB = 90°$. Thus $\overline{BX}$ is the perpendicular bisector of AC.

34. It is given that $\overline{AK} \cong \overline{BK}$. Since $\angle KBC \cong \angle KCB$, $\triangle KBC$ is isosceles by the Converse of the Isosceles Triangle Theorem, so $\overline{BK} \cong \overline{CK}$. Since $\triangle KBC \cong \triangle KCD$ by SAS, $\overline{CK} \cong \overline{DK}$ because CPCTC. Also, $\triangle KCD \cong \triangle DKJ$ by SAS, so $\overline{DK} \cong \overline{DJ}$. Since $\triangle DKJ$ is isosceles, $\angle DKJ \cong \angle DJK$ by the Isosceles Triangle Theorem, so $\triangle DKJ \cong \triangle DJI$ by AAS, and so $\overline{DJ} \cong \overline{DI}$ because CPCTC. $\triangle DJI \cong \triangle DHI$ by SAS, so $\overline{DH} \cong \overline{DJ}$ because CPCTC. Since $\overline{HJ} \parallel \overline{DE}$, $\angle IHD \cong \angle EDH$ by the Alternate Interior Angles Theorem, so $\triangle DHI \cong \triangle HDE$ by AAS, and so $\overline{DI} \cong \overline{HE}$ because CPCTC. $\angle EHD \cong \angle HDI$ because CPCTC, so $\triangle HDE \cong \triangle HFE$ by SAS, and so $\overline{HD} \cong \overline{HF}$ because CPCTC. Using the Transitive Property of Congruence, $\triangle HFE \cong \triangle DJK$, so $\overline{EF} \cong \overline{JK}$ because CPCTC, and so $\overline{EF} \cong \overline{GF}$ by the Transitive Property of Congruence. Thus, $\triangle HFE \cong \triangle HFG$ by SAS, so $\overline{HG} \cong \overline{HE}$. Since it is given that $\overline{AK} \cong \overline{HE}$, $\overline{AK} \cong \overline{HG}$ by the Transitive Property of Congruence.

35. The surveyor constructs $\triangle ZYX \cong \triangle TBX$ by SAS. Because CPCTC, the surveyor knows that $YZ = TB$. Thus by measuring $\overline{YZ}$, he can determine the distance across the pond.

36. All angles measure 60° by Corollary 4.4.3.

PAGE 242, LOOK BACK

37. $n = 4$

$$
\begin{aligned}
\text{Sum of interior angles} &= 180°(n - 2) \\
&= 180°(4 - 2) \\
&= 180°(2) \\
&= 360°
\end{aligned}
$$

$$
\begin{aligned}
\text{Sum of exterior angles} &= n\left(180° - \frac{\text{sum of interior angles}}{}\right) \\
&= 4\left(180° - \frac{360°}{4}\right) \\
&= 4(180° - 90°) \\
&= 4(90°) \\
&= 360°
\end{aligned}
$$

38. $n = 6$

$$\begin{aligned}
\text{Sum of interior angles} &= 180°(n - 2) \\
&= 180°(6 - 2) \\
&= 180°(4) \\
&= 720°
\end{aligned}$$

$$\begin{aligned}
\text{Sum of exterior angles} &= n\left(180° - \frac{\text{sum of interior angles}}{}\right) \\
&= 6\left(180° - \frac{720°}{6}\right) \\
&= 6(180° - 120°) \\
&= 6(60°) \\
&= 360°
\end{aligned}$$

39. $n = 12$

$$\begin{aligned}
\text{Sum of interior angles} &= 180°(n - 2) \\
&= 180°(12 - 2) \\
&= 180°(10) \\
&= 1800°
\end{aligned}$$

$$\begin{aligned}
\text{Sum of exterior angles} &= n\left(180° - \frac{\text{sum of interior angles}}{}\right) \\
&= 12\left(180° - \frac{1800°}{12}\right) \\
&= 12(180° - 150°) \\
&= 12(30°) \\
&= 360°
\end{aligned}$$

40. No, an equilateral octagon is not necessarily equiangular.

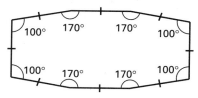

No, an equiangular octagon is not necessarily equilateral.

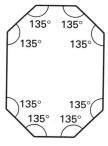

41. Slope of $\overline{AB} = \frac{6 - 0}{3 - 0} = \frac{6}{3} = 2$

If $D(1, 2)$ then

Slope of $\overline{CD} = \frac{2 - 1}{1 - 3} = -\frac{1}{2}$.

$2\left(-\frac{1}{2}\right) = -1$ thus $\overline{AB} \perp \overline{CD}$.

42. Slope of $\overline{AB} = \frac{6 - 0}{3 - 0} = \frac{6}{3} = 2$

If $D(4, 4)$ then

Slope of $\overline{CD} = \frac{4 - 1}{4 - 3} = \frac{3}{1}$.

The slopes are not the same and their product is not -1. $\overline{AB}$ is neither parallel nor perpendicular to $\overline{CD}$.

43. Slope of $\overline{AB} = \frac{6 - 0}{3 - 0} = \frac{6}{3} = 2$

If $D(2, -1)$ then

Slope of $\overline{CD} = \frac{1 - (-1)}{3 - 2} = \frac{2}{1} = 2$.

The slopes are the same thus $\overline{AB} \parallel \overline{CD}$.

44. Given: $\triangle ABC$ is isosceles and $\overline{AC} \cong \overline{BC}$.
Prove: $\angle A \cong \angle B$

Statements	Reasons
$\overline{AC} \cong \overline{BC}$	Given
$\angle C \cong \angle C$	Reflexive Property of Congruence
$\overline{CB} \cong \overline{CA}$	Given
$\triangle ACB \cong \triangle BCA$	SAS
$\angle A \cong \angle B$	CPCTC

45. Given: $\triangle ABC$ with $\angle A \cong \angle B$.
Prove: $\triangle ABC$ is isosceles.

Statements	Reasons
$\angle A \cong \angle B$	Given
$\overline{AB} \cong \overline{BA}$	Reflexive Property of Congruence
$\angle B \cong \angle A$	Given
$\triangle ABC \cong \triangle BAC$	ASA
$\overline{BC} \cong \overline{AC}$	CPCTC
$\triangle ABC$ is isosceles	Definition of isosceles triangle

4.5 **PAGE 245, GUIDED SKILLS PRACTICE**

5. $m\angle WXZ = m\angle XZY = 25°$

6. $m\angle W = m\angle Y = 50°$

7. $XY = WZ = 2.2$

8. The ramp, handrail, and the two upright posts form a parallelogram. Opposite sides of a parallelogram are congruent.

PAGES 246–251, PRACTICE AND APPLY

9. $CD = BA = 5$

10. $DA = CB = 7$

11. $m\angle C = m\angle A = 50°$

12. $m\angle D = 180° - m\angle A = 180° - 50° = 130°$

13. $m\angle Q = m\angle S = 60°$

14. $m\angle RPQ = m\angle PRS = 40°$

15. $m\angle SPR = 180° - m\angle PSR - m\angle PRS$
$\qquad = 180° - 60° - 40° = 80°$

16. $m\angle PRQ = m\angle SPR = 80°$

17. $TU = SV$
$\quad 4a - 2 = a + 7$
$\qquad 3a = 9$
$\qquad a = 3$
$\quad SV = a + 7$
$\qquad = 3 + 7$
$\qquad = 10$

18. $m\angle C = m\angle A$
$\quad (2x)° = \left(\frac{3}{2}x + 30\right)°$
$\qquad 2x = \frac{3}{2}x + 30$
$\qquad \frac{1}{2}x = 30$
$\qquad x = 60$
$\quad m\angle A = \left(\frac{3}{2}(60) + 30\right)°$
$\qquad = 90 + 30$
$\qquad = 120°$

19. $QT = RS$
$6x - 2 = 10$
$6x = 12$
$x = 2$
$QR = TS = x + 4$
$= 2 + 4$
$= 6$

20. $m\angle E = m\angle C$
$(2x + 6)° = 50°$
$2x + 6 = 50$
$2x = 44$
$x = 22$
$CD = FE = x - 7$
$= 22 - 7$
$= 15$

21. $m\angle M + m\angle P = 180°$
$(6x + 16)° + (x - 4)° = 180°$
$6x + 16 + x - 4 = 180$
$7x + 12 = 180$
$7x = 168$
$x = 24$
$m\angle N = m\angle P = (x - 4)°$
$= (24 - 4)°$
$= 20°$

22. $m\angle EHG + m\angle G = 180°$
$2(27°) + m\angle G = 180°$
$54° + m\angle G = 180°$
$m\angle G = 126°$

23. a. Yes; The opposite sides are parallel by the Alternate Interior Angles Theorem.

b. No; The Triangles fit together to form a kite.

c. No; the sides of the triangles will not match up.

24. Definition of a parallelogram

25. Alternate Interior Angles Theorem

26. $\overline{AD} \parallel \overline{BC}$

27. Definition of a parallelogram

28. $\angle 2 \cong \angle 4$

29. Alternate Interior Angles Theorem

30. $\overline{DB} \cong \overline{DB}$

31. ASA

32. Given

33. $\triangle ABD \cong \triangle CDB$

34. CPCTC

35. $\triangle CDB$

36. A diagonal of a parallelogram divides the parallelogram into two congruent triangles.

37. $\triangle DCA$

38. A diagonal of a parallelogram divides the parallelogram into two congruent triangles.

39. $\angle BAD \cong \angle DCB$

40. $\angle ABC \cong \angle CDA$

41. Given: $ABCD$ is a parallelogram.
Prove: $\angle A$ and $\angle B$ are supplementary,
$\angle B$ and $\angle C$ are supplementary,
$\angle C$ and $\angle D$ are supplementary, and
$\angle D$ and $\angle A$ are supplementary.

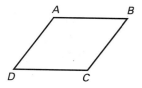

By definition of a parallelogram, $\overline{AB} \parallel \overline{DC}$ and $\overline{AD} \parallel \overline{BC}$. By Same-Side Interior Angles Theorem, $\angle A$ and $\angle B$ are supplementary, $\angle B$ and $\angle C$ are supplementary, $\angle C$ and $\angle D$ are supplementary, and $\angle D$ and $\angle A$ are supplementary.

42. Definition of a parallelogram

43. alternate interior

44. opposite

45. congruent

46. ASA

47. CPCTC

48. CPCTC

49. Definition of a rhombus

50. Reflexive Property of Congruence

51. SSS

52. Converse of the Alternate Interior Angles Theorem

53. CPCTC

54. Converse of the Alternate Interior Angles Theorem

55. Definition of a rectangle

56. Addition Property

57. Converse of the Same-Side Interior Angles Theorem

58. Definition of a rectangle

59. Converse of the Same-Side Interior Angles Theorem

60. Definition of a rhombus

61. Diagonals of a parallelogram bisect each other

62. Reflexive Property of Congruence

63. SSS

64. CPCTC

65. 180°

66. 90°

67. Given: $RSTU$ is a rectangle with diagonals $\overline{RT}$ and $\overline{US}$ intersecting at V.
Prove: $\overline{RT} \cong \overline{US}$

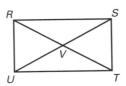

Statements	Reasons
$RSTU$ is a rectangle.	Given
$RSTU$ is a parallelogram.	Theorem 4.5.7
$\overline{ST} \cong \overline{UR}$	Theorem 4.5.2
$\overline{TU} \cong \overline{TU}$	Reflexive Property of Congruence
m∠RUT = 90° m∠STU = 90°	Definition of a rectangle
m∠RUT = m∠STU (∠RUT ≅ ∠STU)	Transitive Property or Substitution Property
$\triangle RTU \cong \triangle SUT$	SAS
$\overline{RT} \cong \overline{SU}$	CPCTC

68. Given: Kite $WXYZ$ with diagonals $\overline{WY}$ and $\overline{XZ}$ intersecting at point A
Prove: $\overline{WY} \perp \overline{XZ}$

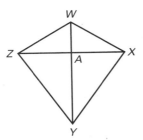

StatemenLts	Reasons
$\overline{WX} \cong \overline{WZ}$ $\overline{XY} \cong \overline{ZY}$	Given
$\overline{WY} \cong \overline{WY}$	Reflexive Property of Congruence
$\triangle WXY \cong \triangle WZY$	SSS
∠ZWY ≅ ∠XWY	CPCTC
$\overline{WA} \cong \overline{WA}$	Reflexive Property of Congruence
$\triangle WAX \cong \triangle WAZ$	SAS
∠WAX ≅ ∠WAZ	CPCTC
m∠WAX + m∠WAZ = 180°	Linear Pair Property
m∠WAX + m∠WAX = 180°	Substitution Property
m∠WAX = 90°	Division Property
$WY \perp ZX$	Definition of perpendicular

69. Theorem: A square is a rectangle

A square is a quadrilateral with the property that all of its sides are equal and every angle is a right angle. Since every angle is a right angle, a square is a rectangle by the definition of a rectangle.

70. Theorem: A square is a rhombus.

A square is a quadrilateral with the property that all of its sides have the same length. This is the definition of a rhombus.

71. Theorem: The diagonals of a square are congruent and are the perpendicular bisectors of each other.

A square is a rhombus and hence a parallelogram by theorems 4.5.12 and 4.5.6, respectively. Therefore its diagonals are perpendicular bisectors of each other by theorems 4.5.8 and 4.5.5. A square is also a rectangle by theorem 4.5.12 and thus its diagonals are congruent by theorem 4.5.9.

72. Given: *ABCD* is a parallelogram.
Prove: Quadrilateral *EFGH* is a parallelogram.

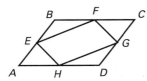

Given: *ABCD* is a parallelogram, and *E*, *F*, *G*, and *H* are the midpoints of sides $\overline{AB}$, $\overline{BC}$, $\overline{CD}$, and $\overline{DA}$.

Prove: *EFGH* is a parallelogram.

Proof: By Theorem 4.5.2, $\overline{AB} \cong \overline{CD}$ and $\overline{BC} \cong \overline{DA}$. By definition of midpoint, $AE = BE = \frac{1}{2}AB$, and $CG = DG = \frac{1}{2}CD$, so $\overline{AE} \cong \overline{CG}$ and $\overline{BE} \cong \overline{DG}$. Also by definition of midpoint, $BF = CF = \frac{1}{2}BC$, and $AH = DH = \frac{1}{2}AD$, so $\overline{BF} \cong \overline{DH}$ and $\overline{CF} \cong \overline{AH}$. By Theorem 4.5.3, $\angle A \cong \angle C$ and $\angle B \cong \angle D$, so $\triangle BEF \cong \triangle DGH$ and $\triangle AEH \cong \triangle CGF$ by SAS. $\overline{EF} \cong \overline{GH}$ and $\overline{FG} \cong \overline{HE}$ because CPCTC, so $\triangle EFG \cong \triangle GHE$ by SSS. Because CPCTC, $\angle FEG \cong \angle EGH$, so $\overline{EF} \parallel \overline{GH}$ by the Converse of the Alternate Interior angles Theorem. Also because CPCTC, $\angle FGE \cong \angle HEG$, so $\overline{FG} \parallel \overline{HE}$ by the Converse of the Alternate Interior Angles Theorem. Thus, *EFGH* is a parallelogram by the definition of a parallelogram.

73. a. 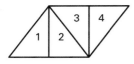 **b.** 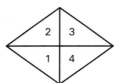 **c.**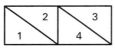

d. 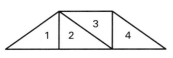 **e.**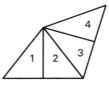

74. a. True because the diagonals of a rectangle are congruent.

b. False because the diagonals of a parallelogram bisect each other.

c. True because opposite sides of a parallelogram are parallel.

PAGES 251–252, LOOK BACK

---

75. No, the triangles are not necessarily congruent. One triangle could be an enlargement of the other.

76. $\triangle ABC \cong \triangle EDF$, SSS

77. $\triangle GHI \cong \triangle JKL$, Hypotenuse-Leg Congruence Theorem

78. $\triangle ONM \cong \triangle PQR$, SAS

79.
$$m\angle R = m\angle K$$
$$(3x + y)° = 65°$$
$$3x + y = 65$$
$$y = 65 - 3x$$

$$m\angle S = m\angle L$$
$$m\angle S = (x + y)°$$

$$m\angle R + m\angle S + m\angle T = 180°$$
$$(3x + y)° + (x + y)° + 80° = 180°$$
$$3x + y + x + y + 80 = 180$$
$$4x + 2y = 100$$
$$4x + 2(65 - 3x) = 100$$
$$4x + 130 - 6x = 100$$
$$30 = 2x$$
$$15 = x$$

$$y = 65 - 3x$$
$$= 65 - 3(15)$$
$$= 65 - 45$$
$$= 20$$

$$x = 15; y = 20$$

PAGE 252, LOOK BEYOND

80. Given: $ABCD$ is an isosceles trapezoid and $\overline{AE}$ and $\overline{BF}$ are perpendicular to $\overline{CD}$.
Prove: $\angle ADE \cong \angle BCE$

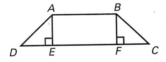

Statements	Reasons
$\overline{AB} \parallel \overline{CD}$	Definition of a trapezoid
$\overline{AE} \perp \overline{CD}$ $\overline{BF} \perp \overline{CD}$	Given
$\overline{AE} \parallel \overline{BF}$	Converse of the Corresponding Angles Postulate
$ABFE$ is a parallelogram.	Definition of parallelogram
$\overline{AE} \cong \overline{BF}$	Theorem 4.5.2
$\overline{AD} \cong \overline{BC}$	Given
$\angle AED \cong \angle BFC$	Right angles are congruent.
$\triangle AED \cong \triangle BFC$	HL
$\angle ADE \cong \angle BCE$	CPCTC

81. Given: $ABCD$ is an isosceles trapezoid.
  Prove: $\overline{AC} \cong \overline{BD}$

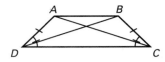

Statements	Reasons
$ABCD$ is an isosceles trapezoid.	Given
$\angle ADC \cong \angle BCD$	Exercise 80
$\overline{AD} \cong \overline{BC}$	Definition of isosceles trapezoid
$\overline{DC} \cong \overline{CD}$	Reflexive Property of Congruence
$\triangle ADC \cong \triangle BCD$	SAS
$\overline{AC} \cong \overline{BD}$	CPCTC

4.6 | **PAGE 257, GUIDED SKILLS PRACTICE**

5. none

6. parallelogram

7. parallelogram, rectangle

8. parallelogram, rhombus

9. The diagonals must be congruent.

PAGES 257–260, PRACTICE AND APPLY

10. No. The figure could be a trapezoid.

11. Yes, Theorem 4.6.3

12. No. The figure could be a trapezoid.

13. Yes, Theorem 4.6.2

14. Rectangle, Theorem 4.6.5

15. Rhombus, Theorem 4.6.6

16. Rectangle, Theorem 4.6.4

17. Neither

18. Rhombus, Theorem 4.6.8

19. Neither

20. $KLMN$ is a square.
  The diagonals are perpendicular, so it is a rhombus.
  $\angle KLM$ is a right angle so $KLMN$ is a rectangle.
  Therefore $KLMN$ is a square.

21. No, $KLMN$ is a rhombus by Theorem 4.6.6.
  But it is possible that $\angle KLM > 90°$. Thus,
  $KLMN$ is not necessarily a square.

22. $KLMN$ is a rhombus by Theorem 4.6.7.
  $KLMN$ is a rectangle since m$\angle LKN = 90°$.
  Therefore $KLMN$ is a square.

23. $KLMN$ is a rhombus by Theorem 4.6.6.
  $KLMN$ is a rectangle by Theorem 4.6.5.
  Therefore $KLMN$ is a square.

24. No. $KLMN$ is a rhombus by Theorem 4.6.7.
  But it is possible that $\angle KLM > 90°$.
  Thus $KLMN$ is not necessarily a square.

25. $KLMN$ is a rhombus by Theorem 4.6.6,
  since $\overline{NK} \cong \overline{KL}$ by CPCTC. Also by CPCTC,
  $\overline{LN} \cong \overline{MK}$, so $KLMN$ is a rectangle by
  Theorem 4.6.5. Thus, $KLMN$ is a square by the
  definition of a square.

26. Given

27. Reflexive Property

28. SSS

29. CPCTC

30. Converse of the Alternate Interior Angles Theorem

31. Definition of a parallelogram

32. $\overline{FG}$

33. SSS

34. CPCTC

35. same-side interior angles

36. They are congruent and supplementary.

37. Opposite angles in a parallelogram are congruent.

38. Definition of a rectangle

39.

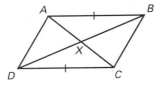

Given: Quadrilateral $ABCD$, $\overline{AB} \cong \overline{CD}$, $AB \parallel CD$
Prove: $ABCD$ is a parallelogram.

Statements	Reasons
$\overline{AB} \parallel \overline{CD}$ $\overline{AB} \cong \overline{CD}$	Given
$\angle XAB \cong \angle XCD$ $\angle XBA \cong \angle XDC$	Alternate Interior Angles Theorem
$\triangle AXB \cong \triangle CXD$	ASA
$\overline{XB} \cong \overline{XD}$ $\overline{XC} \cong \overline{XA}$	CPCTC
$\angle AXD \cong \angle BXC$	Vertical Angles Theorem
$\triangle AXD \cong \triangle BXC$	SAS
$\angle DAX \cong \angle BCX$	CPCTC
$\overline{AD} \parallel \overline{BC}$	Converse of the Alternate Interior Angles Theorem
$ABCD$ is a parallelogram.	Definition of a parallelogram

40.

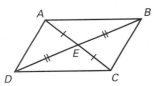

Given: Quadrilateral $ABCD$ and AC, BD bisect each other.
Prove: $ABCD$ is a parallelogram.

Statements	Reasons
$\overline{AC}$ bisects $\overline{BD}$ $\overline{BD}$ bisects $\overline{AC}$	Given
$\overline{AE} \cong \overline{EC}$ $\overline{DE} \cong \overline{EB}$	Definition of bisector
$\angle AED \cong \angle BEC$	Vertical Angles Theorem
$\triangle AED \cong \triangle CEB$	SAS
$\angle EAD \cong ECB$	CPCTC
$\overline{AD} \parallel \overline{BC}$	Converse of the Alternate Interior Angles Theorem
$\angle AEB \cong \angle DEC$	Vertical Angles
$\triangle ABE \cong \triangle CDE$	SAS
$\angle EAB \cong \angle ECD$	CPCTC
$\overline{AB} \parallel \overline{CD}$	Converse of the Alternate Interior Angles Theorem
$ABCD$ is a parallelogram.	Definition of parallelogram

41. Given parallelogram $ABCD$ with $m\angle A = 90°$
Prove: $ABCD$ is a rectangle

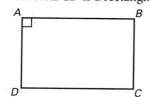

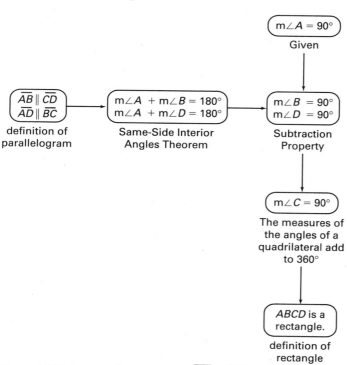

42. Given: $ABCD$ is a parallelogram and $\overline{AB} \cong \overline{BC}$.
Prove: $ABCD$ is a rhombus.

Statements	Reasons
$ABCD$ is a parallelogram.	Given
$\overline{CD} \cong \overline{AB}$ $\overline{AD} \cong \overline{BC}$	Opposite sides of a parallelogram are congruent.
$\overline{AB} \cong \overline{BC}$	Given
$\overline{AB} \cong \overline{BC} \cong \overline{CD} \cong \overline{AD}$	Transitive Property of Congruence
$ABCD$ is a rhombus.	Definition of rhombus

43. Given: $ABCD$ is a parallelogram, $\overline{AC}$ bisects $\angle A$ and $\angle C$,
and $\overline{BD}$ bisects $\angle B$ and $\angle D$.

Prove: $ABCD$ is a rhombus.

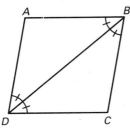

Statements	Reasons
$ABCD$ is a parallelogram.	Given
$m\angle B = m\angle D$	Opposite angles of a parallelogram are congruent.
$\overline{BD}$ bisects $\angle B$, $\angle D$	Given
$m\angle ABD = \frac{1}{2}m\angle B$ $m\angle ADB = \frac{1}{2}m\angle D$	Definition of angle bisector
$m\angle ABD = m\angle ADB$	Substitution
$\overline{AD} \cong \overline{AB}$	Converse to Isoceles Triangle Theorem
$ABCD$ is a rhombus.	Theorem 4.6.6

44. Given: $ABCD$ is a parallelogram.
$\overline{AD} \perp \overline{BC}$; AC intersects BD at X.

Show: $ABCD$ is a rhombus.

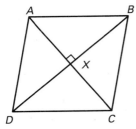

Statements	Reasons
$ABCD$ is a parallelogram.	Given
$\overline{AX} \cong \overline{XC}$	Diagonals of a parallelogram bisect each other.
$\overline{XB} \cong \overline{XB}$	Reflexive Property of Congruence
$\overline{AC} \perp \overline{BD}$	Given
$m\angle AXB = 90°$ $m\angle CXB = 90°$	Definition of perpendicular
$m\angle AXB = m\angle CXB$ $(\angle AXB \cong \angle CXB)$	Transitive Property of Equality
$\triangle AXB \cong \triangle CXB$	SAS
$\overline{AB} \cong \overline{CB}$	CPCTC
$ABCD$ is a rhombus.	Theorem 4.6.6

45. Vertical Angles Theorem

46. SAS

47. CPCTC

48. Converse of the Alternate Interior Angles Theorem

49. Theorem 4.6.2

50. Definition of a parallelogram

51. Opposite sides of a parallelogram are congruent.

52. Segment Addition Postulate

53. Division Property

54. Substitution Property of Equality

55. The boards that are 2 feet long must be opposite each other, and the boards that are 3 feet long must be opposite each other. She also should make sure that the diagonals have the same measure.

56. The spacing of the rungs must be the same on both side pieces. In theory, only one corner brace is needed. When one right angle is established, all corresponding angles must also be right angles. By Theorem 4.6.4, each parallelogram must be a rectangle. In practice, more than one brace would be used, because of the flexibility of the wood.

PAGE 260, LOOK BACK

57. $\dfrac{360°}{72°} = 5$ sides

58. $180° - 72° = 108°$

59. $72°$

60. Sample answer:
A postulate is something that is accepted as true, without proof.
A theorem is something you can prove using postulates and definitions.
A conjecture is something that you think is true but have not yet proven.

61. Given: $\overline{BA} \cong \overline{BC}$ and $\overline{BD} \cong \overline{BE}$
Prove: $\triangle ADC \cong \triangle CEA$ ($\triangle ABC$ is isoceles)

Since $\overline{BA} \cong \overline{BC}$, $\triangle ABC$ is isosceles, so $\angle BAC \cong \angle BCA$ by the Isosceles Triangle Theorem. Also, since $\overline{BD} \cong \overline{BE}$, by the Segment Addition Postulate $\overline{AD} \cong \overline{CE}$. By the Reflexive Property of Congruence $\overline{AC} \cong \overline{AC}$, so $\triangle ADC \cong \triangle CEA$ by SAS.

PAGE 260, LOOK BEYOND

62. parallelogram

63. midsegments

64.

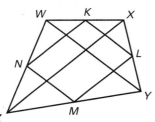

Given: quadrilateral $WXYZ$ with K, L, M, and N the midpoints of $\overline{WX}$, $\overline{XY}$, $\overline{YZ}$, and $\overline{ZW}$.
Prove: $KLMN$ is a parallelogram.
Proof: $\overline{KL}$ is a midsegment of $\triangle WXY$, so by the Triangle Midsegment Theorem, $\overline{KL} \parallel \overline{WY}$. $\overline{MN}$ is a midsegment of $\triangle WZY$, so by the Triangle Midsegment Theorem, $\overline{MN} \parallel \overline{WY}$. Two segments that are parallel to the same segment are parallel to each other, so $\overline{KL} \parallel \overline{MN}$. Also by the Triangle Midsegment Theorem, $KL = \frac{1}{2}WY$ and $MN = \frac{1}{2}WY$, so by the Transitive Property of Equality $KL = MN$ ($\overline{KL} \cong \overline{MN}$). Since $KLMN$ has one pair of opposite sides that are parallel and congruent, $KLMN$ is a parallelogram by Theorem 4.6.2.

5. Trace the line segment and label endpoints A and B.

Draw a line and label point C on the line.

Set the compass to the distance of AB.

Place the point of the compass on C and draw an arc that intersects the line. Label the intersection point D.

$\overline{CD} \cong \overline{AB}$

6. Trace the triangle and label vertices A, B, and C.

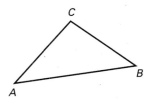

Draw a line and label point D on the line.

Set the compass to the distance AB in the triangle.

Place the point of the compass on D and draw an arc that interesects the line. Label the intersection point E.

Set the compass to the distance AC in the triangle. Place the point of the compass on D and draw an arc above $\overline{DE}$.

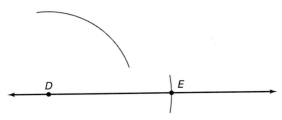

Set the compass to the distance BC in the triangle. Place the point of the compass on E and draw an arc that intersects the other arc. Label the intersection point F.

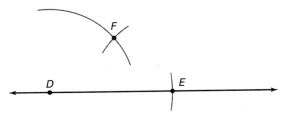

Connect points *D*, *E*, and *F*.

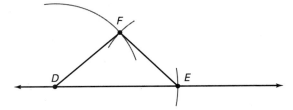

$\triangle DEF \cong \triangle ABC$

7. Trace $\triangle ABC$.

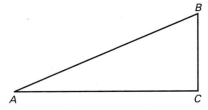

Place the point of the compass on *A* and draw an arc through the rays of the angle. Label the points of intersection *D* and *E*.

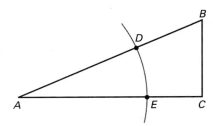

Using the same compass setting, draw intersecting arcs from points *D* and *E*. Label the interesection point *F*.

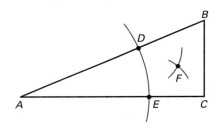

Draw the ray $\overrightarrow{AF}$.

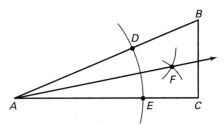

$\overrightarrow{AF}$ is the angle bisector of $\angle A$.

8. Trace △*ABC*

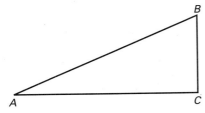

Place the point of the compass on *B* and draw an arc through the rays of the angle. Label the points of intersection *D* and *E*.

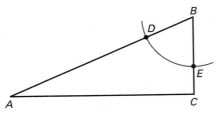

Using the same compass setting, draw intersecting arcs from points *D* and *E*. Label the intersection point *F*.

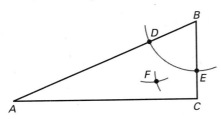

Draw the ray $\overrightarrow{BF}$.

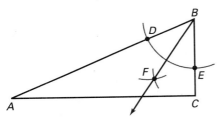

$\overrightarrow{BF}$ is the angle bisector of ∠*B*.

9. Trace △*ABC*

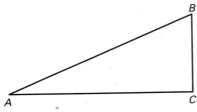

Place the point of the compass on *C* and draw an arc through the rays of the angle. Label the points of intersection *D* and *E*.

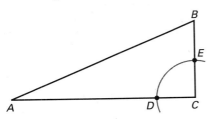

Using the same compass setting, draw intersecting arcs from points D and E. Label the intersection point F.

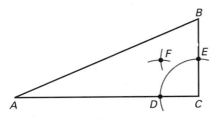

Draw the ray $\overrightarrow{CF}$.

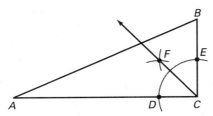

$\overrightarrow{CF}$ is the angle bisector of $\angle C$.

10. Draw a line and label point A on the line

Set the compass to the distance of the given line segment. Place the point of the compass on A and draw an arc that intersects the line. Label the intersection point B.

$\overline{AB}$ is congruent to the given line segment.

11. Draw a line and label point A on the line

Set compass to the distance of the given line segment. Place the point of the compass on A and draw an arc that intersects the line. Label the intersection point B.

$\overline{AB}$ is congruent to the given line segment.

12. Draw a line and label point A on the line.

Set the compass to the distance of the bottom edge of the triangle. Place the point of the compass on *A* and draw an arc that intersects the line. Label the intersection point *B*.

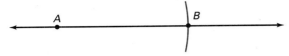

Set the compass to the distance of the left edge of the triangle. Place the point of the compass on *A* and draw an arc above *AB*.

Set the compass to the distance of the right edge of the triangle. Place the point of the compass on *B* and draw an arc that intersects the other arc. Label the intersection point *C*.

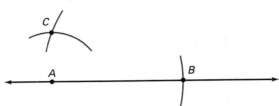

Connect points *A*, *B*, and *C*.

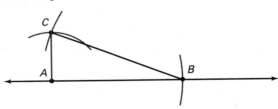

△*ABC* is congruent to the given triangle.

13. Draw a line and label point *A* on the line.

Set the compass to the distance of the bottom edge of the triangle. Place the point of the compass on *A* and draw an arc that intersects the line. Label the intersection point *B*.

Set the compass to the distance of the left edge of the triangle. Place the point of the compass on *A* and draw an arc above *AB*.

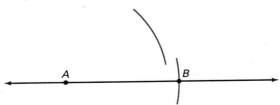

Set the compass to the distance of the right edge of the triangle. Place the point of the compass on *B* and draw an arc that intersects the other arc. Label the intersection point *C*.

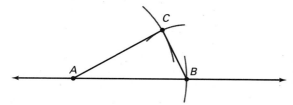

Connect points *A*, *B*, and *C*.

$\triangle ABC$ is congruent to the given triangle.

14. Draw a line and label point *A* on the line.

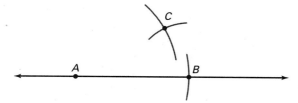

Set the compass to the distance of the bottom edge of the triangle. Place the point of the compass on *A* and draw an arc that intersects the line. Label the intersection point *B*.

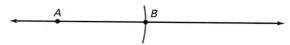

Set the compass to the distance of the left edge of the triangle. Place the point of the compass on *A* and draw an arc above *AB*.

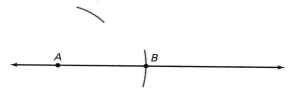

Set the compass to the distance of the right edge of the triangle. Place the point of the compass on *B* and draw an arc that intersects the other arc. Label the intersection point *C*.

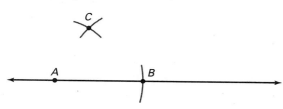

Connect points *A*, *B*, and *C*.

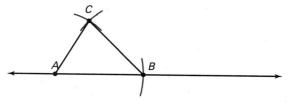

$\triangle ABC$ is congruent to the given triangle.

15. Draw a line and label point *A* on the line.

Set the compass to the distance of the bottom edge of the triangle. Place the point of the compass on *A* and draw an arc that intersects the line. Label the intersection point *B*.

Set the compass to the distance of the left edge of the triangle. Place the point of the compass on *A* and draw an arc above *AB*.

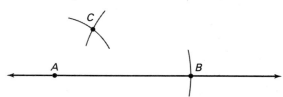

Set the compass to the distance of the right edge of the triangle. Place the point of the compass on *B* and draw an arc that intersects the other arc. Label the intersection point *C*.

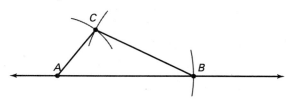

Connect points *A*, *B*, and *C*.

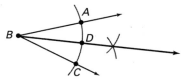

△*ABC* is congruent to the given triangle.

16. Trace the angle. Label the vertex *B*. Draw an arc with its center at *B*. Label the intersection points *A* and *C*. Draw two intersecting arcs having the same radius from *A* and *C*. Label the intersection point *D*. Draw $\overrightarrow{BD}$.

$\overrightarrow{BD}$ is the angle bisector of ∠*ABC*.

17. Trace the angle. Label the vertex *B*. Draw an arc with its center at *B*. Label the intersection points *A* and *C*. Draw two intersecting arcs having the same radius from *A* and *C*. Label the intersection point *D*. Draw $\overrightarrow{BD}$.

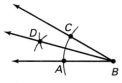

$\overrightarrow{BD}$ is the angle bisector of ∠*ABC*.

18. Trace the original angle.

Use a straightedge to draw a ray with endpoint B.

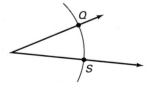

Place the compass point on the vertex of the traced angle and draw an arc. Label the intersection points Q and S.

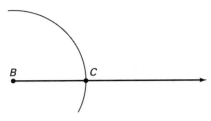

Without adjusting the compass, place the compass point at B. Draw an arc that crosses the ray. Label the intersection point C.

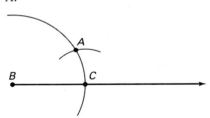

Set the compass equal to the distance QS. Without adjusting the compass, place the compass at C and draw an arc that crosses the first arc. Label the intersection point A.

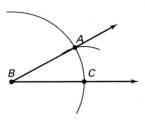

Draw $\overrightarrow{BA}$ to form ∠B.

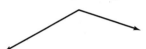

∠B is congruent to the given angle.

19. Trace the original angle.

Use a straightedge to draw a ray with endpoint B.

Place the compass point on the vertex of the traced angle and draw an arc. Label the intersection points *Q* and *S*.

Without adjusting the compass, place the compass point at *B*. Draw an arc that crosses the ray. Label the intersection point *C*.

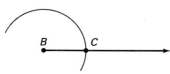

Set the compass equal to the distance *QS*. Without adjusting the compass, place the compass at *C* and draw an arc that crosses the first arc. Label the intersection point *A*.

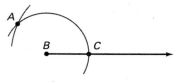

Draw $\overrightarrow{BA}$ to form ∠*B*.

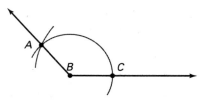

∠*B* is congruent to the given angle.

20. Trace the original angle.

Use a straightedge to draw a ray with endpoint *B*.

Place the compass point on the vertex of the traced angle and draw an arc. Label the intersection points *Q* and *S*.

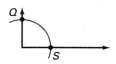

Without adjusting the compass, place the compass point at *B*. Draw an arc that crosses the ray. Label the intersection point *C*.

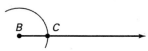

Set the compass equal to the distance QS. Without adjusting the compass, place the compass at *C* and draw an arc that crosses the first arc. Label the intersection point *A*.

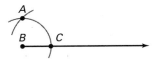

Draw $\overrightarrow{BA}$ to form ∠*B*.

∠*B* is congruent to the given angle.

21. Trace the original angle.

Use a straightedge to draw a ray with endpoint *B*.

Place the compass point on the vertex of the traced angle and draw an arc. Label the intersection points *Q* and *S*.

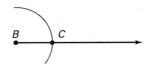

Without adjusting the compass, place the compass point at *B*. Draw an arc that crosses the ray. Label the intersection point *C*.

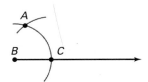

Set the compass equal to the distance QS. Without adjusting the compass, place the compass at *C* and draw an arc that crosses the first arc. Label the intersection point *A*.

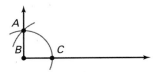

Draw $\overrightarrow{BA}$ to form ∠*B*.

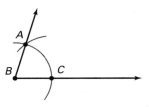

∠*B* is congruent to the given angle.

22. Trace the original segment.

Set the compass equal to a distance greater than half the length of the segment. Place the compass point at one endpoint of the segment and draw an arc which intersects the segment.

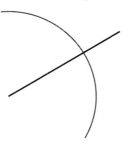

Without adjusting the compass, place the compass at the other endpoint of the segment and draw an arc which intersects the first arc at two points. Label the intersection points A and B.

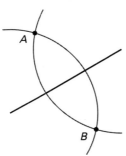

Use a straightedge to draw $\overleftrightarrow{AB}$. Label the point where $\overleftrightarrow{AB}$ intersects the line segment C.

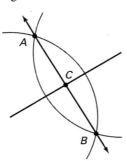

$\overleftrightarrow{AB}$ is the perpendicular bisector of the segment and C is its midpoint.

23. Trace the original segment.

Set the compass equal to a distance greater than half the length of the segment. Place the compass point at one endpoint of the segment and draw an arc which intersects the segment.

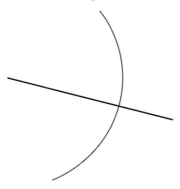

Without adjusting the compass, place the compass at the other endpoint of the segment and draw an arc which intersects the first arc at two points. Label the intersection points A and B.

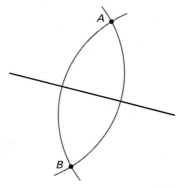

Use a straightedge to draw $\overleftrightarrow{AB}$. Label the point where $\overleftrightarrow{AB}$ intersects the line segment C.

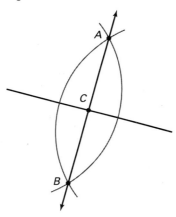

$\overleftrightarrow{AB}$ is the perpendicular bisector of the segment and C is its midpoint.

24. Trace the original triangle. Label the vertices A, B, and C.

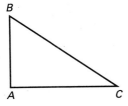

Construct perpendicular bisector to $\overline{AB}$.

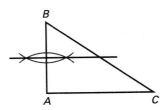

Construct perpendicular bisector to $\overline{BC}$.

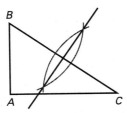

Construct perpendicular bisector to $\overline{AC}$.

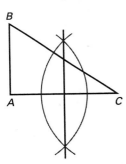

Notice that the perpendicular bisectors intersect at a common point. Label this point D. Set the compass equal to the distance of $\overline{AD}$. Place the compass point on D and draw a circle.

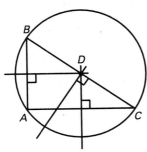

The circle drawn is the circumscribed circle of the triangle.

25. Trace the original triangle. Label the vertices *A*, *B*, and *C*.

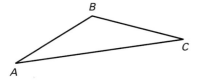

Construct perpendicular bisector to $\overline{AB}$.

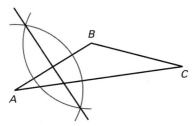

Construct perpendicular bisector to $\overline{BC}$.

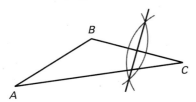

Construct perpendicular bisector to $\overline{AC}$.

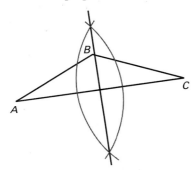

Notice that the perpendicular bisectors intersect at a common point. Label this point *D*. Set the compass equal to the distance of $\overline{AD}$. Place the compass point on *D* and draw a circle.

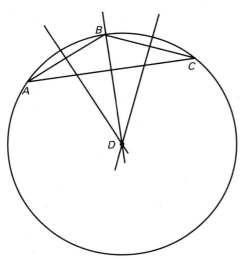

The circle drawn is the circumscribed circle of the triangle.

26. Trace the original line and point. Label the point *A*.

Place the compass point at *A* and make an arc that intersects the line at two points. Label the points of intersection *D* and *B*.

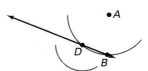

Place the compass point at *D* and make an arc on one side of the line.

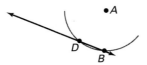

Using the same setting, place the compass point at *B* and draw an arc which intersects the arc drawn from point *D*. Label the intersection point *C*.

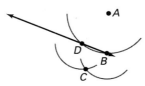

Use a straightedge to draw $\overleftrightarrow{AC}$.

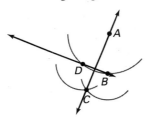

$\overleftrightarrow{AC}$ is perpendicular to the traced line.

27. Trace the original line and point. Label the point *A*.

Place the compass point at *A* and make an arc that intersects the line at two points. Label the points of intersection *D* and *B*.

Place the compass point at *D* and make an arc on one side of the line.

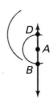

Using the same setting, place the compass point at B and draw an arc which intersects the arc drawn from point D. Label the intersection point C.

Use a straightedge to draw $\overleftrightarrow{AC}$.

$\overleftrightarrow{AC}$ is perpendicular to the traced line.

28. a. Trace the given triangle. Construct a line through A which is perpendicular to $\overleftrightarrow{BC}$.

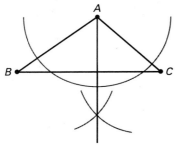

Construct a line through B which is perpendicular to $\overleftrightarrow{AC}$.

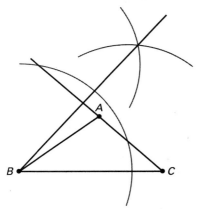

Construct a line through C which is perpendicular to $\overleftrightarrow{AB}$.

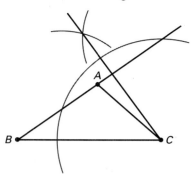

b. Here is $\triangle ABC$ with its 3 altitudes.

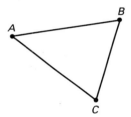

The 3 altitudes intersect at one point.

29. a. Sample answer:
Draw a triangle and label the vertices A, B, and C.

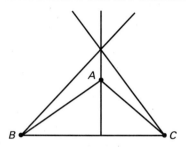

Bisect $\angle A$.

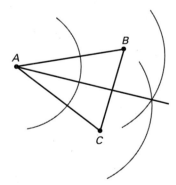

Bisect ∠*B*.

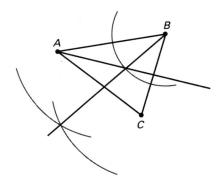

Bisect ∠*C*.

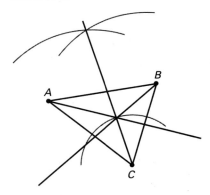

b. Notice that the angle bisectors intersect at a common point. Label this point *I*. Construct a line through *I* that is perpendicular to $\overline{AB}$. Label the point where the perpendicular line intersects $\overline{AB}$.

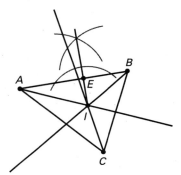

c. Set the compass to the distance *EI*. Place the compass point at *I* and draw a circle.

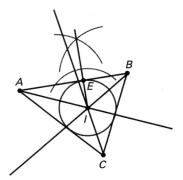

The circle drawn is the inscribed circle of the triangle.

30. Trace the original figure. Label the point *M*.

Draw a line through *M* that intersects the given line. Label the intersection point *P*.

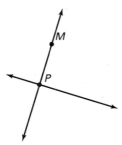

Draw an arc with center at *P*. Label new intersection points *R* and *T*.

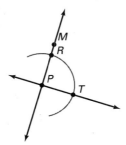

Without adjusting the compass setting, place the compass point at *M* and draw a new arc. Label the new intersection point *N*.

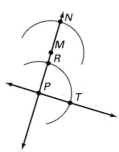

Set the compass to the distance *RT*. Place the compass point on *N* and draw an arc. Label the point of intersection *O*.

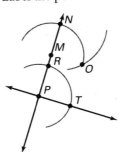

Draw the line containing *M* and *O*.

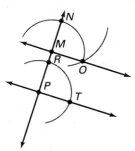

The line $\overleftrightarrow{MO}$ is parallel to the given line.

31. Trace the original figure. Label the point *M*.

Draw a line through *M* that intersects the given line. Label the intersection point *P*.

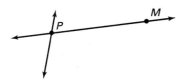

Draw an arc with its center at *P*. Label new intersection points *R* and *T*.

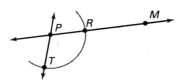

Without adjusting the compass setting, place the compass point at *M* and draw a new arc. Label the new intersection point *N*.

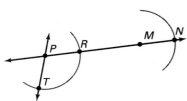

Set the compass to the distance *RT*. Place the compass point on *N* and draw an arc. Label the point of intersection *O*.

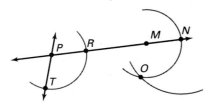

Draw the line containing M and O.

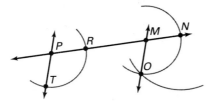

The line $\overleftrightarrow{MO}$ is parallel to the given line.

32. Sample answer: A trapezoid has exactly one pair of parallel lines. To construct one, construct a pair of parallel line segments with different lengths. Begin by drawing an intersecting line and line segment. Label the intersection point P, the other endpoint of the segment O and a point on the line M.

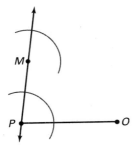

Using the same compass setting, make an arc with center at P and an arc with center at M.

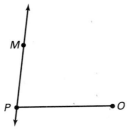

Set the compass to the distance between the intersection points of the lower arc. Use this setting to draw an arc from the upper arc intersection point of $\overleftrightarrow{PM}$.

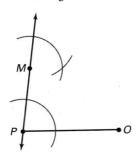

Draw a line segment with endpoint M which contains the intersection point formed by the two upper arcs and is not the same length as $\overline{PO}$. Label its endpoint N. Connect N and O.

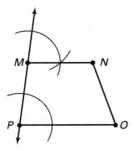

Quadrilateral $MNOP$ is a trapezoid.

33. Sample answer: A parallelogram has two points of parallel lines. Begin by drawing a pair of intersecting lines. Label the intersection point P and a point on each line M and O.

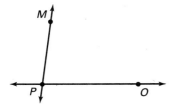

Using the same compass setting, make an arc with center at P, an arc with center at M, and an arc with center at O.

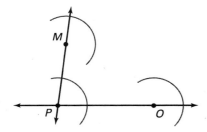

Set the compass to the distance between the intersection points of the arc at P. Use this setting to draw an arc from the intersection of the arc at M and the line $\overleftrightarrow{PM}$ and to draw an arc from the intersection of the arc at O and the line $\overleftrightarrow{PO}$.

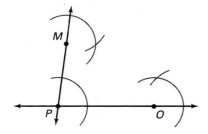

Draw the line containing M and the intersection of the two arcs near M and the line containing O and the intersection of the two arcs near O. Label the intersection point of these lines N.

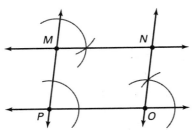

Quadrilateral $MNOP$ is a parallelogram.

34. Sample answer: A rectangle is a parallelogram with 90° interior angles. Begin with a line and a point not on the line. Label the point M.

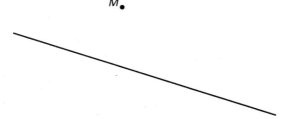

Construct the perpendicular to the line through M. Label the intersection point P.

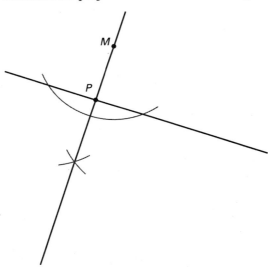

Construct the line containing *M* that is parallel to the original line. Label a point on the line *N*.

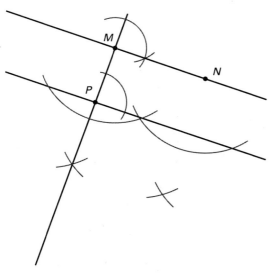

Construct the perpendicular to the original line through *N*. Label the intersection point *O*.

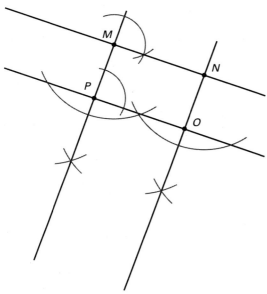

Quadrilateral *MNOP* is a rectangle.

35. Sample answer: A rhombus is a parallelogram with congruent line segments. Its diagonals are perpendicular. Begin by drawing a line and a point not on the line.

•

Construct the line through the point that is perpendicular to the original line.

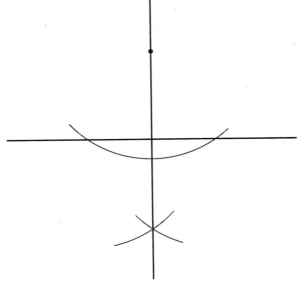

Put the point of the compass at the intersection of the perpendicular line and draw two arcs that intersect the original line. Label the intersection points *M* and *O*.

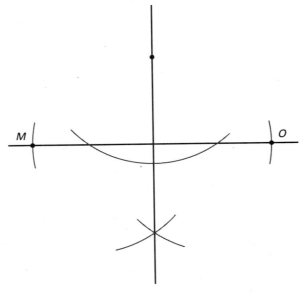

Change the compass setting. Put the point of the compass at the intersection of the perpendicular lines and draw two arcs that intersect the line that was constructed. Label the intersection points *N* and *P*. Connect the labeled points to form a quadrilateral.

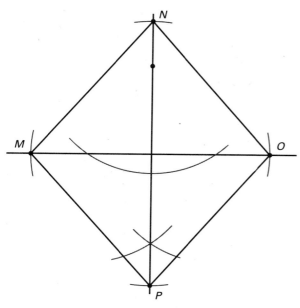

Quadrilateral *MNOP* is a rhombus.

36. Sample answer: A square is a parallelogram with congruent line segments and congruent interior angles. Its diagonals are perpendicular and congruent. Begin by drawing a line and a point not on the line.

Construct the line through the point that is perpendicular to the original line.

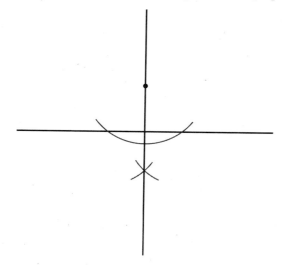

Put the point of the compass at the intersection of the perpendicular lines and draw a circle. Label the four intersection points *M, N, O,* and *P.*

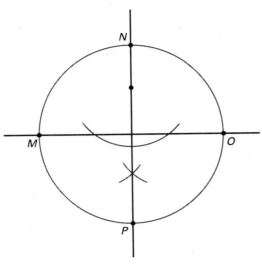

Connect the labeled points to form a quadrilateral.

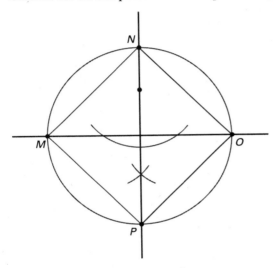

Quadrilateral *MNOP* is a rhombus.

37. Sample answer: A kite is a quadrilateral which has 2 pairs of adjacent sides which are congruent. Its diagonals are perpendicular. Begin by drawing a line and a point not on the line. Label the point *N.*

N •

Construct the line through *N* that is perpendicular to the original line. Label a point *P* on the perpendicular line and below the point of intersection.

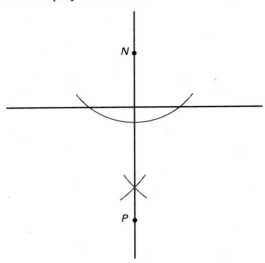

Put the point of the compass at the intersection of the perpendicular lines and draw two arcs that intersect the original line. Label the intersection point *M* and *O*.

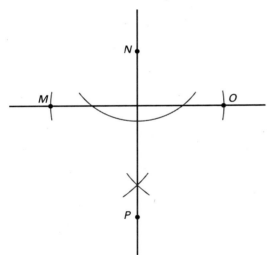

Connect the labeled points to form a quadrilateral.

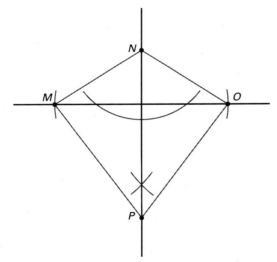

Quadrilateral *MNOP* is a kite.

38. Same compass setting used   **39.** Same compass setting used

40. SSS   **41.** CPCTC

42. The same compass setting was used to create $\overline{AB}$, $\overline{AD}$, $\overline{CB}$, and $\overline{CD}$, so the segments are congruent. Thus, $ABCD$ is a rhombus by definition. By Theorem 4.5.8, the diagonals of a rhombus are perpendicular, so $\overleftrightarrow{BD} \perp \overline{AC}$. By Theorem 4.5.6, $ABCD$ is a parallelogram, so by Theorem 4.5.5, the diagonals of $ABCD$ bisect each other. Thus, $\overleftrightarrow{BD}$ bisects $\overline{AC}$.

43.

Statements	Reasons
$\overline{AB} \cong \overline{AD}$ $\overline{CB} \cong \overline{CD}$	Same compass setting
$ABCD$ is a kite.	Definition of kite
$\overline{AC} \perp \overline{BD}$	Diagonals of a kite are perpendicular.
$BD \parallel \ell$	ℓ is an extension of line segment BD.
$\overline{AC} \perp \ell$	A line which is perpendicular to a line which is parallel to the given line is also perpendicular to the given line.

44.

Statements	Reasons
$PR \cong MN$ $PT \cong MO$	Same compass setting
$RT \cong NO$	Same compass setting
$\triangle PRT \cong \triangle MNO$	SSS
$\angle NMO \cong \angle RPT$	CPCTC
$\overleftrightarrow{MO} \parallel \ell$	Converse of the Corresponding Angles Postulate

45. The shortest path is the line through point X that is perpendicular to the road.

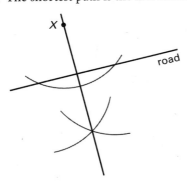

46. Sample answer:

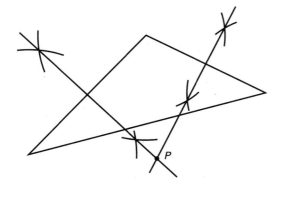

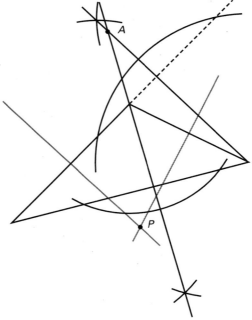

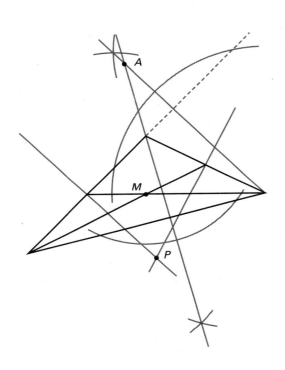

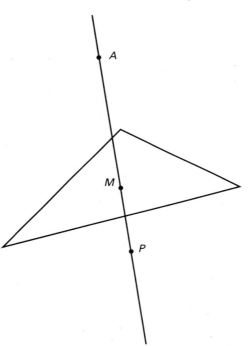

47. rotation

48. reflection

49. translation

50. trapezoid

51. parallelogram and rhombus

52. kite

53. parallelogram, rectangle, square, and rhombus

54.

55. Sample answer:

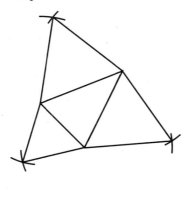

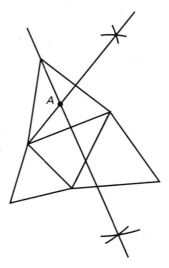

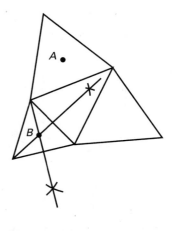

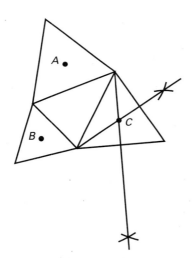

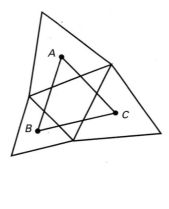

$\triangle ABC$ is equilateral.

5.

6.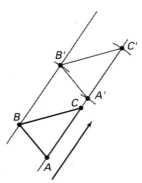

7.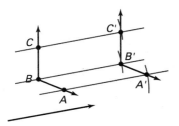

The angles have the same measure.

PAGES 275-279, PRACTICE AND APPLY

8.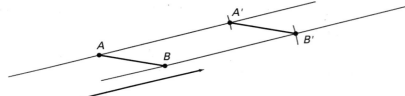

9.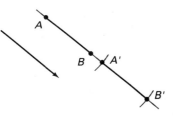

10.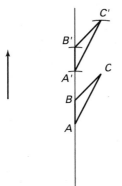

11.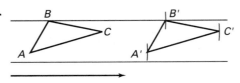

12. Follow the steps given.

a.

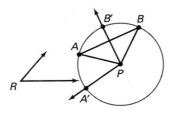

b.

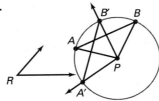

c.

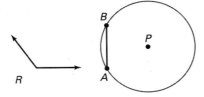

d.

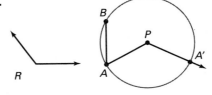

13. Follow the steps given.

a.

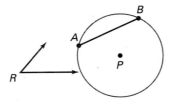

b.

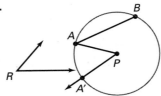

c.

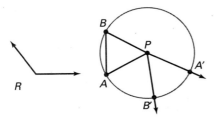

d.

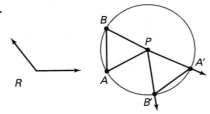

14. Follow the steps given.

a.

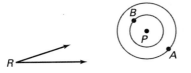

b.

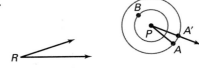

c.

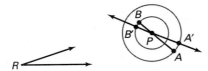

d.

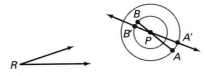

15. Follow the steps given.

a.

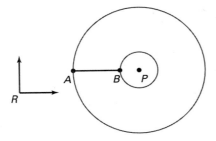

b.

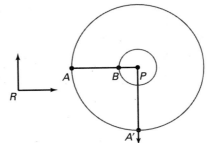

c.

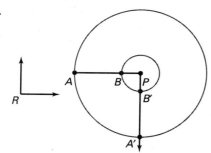

d.

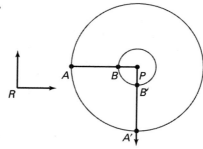

16. Follow the steps given.

a.

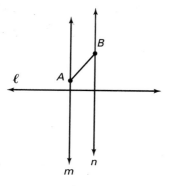

b.

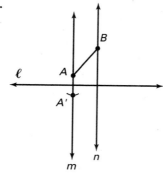

c.

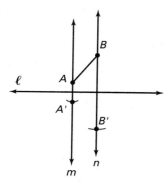

d.

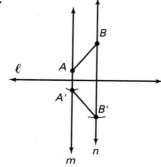

17. Follow the steps given.

a.

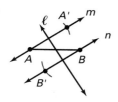

b.

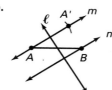

c.

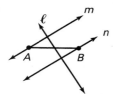

d.

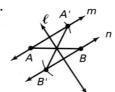

18. Follow the steps given.

a.

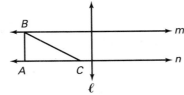

b.

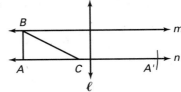

c.

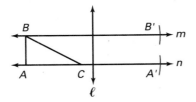

d.

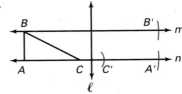

e.

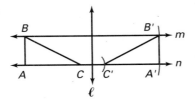

19. Follow the steps given.

a.

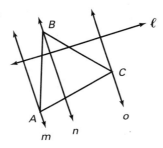

b.

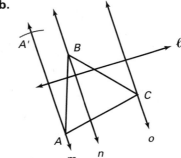

c.

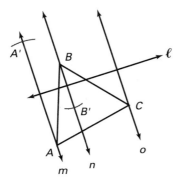

d.

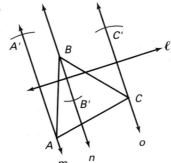

e.

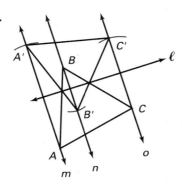

20. $\triangle ABC$ is possible.

21. $\triangle DEF$ is impossible since $DE + EF < DF$.

22. $\triangle GHI$ is impossible since, G, H, and I are collinear.

23. $\triangle JKL$ is possible.

24. a. $10 + 8 > x$ **b.** $x + 8 > 10$ **c.** $x + 10 > 8$

25. $\left. \begin{array}{l} x + 8 > 10 \Rightarrow x > 2 \\ x + 10 > 8 \Rightarrow x > -2 \\ 10 + 8 > x \Rightarrow x < 18 \end{array} \right\} \Rightarrow 2 < x < 18$

26. Same compass setting used

27. Overlapping Angles Theorem

28. SAS

29. CPCTC

30. Corresponding Angles Postulate

31. definition of a rectangle

32. $\overline{A'D}$

33. opposite sides of a rectangle are congruent

34. Transitive Property of Congruence

35. Subtraction Property

36. SAS

37. CPCTC

38.

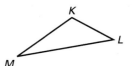

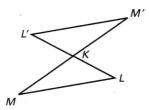

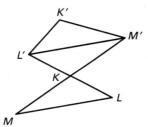

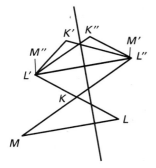

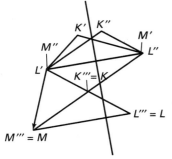

They are the same triangle.

39. Sample answers: They have to walk the same distance in the same direction.

40.

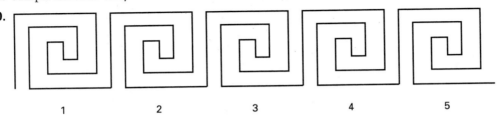

1	2	3	4	5

Sample answer: a rotation of 90° about *B*

PAGES 279–280, LOOK BACK

41. reflectional symmetry across a vertical axis

42. 180° rotational symmetry

43. reflectional symmetry across a vertical and horizontal axis and 180° rotational symmetry

44. reflectional symmetry across a vertical, horizontal, and two diagonal axes and 90°, 180°, and 270° rotational symmetry

45. a, b, and d

46.

Statements	Reasons
$BX \perp AD$ $CY \perp AD$ $AD \parallel BC$	Given
$m\angle BCY = 90°$ $m\angle XBC = 90°$	Alternate Interior Angles
$BCYX$ is a rectangle.	Definition of rectangle
$\overline{BX} \cong \overline{CY}$	Opposite sides of a Rectangle are congruent
$\overline{BA} \cong \overline{CD}$	Given
$\angle BXA \cong \angle CYD$	All right angles are congruent
$\triangle BXA \cong \triangle CYD$	HL
$\angle A \cong \angle D$	CPCTC

47.

Statements	Reasons
$\angle XZY \cong \angle YZS$	Vertical Angles Theorem
$\angle XQZ \cong \angle YZS$	Alternate Interior Angles Theorem
$\overline{QZ} \cong \overline{SZ}$	Diagonals of a parallelogram bisect each other.
$\triangle QXZ \cong \triangle SYZ$	ASA
$\overline{XZ} \cong \overline{YZ}$	CPCTC

48. The opposite sides of the quadrilateral formed by the parallel ruler are congruent, so by Theorem 4.6.1 it is a parallelogram. By definition of a parallelogram, the sides are always parallel.

PAGE 280, LOOK BEYOND

49. Check student's drawing.

50. They are all equilateral triangles, because any triangle chosen has all sides with the same length.

51. $P_1 = 3$ units

$P_2 = \frac{3}{2}$ units

$P_3 = \frac{3}{4}$ units

52. $\frac{3}{2} + \frac{3}{4} + \frac{3}{8} + \frac{3}{16} = 2\frac{13}{16}$. The perimeter of the first triangle is 3. As the perimeters of the additional triangles are added to the sum, the total approaches the perimeter of the first triangle.

CHAPTER REVIEW AND ASSESSMENT

1. $\overline{QR}$ **2.** $\overline{MK}$ **3.** $\angle K$ **4.** $\triangle LKM$

5. Yes; SAS   **6.** Yes; SSS   **7.** Yes; SSS   **8.** Yes; ASA

9. Yes; AAS   **10.** Yes; HL

11. Yes; AAS   **12.** Converse of the Isosceles Triangle Theorem

13. Definition of isosceles triangle   **14.** ASA

15. CPCTC

16. Given: *RSTU* is a parallelogram
Prove: m$\angle R$ + m$\angle S$ = 180°

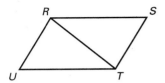

Statements	Reasons
$\overline{RU} \parallel \overline{ST}$	Definition of a parallelogram
m$\angle RTS$ = m$\angle URT$	Alternate Interior Angles Theorem
m$\angle RTS$ + m$\angle TRS$ + m$\angle S$ = 180°	Triangle Sum Theorem
m$\angle URT$ + m$\angle TRS$ + m$\angle S$ = 180°	Substitution Property
m$\angle URT$ + m$\angle TRS$ = m$\angle R$	Angle Addition Postulate
m$\angle R$ + m$\angle S$ = 180°	Substitution Property

17. Given: *CDEF* is a rectangle.
$\overline{CE}$ and $\overline{DF}$ intersect at *G*.
Prove: $\triangle CDG \cong \triangle EFG$

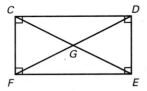

Statements	Reasons
CDEF is a rectangle.	Given
CDEF is a parallelogram.	Theorem 4.5.7
$\overline{CG} \cong \overline{GE}$ $\overline{DG} \cong \overline{GF}$	Theorem 4.5.5
$\overline{CD} \cong \overline{EF}$	Theorem 4.5.2
$\triangle CDG \cong \triangle EFG$	SSS

18. Given: *NOPQ* is a rhombus.
  Prove: $\angle NOQ \cong \angle POQ$

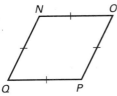

Statements	Reasons
NOPQ is a rhombus.	Given
$\overline{NO} \cong \overline{PO}$	Definition of a rhombus
$\overline{OQ} \cong \overline{OQ}$	Reflexive Property of Congruence
$\overline{QN} \cong \overline{QP}$	Definition of a rhombus
$\triangle NOQ \cong \triangle POQ$	SSS
$\angle NOQ \cong \angle POQ$	CPCTC

19.

Given: *VWXY* is a square.
Prove: $\triangle VWX \cong \triangle WXY$

Statements	Reasons
VWXY is a square.	Given
$\overline{VW} \cong \overline{WX}$ $\overline{WX} \cong \overline{XY}$	Definition of a square
VWXY is a rectangle.	Theorem 4.5.11
$m\angle W = 90°$ $m\angle X = 90°$	Definition of rectangle
$m\angle W = m\angle X$	Transitive Property or Substitution Property
$\triangle VWX \cong \triangle WXY$	SAS

20.

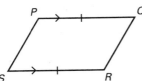

One pair of opposite sides are parallel and congruent.
parallelogram by Theorem 4.6.2

21.

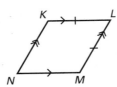

Opposite sides are parallel and one pair of adjacent sides are congruent.
parallelogram by definition
rhombus by Theorem 4.6.6

22.

Diagonals are perpendicular and bisect each other, and one angle is a right angle.
parallelogram by Theorem 4.6.3
rhombus by Theorem 4.6.8
rectangle by Theorem 4.6.4
square by definition

23.

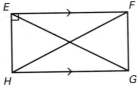

One pair of opposite sides are parallel, diagonals are congruent, and one angle is a right angle.
parallelogram by HL and Theorem 4.6.2
rectangle by Theorem 4.6.4 or 4.6.5

24.

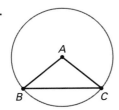

25.

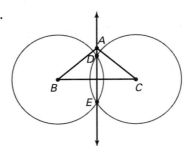

26.

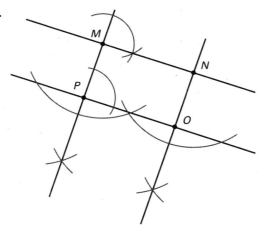

27.

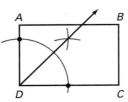

28.

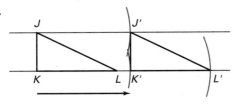

29.

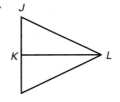

30.

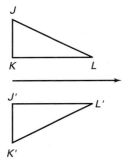

31.

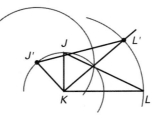

32. Because the length of the lake is part of a triangle which is congruent to a second triangle by SAS, and because CPCTC, we can determine the length of the lake by measuring the distance of the side of the second triangle which corresponds to the length of the lake.

33. Sample construction:
Open your compass to the horizontal length of one of the rhombus' of the figure. Make circle with this compass opening.

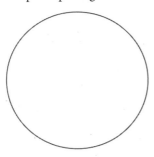

Open your compass to the distance of the length of a side of one of the rhombus'.
Using the same center as before, make a circle with this compass opening.

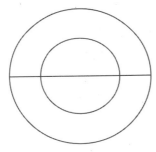

Draw the horizontal diameter of the big circle.

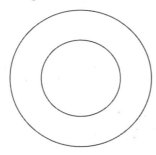

Draw the perpendicular bisector of the each radius of the diameter drawn for the large circle.

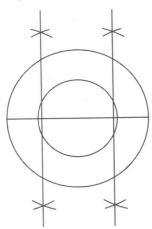

Find the points where the perpendicular bisectors intersect the small circle.

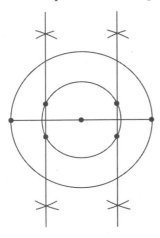

Connect the points as shown in the original figure.

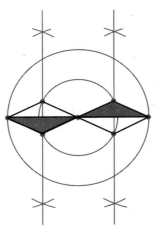

CHAPTER TEST

1. $\overline{FG}$

2. $\overline{HG}$

3. $\overline{BA}$

4. $\angle G$

5. $\angle D$

6. $GFEH$

7. yes; SSS

8. yes; ASA

9. yes; SAS

10. yes; HL

11. $m\angle Y = 50°$ (Isosceles Triangle Theorem)

12. $m\angle Q = 60°$ (Corollary: The measure of each angle of an equilateral triangle is 60°.)

13. $LN = 30$ (Corollary: The bisector of the vertex angle of an isosceles triangle is the perpendicular bisector of the base.)

14. $3x - 20 = 2x$
$x = 20$
$m\angle A = 2(20) = 40$ and $m\angle C = 3(20) - 20 = 40$
$m\angle B = 180° - 80° = 100°$

15. $JK = 85$ ft; By SAS, the triangles are congruent, so $JK = 85$ ft by CPCTC.

16. Opposite sides of a parallelogram are congruent.
$$2x + 10 = 4x - 24$$
$$2x = 34$$
$$x = 17$$
$$RU = 2(17) + 10 = 44$$

17. $\triangle XWZ \cong \triangle XYZ$ by SAS, so by CPCTC, $\angle XZY \cong \angle XZW$.
Therefore, $m\angle XZY = 35°$ and $m\angle WZY = 35° + 35° = 70°$.
Adjacent sides of a parallelogram are supplementary, so $m\angle XYZ = 180° - 70° = 110°$.

18. Opposite sides of a parallelogram are congruent.

19. Given

20. Transitive Property of Congruence

21. Definition of Rhombus

22. Sample answer:

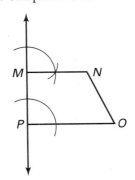

23. Sample answer:

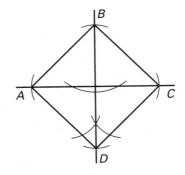

24. $\triangle ABC$ is possible because $17 + 20 > 22$.

25. $\triangle DEF$ is not possible because $25 + 10 < 40$.

26. $\triangle GHI$ is possible because $9 + 9 > 14$.

27. $\triangle XYZ$ is impossible because $13 + 11 = 24$.

CHAPTERS 1–4 CUMULATIVE ASSESSMENT

1. slope of $\overline{DE} = \dfrac{7 - (-3)}{-7 - (-3)} = \dfrac{10}{-4} = -\dfrac{5}{2}$
slope of $\overline{AB} = \dfrac{6 - (-5)}{5 - 1} = \dfrac{11}{4}$
$\dfrac{11}{4} > -\dfrac{5}{2}$
The answer is B.

2. $m\angle R = m\angle O \qquad m\angle Q = m\angle N$
$(3x + 2)° = 56° \qquad (2y + 14)° = 64°$
$3x = 54 \qquad\qquad 2y = 50$
$x = 18 \qquad\qquad y = 25$
$y > x$
The answer is B.

3. SSA does not imply congruence. The answer is D.

4. slope of $\overline{MN} = \dfrac{-4 - 7}{4 - (-4)} = -\dfrac{11}{8}$

A line $\perp$ to MN has slope $\dfrac{8}{11}$

a. $m = \dfrac{-4 - 7}{8 - 0} = \dfrac{-11}{8}$

b. $m = \dfrac{4 - (-7)}{-4 - 4} = \dfrac{11}{-8}$

c. $m = \dfrac{8 - 0}{4 - (-7)} = \dfrac{8}{11}$

d. $m = \dfrac{4 - (-4)}{-4 - 7} = \dfrac{8}{-11}$

The only pair of points given that defines a line perpendicular to $\overline{MN}$ is the pair given in c.

5. $m\angle ACB = \dfrac{1}{2}(180° - 54°)$
$= \dfrac{1}{2}(126°) = 63°$

Choose c.

6. $720° - (120° + 140° + 140° + 130° + 70°) =$
$720° - (600°) = 120°$
Choose a.

7. The two triangles have 2 sides and their included angles congruent.
Choose d.

8. A rhombus is a parallelogram with four congruent sides thus opposite sides are parallel, diagonals are congruent and four sides are congruent. The diagonals are not necessarily perpendicular.
Choose b.

9. If trees are conifers, then the trees bear cones.

10. Translation with points $(3, 0)$, $(6, 3)$, $(7, 0)$

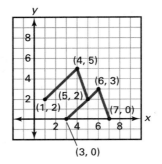

11. $m\angle 2$ is an alternate interior angle to the supplementary angle of the one marked $130°$. Therefore $m\angle 2 = 50°$.

12. Yes; AAS or ASA

13. In **c**, $90° + 65° + 15° = 170°$ but the angles of a triangle sum to $180°$. Thus, it is not a triangle. In **d**, $BC + AC < AB$ but the sum of the lengths of any two sides of a triangle is always greater than the length of the other side. Thus, it is not a triangle.

14. $x = 3x - 8$
$2x = 8$
$x = 4$

15. $\dfrac{(n - 2)180°}{n} = \dfrac{(6 - 2)180°}{6} = \dfrac{2}{3}180° = 120°$

16. $\dfrac{BC + FE}{2} = AD$
$\dfrac{BC + 50}{2} = 46$
$BC + 50 = 92$
$BC = 42$

5. $P = 2 + 1.5 + 2 + 4 + 2.5 + 2.5 + 1.5 + 5 = 21$ units

6. The area can be found by adding the areas of three rectangles:
$A = (1.5)(2) + (2.5)(4) = (2.5)(1.5) = 16.75$ units^2

7. $P = 2b + 2h$
$48 = 2b + 2h$
$24 = b + h$
The maximum area is obtained when the rectangle is a square, or $b = h$. So $b = 12$ and $h = 12$. The maximum area is $A = bh = (12)(12) = 144$ ft.^{2}.

8. $A = bh$
$900 = bh$
The minimum perimeter is obtained when the rectangle is a square, or $b = h$. So $b = 30$ and $h = 30$. The minimum perimeter is $P = 2b + 2h = 2(30) = 120$ feet.

9. Perimeter $ADLI = 2(15) + 2(11) = 52$ in.

10. Area $ADLI = (15)(11) = 165$ in^2

11. $CD = AD - AC = 15 - 13 = 2$
Perimeter $CDLK = 2(2) + 2(11) = 26$ in.

12. Area $CDLK = 2(11) = 22$ in^2

13. $GH = BD - CD = 10 - 2 = 8$
$LJ = BD = 10$
$GJ = DL - CH = 11 - 4 = 7$
Perimeter $GHCDLJ = 8 + 4 + 2 + 11 + 10 + 7 = 42$ in.

14. Area $GHCDLJ =$ Area $GHKJ +$ Area $CDLK = (8)(7) + 22 = 78$ in^2

15. Perimeter $BCHG = 2BC + 2CH = 2(8) + 2(4) = 24$ in.

16. Area $BCHG = (BC)(CH) = (8)(4) = 32$ in^2

17. Area $BHG = \frac{1}{2}$ Area $BCHG = \frac{1}{2}(32) = 16$ in^2

18. Area $ADI = \frac{1}{2}$ Area $ADLI = \frac{1}{2}(165) = 82.5$ in^2

19. Let b be the base and h be the height of a rectangle. Since opposite sides of a rectangle are the same length, two sides have length b and two sides have length h. The perimeter is $b + b + h + h = 2b + 2h$.

20. The base is the length of the segment between the points $(0, 0)$ and $(5, 0)$. Its length is $|5 - 0| = 5$ units. The height is the length of the segment between $(0, 0)$ and $(0, 2)$, $|0 - 2| = |-2| = 2$ units. The area is: $A = bh = 5 \cdot 2 = 10$ units^2.

21. The base is the length of the segment between the points $(3, 1)$ and $(9, 1)$, $|9 - 3|$ $= 6$ units. The height is the length of the segment between $(3, 1)$ and $(3, 7)$, $|7 - 1| = 6$ units. The area is: $A = bh = 6 \cdot 6 = 36$ units^2.

22. The base is the length of the segment between the points $(-2, -5)$ and $(4, -5)$, $|-2 - 4| = |-6| = 6$ units. The height is the length of the segment between $(-2, -5)$ and $(-2, 3)$, $|-5 - 3| = |-8| = 8$ units. The area is $A = bh = 6 \cdot 8 = 48$ units^2.

23. The base is the length of the segment between $(0, 0)$ and $(3, -3)$:
$d = \sqrt{(3 - 0)^2 + (-3 - 0)^2} = \sqrt{9 + 9} = \sqrt{18} = 3\sqrt{2}$. The height is the length of the segment between $(0, 0)$ and $(3, 3)$: $d = \sqrt{(3 - 0)^2 + (3 - 0)^2} = \sqrt{9 + 9} = \sqrt{18} = 3\sqrt{2}$. The area is $A = bh = (3\sqrt{2}) \cdot (3\sqrt{2}) = 9 \cdot 2 = 18$ units^2.

24. Let h be the height.
Then $b = 3h$ and $P = 2(3h) + 2(h) = 72$
$8h = 72$
$h = 9$ cm
$b = 3(9) = 27$ cm
$A = bh = (9)(27) = 243$ cm^2

25. Let h be the height.
Then $b = 2h + 3$ and
$A = (2h + 3)h = 2h^2 + 3h = 27$
$2h^2 + 3h - 27 = 0$
$(h - 3)(2h + 9) = 0$
$h = 3$ or $h = -\dfrac{9}{2}$
The height cannot be negative, so $h = 3$.
$b = 2(3) + 3 = 9$ cm
$P = 2b + 2h = 18 + 6 = 24$ cm^2

26. Let h be the height.
Then $b = 7h$ and $2(7h) + 2h = 80x$
$14h + 2h = 80x$
$16h = 80x$
$h = 5x$
$b = 7(5x) = 35x$
$A = bh = (5x)(35x) = 175x^2$

27. Let b be the base.
Then $h = b + 3$ and $2b + 2h = bh$.
$2b + 2(b + 3) = b(b + 3)$
$2b + 2b + 6 = b^2 + 3b$
$b^2 - b - 6 = 0$
$(b - 3)(b + 2) = 0$
$b = 3$ or $b = -2$
The base cannot be negative, so
$b = 3$
$h = b + 3 = 6$
$A = bh = 3(6) = 18$

28. $2b + 2h = 100$
$b + h = 50$
$b = 50 - h$
or $h = 50 - b$

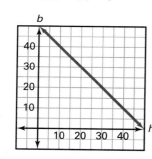
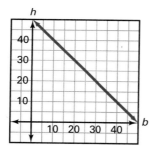

The relationship is a linear function. Any function value that produces a zero or negative value for b or h doesn't make sense for the perimeter example—specifically:
$b \leq 0$, $b \geq 50$ or $h \leq 0$, $h \geq 50$.

29. $bh = 100$

$$b = \frac{100}{h}$$

$$\text{or } h = \frac{100}{b}$$

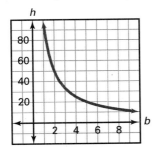 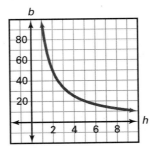

The relationship is a nonlinear function. Any value which causes division by zero will make the function undefined, and a negative value for b or h doesn't make sense for a side length. That is, $b \leq 0$ or $h \leq 0$ are values that don't make sense in the equation.

30. Sample answer:
Area of one square $= 0.5(0.5) = 0.25$
There are about 14 squares in the figure.
$A \approx 14(0.25) = 3.5 \text{ cm}^2$

31. Sample answer:
Area of one square $= 0.5(0.5) = 0.25$
There are about 30 squares in the figure.
$A \approx 30(0.25) = 7.5 \text{ cm}^2$

32. Area of roof $= 2(25)(42) = 2100 \text{ ft}^2$
Area of one sheet of plywood $= 8(4) = 32 \text{ ft}^2$
Number of sheets needed $= \frac{2100}{32} \approx 66$

33. Sample answer:

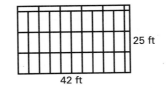

12 ft^2 of plywood is left over after covering the whole roof.

34. $(6)(15) = 90 \text{ ft}^2$ of panel are needed. Each panel must cover an area of $90 \div 2 = 45 \text{ ft}^2$ and be $45 \div 10 = 4.5$ ft wide.

35.

base	height	perimeter	area
1	99	200	99
2	98	200	196
5	95	200	475
20	80	200	1600
30	70	200	2100
40	60	200	2400
50	50	200	2500
60	40	200	2400

Maximum area $= 2500 \text{ ft}^2$

36.

base	height	perimeter	area
1	5625	11,252	5625
2	2812.5	5629	5625
5	1125	2260	5625
20	281.25	602.5	5625
30	187.5	435	5625
40	140.625	361.25	5625
50	112.5	325	5625
60	93.75	307.50	5625
70	80.36	300.71	5625
75	75	300	5625
80	70.31	300.63	5625
90	62.5	305	5625

The minimum amount of fencing needed is 300 feet.

37.

$$P = b + 2h$$
$$200 = b + 2h$$
$$200 - b = 2h$$
$$h = \frac{200 - b}{2}$$

base	height	perimeter	area
1	99.5	200	99.5
2	99	200	198
5	97.5	200	487.5
20	90	200	1800
30	85	200	2550
40	80	200	3200
50	75	200	3750
60	70	200	4200
70	65	200	4550
80	60	200	4800
90	55	200	4950
100	50	200	5000
110	45	200	4950

Maximum area = 5000 ft^2

38.

$$P = b + 2h$$
$$A = bh = 5625$$
$$h = \frac{5625}{b}$$

base	height	perimeter	area
1	5625	11,251	5625
2	2812.5	5627	5625
5	1125	2255	5625
20	281.25	582.5	5625
30	187.5	405	5625
40	140.63	321.25	5625
50	112.5	275	5625
60	93.75	247.5	5625
70	80.357	230.7	5625
80	70.313	220.63	5625
90	62.5	215	5625
100	56.25	212.5	5625
106	53.066	212.13	5625
110	51.136	212.27	5625
120	46.875	213.75	5625

Minimum fencing needed ≈ 212.13 feet.

39. Divide the lawn into two rectangles.
Area of rectangle 1 = $(40 - 30)(21 - 15)$
$$= 10(6) = 60 \text{ ft}^2$$
Area of rectangle 2 = $30(21) = 630 \text{ ft}^2$
Total area = $60 + 630 = 690 \text{ ft}^2$

40. The area of the garden is $10 \times 16 = 160 \text{ ft}^2$. Since the rows are 10 feet long, and his marigold row will be 1 ft wide, it will take up 10 ft^2 of space, leaving $160 - 10 = 150 \text{ ft}^2$. The five vegetables need an equal amount of space, $150 \div 5 = 30 \text{ ft}^2$ each.

41. $A = 2[(16)(10) + (14)(10)] + (14)(16) - [(7)(3) + (4)(6)]$
$$= 2(160 + 140) + 224 - (21 + 24)$$
$$= 2(300) + 224 - 45$$
$$= 779 \text{ ft}^2$$
Since the paints are available only in gallon cans, she needs 2 gallons of base paint at $10/gal and 4 gallons of finish paint at $20/gal.
Cost of paint = $2(10) + 4(20) = 20 + 80 = \$100$
Tax = $(100)(0.07) = \$7.00$
Total cost = $\$100 + \$7.00 = \$107.00$

42.

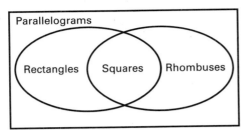

Parallelograms

Rectangles   Squares   Rhombuses

43. $\frac{360}{3} = 120°$

44. Sum of the interior angle measures
$= (4 - 2)(180°) = 2(180) = 360°;$
fourth angle measure $= 360 - 300 = 60°$

45. $(n - 2)180°$

46. $\frac{(6 - 2)(180°)}{6} = \frac{720°}{6} = 120°$

47. $m = \frac{y_2 - y_1}{x_2 - x_1} = \frac{3 - (-1)}{2 - 4} = \frac{3 + 1}{-2} = -2$

48. The slope of the line in Exercise 46 is -2, so the slope of a line perpendicular to the segment is $-\frac{1}{-2} = \frac{1}{2}$.

49. 0

50. undefined

51. $\left(\frac{-4 + 6}{2}, \frac{6 + 4}{2}\right) = (1, 5)$

52. A circle with a diameter equal to the length of the side of the square can be inscribed inside the square as shown. Therefore, the square has the greater area.

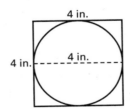

4 in.

4 in.   4 in.

53. Divide the square into smaller squares. Then estimate the number of squares outside the circle and subtract from the total number of squares. The area of the remaining squares approximates the area of the circle.

5.2 **PAGE 308, GUIDED SKILLS PRACTICE**

6. $A = \frac{1}{2}bh = \frac{1}{2} \cdot (40)(15) = 300$ units^2

7. $A = bh = (7)(5) = 35$ units^2

8. Trapezoid: $A = \frac{1}{2}(b_1 + b_2)h = \frac{1}{2}(10 + 7)(6) = 51$ units^2
Parallelogram: $A = bh = (10 + 7)(6) = 102$ units^2

9. a. $A = \frac{1}{2}bh = \frac{1}{2}(17)(24) = 204$ units^2

b. $A = bh = (20)(24) = 480$ units^2

c. $DC = DE + EC = 17 + 20 = 37$
$A = \frac{1}{2}(b_1 + b_2)h = \frac{1}{2}(37 + 20)(24) = 684$ units^2

10. $A = \frac{1}{2}(4)(7) = 14$ units^2

11. $A = \frac{1}{2}(8)(3) = 12$ units^2

12. $A = \frac{1}{2}(14)(6) = 42$ units^2

13. $A = (24)(13) = 312$ units^2

14. $A = (2)(11) = 22$ units^2

15. $A = 8(6) = 48$ units^2

16. $A = \frac{1}{2}(9 + 7)(3) = 24$ units^2

17. $A = \frac{1}{2}(4 + 2)(2) = 6$ units^2

18. $A = \frac{1}{2}(25 + 40)(20) = 650$ units^2

19. height $= BK = 14$.
$A = \frac{1}{2}(KL)(BK) = \frac{1}{2}(14)(14) = 98$ units^2

20. $JK = IL - KL - IJ = 32 - 14 - 8 = 10$.
$A = \frac{1}{2}(10)(14) = 70$ units^2

21 Since $KC \parallel JB$, $BC = JK = 10$.
Thus $A = \frac{1}{2}(10)(14) = 70$ units^2

22. Since $\overline{DH} \parallel \overline{IL}$, the altitude of $\triangle DIJ$ is congruent to $\overline{FK}$. So $h = 7$.
$A = \frac{1}{2}(8)(7) = 28$ units^2

23. $A = (JK)(BK) = (10)(14) = 140$ units^2

24. $A = (JK)(FK) = (10)(7) = 70$ units^2

25. Since $\overline{AC} \parallel \overline{IL}$ and $\overline{BK} \perp \overline{IL}$, $\overline{BK} \perp \overline{AC}$. Therefore $\overline{BF} \perp \overline{AC}$. $BF = BK - FK = 14 - 7 = 7$. So $h = 7$.
$A = 8 \cdot 7 = 56$ units^2

26. $KI = IL - KL = 32 - 14 = 18$.
$A = (KI)(BK) = (18)(14) = 252$ units^2

27. $JL = IL - IJ = 32 - 8 = 24$.
$A = \frac{1}{2}(EH + JL)(FK) = \frac{1}{2}(17 + 24)(7)$
$= 143.5$ units^2

28. $BC = 10$, $BF = BK - FK = 14 - 7 = 7$
$A = \frac{1}{2}(BC + EH)(BF) = \frac{1}{2}(10 + 17)(7)$
$= 94.5$ units^2

29. $A = \frac{1}{2}(BC + JL)(BK) = \frac{1}{2}(10 + 24)(14)$
$= 238$ units^2

30. $AC = IK = IL - KL = 32 - 14 = 18$.
$A = \frac{1}{2}(AC + IL)(BK) = \frac{1}{2}(18 + 32)(14) = 350$ units^2

31. $b = |4 - 0| = 4$, $h = |3 - 0| = 3$
$A = \frac{1}{2}(4)(3) = 6$ units^2

32 $b = |5 - 1| = 4$. $h = 7 - 2 = 5$.
$A = \frac{1}{2}(4)(5) = 10$ units^2

33. $b = |3 - 1| = 2$. $h = |6 - 0| = 6$.
$A = \frac{1}{2}(2)(6) = 6$ units^2

34. $b = |-2 - 4| = 6$. $h = |-3 - 1| = 4$.
$A = \frac{1}{2}(6)(4) = 12$ units^2

35. $b = |4 - 0| = 4$. $h = |2 - 0| = 2$.
$A = 4 \cdot 2 = 8$ units^2

36. $b = |3 - 1| = 2$. $h = |3 - 0| = 3$.
$A = 2 \cdot 3 = 6$ units^2

37. $b = |-1 - 3| = 4$. $h = |-1 - 3| = 4$.
$A = 4 \cdot 4 = 16$ units^2

38 $b = |1 - 2| = 1$. $h = |1 - (-2)| = 3$.
$A = 1 \cdot 3 = 3$ units^2

39. $b_1 = |5 - (-1)| = 6$, $b_2 = |3 - 0| = 3$,
$h = |2 - 0| = 2$.
$A = \frac{1}{2}(6 + 3) \cdot 2 = 9$ units^2

40. $b_1 = |4 - 2| = 2$, $b_2 = |6 - (-1)| = 7$,
$h = |4 - 1| = 3$.
$A = \frac{1}{2}(2 + 7) \cdot 3 = 13.5$ units^2

41. $b_1 = |3 - 0| = 3$, $b_2 = |5 - 0| = 5$, $h = |5 - 1| = 4$.
$A = \frac{1}{2}(3 + 5) \cdot 4 = 16$ units^2

42. $b_1 = |6 - (-1)| = 7$, $b_2 = |3 - (-2)| = 5$,
$h = |1 - (-1)| = 2$.
$A = \frac{1}{2}(7 + 5) \cdot 2 = 12$ units^2

43. $100 = \frac{1}{2}(10)b$
$100 = 5b$
$b = 20$ cm

44. $123 = 15h$

$h = 8.2$ cm

45. The diagram shows that the area of the trapezoid is the same as the area of a rectangle that has a base equal to the length of the midsegment of the trapezoid, m, and the same height as the trapezoid, h. Thus, the area of the trapezoid is $A = mh$.

46. Left to right:

$A_1 = \frac{1}{2}(8)(6) = 24$ units^2,

$A_2 = \frac{1}{2}(8)(6.93) = 27.72$ units^2,

$A_3 = \frac{1}{2}(8)(4.95) = 19.8$ units^2.

The triangle with the largest area is the center one, which is an equilateral triangle. Conjecture: For triangles with a given perimeter, an equilateral triangle has the largest area.

Sample answer: This is similar to the fact that the largest area enclosed by a rectangle is enclosed by a square.

47. Left to right:

$A_1 = (7)(5) = 35$ units^2,

$A_2 = (7)(6) = 42$ units^2,

$A_3 = (7)(2) = 14$ units^2.

The parallelogram with the largest area is the rectangle. Conjecture: For parallelograms with a given perimeter, the parallelogram with the largest area is a rectangle.

Sample answer: The shortest distance between a point and a line is along the perpendicular. This perpendicular distance is larger in the rectangle than in the other two parallelograms.

48. Given a right or acute triangle ABC with altitude AF of length h and base b. Let $DEBC$ be a rectangle with base b and height h. The altitude AF forms two rectangles, $DAFC$ and $AEBF$. Since $\overline{AC}$ divides rectangle $DAFC$ in half and $\overline{AB}$ divides rectangle $AEBF$ in half, the area of the triangle is half the sum of the areas of the two smaller rectangles. The sum of the areas of the rectangles is bh, so the area of the triangle is $\frac{1}{2}bh$.

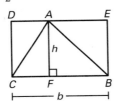

49. Given a parallelogram $ABCD$ with base b and height h, let $\overline{AE}$ be a segment perpendicular to the base, so that $\triangle AED$ is formed. Translate $\triangle AED$ by b units so that points D and C coincide, forming a rectangle with base b and height h. Its area is bh, so the area of the parallelogram is bh.

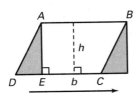

50. Given a trapezoid $ABCD$ with bases b_1 and b_2 and height h, let $EFGH \cong ABCD$. Rotate $EFGH$ 180° and translate it so that C and F coincide and B and G coincide, forming parallelogram $AHED$. The parallelogram has base $b_1 + b_2$ and height h, so its area is $(b_1 + b_2) \cdot h$. The area of the trapezoid is half the area of the parallelogram, so the area is: $\frac{1}{2}(b_1 + b_2) \cdot h$.

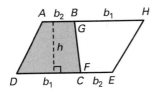

51. Sample answer:

Given: kite $ABCD$ with $\overline{AC} \perp \overline{BD}$,

Prove: Area $ABCD = \frac{1}{2}(BD)(AC)$

Statement	Reasons
$ABCD$ is a kite with $\overline{AC} \perp \overline{BD}$.	Given
Area kite $ABCD =$ Area $\triangle ABD +$ Area $\triangle BCD$	Sum of Areas Postulate
Area kite $ABCD = \frac{1}{2}(AX)(BD) + \frac{1}{2}(XC)(BD)$	Area of a Triangle Formula
Area kite $ABCD = \frac{1}{2}(BD)(AX + XC)$	Distributive Property
Area kite $ABCD = \frac{1}{2}(BD)(AC)$	Segment Addition Postulate

52. Rectangle $KQLN$ is divided into 2 congruent right triangles by side $\overline{KL}$. The shaded area is the area of $\triangle KLN$ minus the area of $\triangle KMN$. The area of a right triangle is $A = \frac{1}{2}bh$, so:

Area of $\triangle KLN = \frac{1}{2}(b + x) \cdot h = \frac{1}{2}bh + \frac{1}{2}xh$

Area of $\triangle KMN = \frac{1}{2}xh$

So Area of $\triangle KLM = \left(\frac{1}{2}bh + \frac{1}{2}xh\right) - \frac{1}{2}xh = \frac{1}{2}bh$

53. $A = \frac{1}{2}bh = \frac{1}{2}(3500)(2800) = 4{,}900{,}000 \text{ ft}^2$

$A = \frac{4{,}900{,}000}{43{,}560} \approx 112.5 \text{ acres}$

$112.5 \text{ acres} \cdot 435 \frac{\text{lb}}{\text{acre}} = 48{,}937.5 \text{ lb}$

PAGE 311, LOOK BACK

54.

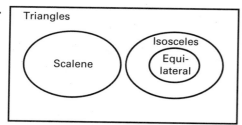

55. Sample proof:

Statements	Reasons
$\angle A \cong \angle D$, $\angle EFD \cong \angle BCA$, $\overline{AF} \cong \overline{CD}$	Given
$\overline{FC} \cong \overline{FC}$	Reflexive Property
$\overline{AC} \cong \overline{FD}$	Overlapping Segments theorem
$\triangle ABC \cong \triangle DEF$	ASA

56. $A = bh = (x + y)(x + y) = x^2 + 2xy + y^2$

57. $A = (x + 2)(x - 3) = x^2 - x - 6$

PAGE 311, LOOK BEYOND

58. $\frac{1}{2}(7 + 8 + 9) = \frac{1}{2}(24) = 12 \text{ cm}$

59. $A = \sqrt{12(12 - 7)(12 - 8)(12 - 9)}$

$= \sqrt{(12)(5)(4)(3)}$

$= \sqrt{720} \approx 26.83 \text{ cm}^2$

60. Check student's work.

	A	B	C	D	E
2	7	8	9	12	26.83
3	3	4	5	6	6
4	2	10	10	11	9.95

5.3 **PAGE 317, GUIDED SKILLS PRACTICE**

The answers given in this lesson are computed using the "π" key on a calculator, unless stated otherwise.

4. $C = 2\pi(3) = 6\pi \approx 18.85$

5. $C = 2\pi r = 25\pi \approx 78.54$

6. $A = \pi \cdot 5^2 = 25\pi \approx 78.54$

7. $A = \pi r^2 = \pi\left(\frac{28}{2}\right)^2 = 196\pi \approx 615.75$

PAGES 317–319, PRACTICE AND APPLY

8. $C = 2\pi(6) \approx 12(3.14) \approx 37.7$
$A = \pi(6)^2 \approx 36(3.14) \approx 113.0$

9. $C = 2\pi(10) = 20\pi \approx 20(3.14) = 62.8.$
$A = \pi(10)^2 = 100\pi \approx 100(3.14) = 314$

10. $C = 2\pi r = \pi d = \pi(18) \approx (3.14)(18) \approx 56.5$
$A = \pi r^2 = \pi\left(\frac{d}{2}\right)^2 = \pi\left(\frac{18}{2}\right)^2 = \pi(9)^2 \approx (3.14)(81)$
≈ 254.3

11. $C = 2\pi(6) = 12\pi \approx 12\left(\frac{22}{7}\right) = \frac{264}{7}$
$A = \pi(6)^2 = 36\pi \approx 36\left(\frac{22}{7}\right) = \frac{792}{7}$

12. $C = 2\pi r = \pi d = \pi \cdot 21 \approx \left(\frac{22}{7}\right) \cdot 21 = 66$
$A = \pi r^2 = \pi\left(\frac{d}{2}\right)^2 = \pi\left(\frac{21}{2}\right)^2 = \pi\left(\frac{441}{4}\right) \approx \left(\frac{22}{7}\right)\left(\frac{441}{4}\right)$
$= \frac{693}{2}$

13. $C = 2\pi r = \pi d = \pi\left(\frac{35}{8}\right) \approx \left(\frac{22}{7}\right)\left(\frac{35}{8}\right) = \frac{55}{4}$
$A = \pi r^2 = \pi\left(\frac{d}{2}\right)^2 \approx \pi\left(\frac{\frac{35}{8}}{2}\right)^2 = \pi\left(\frac{35}{16}\right)^2$
$\approx \left(\frac{22}{7}\right)\left(\frac{1225}{256}\right) = \frac{1925}{128}$

14. $12 = 2\pi r$
$r = \frac{12}{2\pi} = \frac{6}{\pi} \approx 1.9$

15. $62.8 = 2\pi r$
$r = \frac{62.8}{2\pi} = \frac{31.4}{\pi} \approx 10$

16. $50\pi = 2\pi r$
$r = 25$

17. $314 = \pi r^2$
$r^2 = \frac{314}{\pi} \Rightarrow r = \sqrt{\frac{314}{\pi}} \approx 10$

18. $50 = \pi r^2$
$r^2 = \frac{50}{\pi} \Rightarrow r = \sqrt{\frac{50}{\pi}} \approx 4$

19. $100\pi = \pi r^2$
$100 = r^2$
$r = 10$

20. $A = \frac{1}{2}(\pi \cdot 4^2) = 8\pi \approx 25.13$

21. $A = \pi \cdot 9^2 - \pi \cdot 4^2 = 81\pi - 16\pi = 65\pi \approx 204.20$

22. $A = \pi \cdot (1.5)^2 = 2.25\pi \approx 7.07$

23. $A = \pi(2)^2 - (2.8)^2 = 4\pi - 7.84 \approx 4.73$

24. $A = (20)(12) + \frac{1}{2}(\pi \cdot 6^2) = 240 + 18\pi \approx 296.55$

25. $A = \frac{1}{2}(30)(24) + \frac{1}{2}(\pi \cdot 15^2) = 360 + 112.5\pi$
≈ 713.43

26. Since the diameters of the three circles add to 60, each has diameter 20, so $r = 10$. The height of the rectangle is the same as the diameter of a circle, 20.
$A = 60 \cdot 20 - 3(\pi \cdot 10^2) = 1200 - 300\pi \approx 257.52$

27. The shaded area is the area of the trapezoid minus the area of the half circle.
$A = \frac{1}{2}(20 + 8) \cdot 4 - \frac{1}{2}(\pi \cdot 4^2) = 56 - 8\pi \approx 30.87$

28. $C = 2\pi r$, so if r is doubled the circumference is multiplied by 2.

29. $A = \pi r^2$, so if r is doubled the area is multiplied by $2^2 = 4$.

30. The area of the 10-inch pizza $= \pi\left(\frac{10}{2}\right)^2 = 25\pi$ in^2.

The area of the 18-inch pizza $= \pi\left(\frac{18}{2}\right)^2 = 81\pi$ in^2.

Since the area of the 18-inch pizza is more than three times the area of the 10-inch pizza, the 18-inch pizza should feed 6, which is three times the number of people.

31. The area of the 10-in. pizza $= \pi\left(\frac{10}{2}\right)^2 = 25\pi \approx 78.54$ in^2.

The area of the 18-in. pizza $= \pi\left(\frac{18}{2}\right)^2 = 81\pi \approx 254.47$ in^2.

The 18-inch pizza is the better deal. The 18-inch pizza gives about 17 in^2 per dollar; while the 10-inch pizza gives about 16 in^2 per dollar.

32. $d = 5280$ ft; so $r = 2640$ ft, $A = (2640^2)\pi = 6,969,600\pi \approx 21,895,644$ ft^2

33. $C = 2\pi(7) = 14\pi \approx 44$ ft.

34. $C = 2(12.5)\pi \approx 78.5$ ft

35. The inside tires do not have to go as far, so they are not moving as fast.

36. $r = 15$ in. $= 15$ in. $\times \frac{1 \text{ ft}}{12 \text{ in.}} = 1.25$ ft.

In one revolution, the tire covers $2\pi(1.25) = 2.5\pi$ ft. The inside tire must make $\frac{44}{2.5\pi} \approx 5.6$ revolutions

37. The outside tire must make $\frac{78.5}{2.5\pi} \approx 9.99$ revolutions to go around the circle one time.

38. Measuring the crater in the photo with a ruler, the diameter is 0.5 cm. Since the diameter of the moon in the photo is 12.2 cm and it is actually 1077 miles, set up a proportion to find the actual diameter of the crater, d:

$$\frac{0.5}{d} = \frac{12.2}{1077}$$
$$12.2d = 538.5$$
$$d \approx 44.14 \text{ miles}$$

Then the area of the crater is approximately the area of a circle with radius 22.07 miles.

$A = \pi r^2 = \pi(22.07)^2 \approx 1530$ miles^2

So the crater is closest in size to Delaware.

PAGE 319, LOOK BACK

39. $3\sqrt{8} = 3\sqrt{4 \times 2} = 3 \times 2\sqrt{2} = 6\sqrt{2}$

40. $16\sqrt{32} = 16\sqrt{2 \times 16} = 16 \times 4\sqrt{2} = 64\sqrt{2}$

41. $3\sqrt{500} = 3\sqrt{5 \times 100} = 3 \times 10\sqrt{5} = 30\sqrt{5}$

42. $x^2 + 16 = 25$
$x^2 = 9$
$x = 3$ is the positive solution.

43. $x^2 + 144 = 169$
$x^2 = 25$
$x = 5$ is the positive solution.

44. $x^2 + 12.25 = 13.69$
$x^2 = 1.44$
$x = 1.2$ is the positive solution.

45. Sample proof:

Statements	Reasons
$ABCD$ is a parallelogram	Given
$\overline{AB} \cong \overline{CD}$, $\overline{BC} \cong \overline{AD}$	Opposite sides of a parallelogram are congruent.
$\overline{AC} \cong \overline{AC}$	Reflexive Property of Congruence
$\triangle ABC \cong \triangle CDA$	SSS

Alternately, students can use $\angle B \approx \angle D$ (opposite angles of a parallelogram are congruent) in Step 3, then prove that the triangles are congruent by SAS.

46. $A = \frac{1}{2}(9)(7) = 31.5$ in^2

47. $A = 5(3.5) = 17.5$ cm^2

48. $A = \frac{1}{2}(3)(6 + 5) = 16.5$ cm^2

49. Let C_L = circumference of large circle

$\quad\quad d_L$ = diameter of large circle

$C_L = \pi d_L = \pi(d_1 + d_2 + d_3 + d_4)$
$\quad\quad = \pi d_1 + \pi d_2 + \pi d_3 + \pi d_4$
$\quad\quad = C_1 + C_2 + C_3 + C_4$

50. $C_L = \pi d_L = \pi(d_1 + d_2 + d_3 + \dots + d_n)$
$\quad\quad = \pi d_1 + \pi d_2 + \pi d_3 + \dots + \pi d_n$
$\quad\quad = C_1 + C_2 + C_3 + \dots + C_n$

51. Path AB on the outer circle is $\frac{1}{2}$ the circumference of the outer circle:

$ABO = \frac{1}{2}\pi d = \frac{\pi d}{2}$

Path AB through the interior is the circumference of a circle with $\frac{1}{2}$ the diameter:

$ABI = \pi\left(\frac{d}{2}\right) = \frac{\pi d}{2}$

All three paths are the same length.

5.4 **PAGE 326, GUIDED SKILLS PRACTICE**

---

5. $a = 48, c = 80.$ $\quad 48^2 + b^2 = 80^2$
$\quad\quad\quad\quad\quad\quad\quad\quad 2304 + b^2 = 6400$
$\quad\quad\quad\quad\quad\quad\quad\quad\quad\quad b^2 = 4096$
$\quad\quad\quad\quad\quad\quad\quad\quad\quad\quad b = \sqrt{4096} = 64$

6. $a = 7, b = 10, c = 12.$ $\quad 7^2 + 10^2 \stackrel{?}{=} 12^2$
$\quad\quad\quad\quad\quad\quad\quad\quad\quad\quad\quad\quad\quad 149 \neq 144$

No, the triangle is not a right triangle.

7. $a = 8, b = 15, c = 18.$ $\quad 8^2 + 15^2 \underset{\quad}{\overset{?}{}} 18^2$
$\quad\quad\quad\quad\quad\quad\quad\quad\quad\quad\quad\quad 289 < 324$

The triangle is obtuse.

PAGE 326–329, PRACTICE AND APPLY

---

8. $c^2 = 3^2 + 4^2 = 25 \Rightarrow c = \sqrt{25} = 5$

9. $c^2 = 10^2 + 15^2 = 325 \Rightarrow c = \sqrt{325} = 5\sqrt{13}$

10. $c^2 = 46^2 + 73^2 = 2116 + 5329 = 7445$
$\quad\quad c = \sqrt{7445}$

11. $8^2 = a^2 + 6^2$
$\quad 64 = a^2 + 36$
$\quad 28 = a^2$
$\quad\quad a = \sqrt{28} = 2\sqrt{7}$

12. $\quad 53^2 = 27^2 + b^2$
$\quad 2809 = 729 + b^2$
$\quad 2080 = b^2$
$\quad\quad\quad b = \sqrt{2080} = 4\sqrt{130}$

13. $c^2 = 1^2 + 1^2 = 2 \Rightarrow c = \sqrt{2}$

14. $5^2 = 3^2 + b^2$
$\quad 25 = 9 + b^2$
$\quad 16 = b^2$
$\quad\quad b = 4$
$\quad P = 4 + 3 + 5 = 12$ units

15. $c^2 = 50^2 + 24^2$
$\quad c^2 = 2500 + 576$
$\quad c^2 = 3076$
$\quad\quad c = \sqrt{3076} \approx 55.5$
$\quad P \approx 50 + 55.5 + 24 = 129.5$ units

16. The altitude divides the triangle into 2 right triangles. Use the Pythagorean Theorem to find the third side of each, then add to find the third side of the larger triangle.

$17^2 = 8^2 + b^2$ $\quad\quad\quad\quad 10^2 = 8^2 + e^2$
$289 = 64 + b^2$ $\quad\quad\quad\quad 100 = 64 + e^2$
$225 = b^2$ $\quad\quad\quad\quad\quad\quad 36 = e^2$
$\quad b = \sqrt{225} = 15$ $\quad\quad\quad e = \sqrt{36} = 6$

The length of the third side of the outer triangle is: $b + e = 15 + 6 = 21$
The perimeter is: $P = 21 + 17 + 10 = 48$ units.

17. The altitude divides the triangle into 2 right triangles. Use the Pythagorean Theorem to find the third side of each, then add to find the third side of the larger triangle.

$$9.1^2 = 8.4^2 + b^2$$
$$82.81 = 70.56 + b^2$$
$$12.25 = b^2$$
$$b = \sqrt{12.25} = 3.5$$

$$25.9^2 = 8.4^2 + e^2$$
$$670.81 = 70.56 + e^2$$
$$600.25 = e^2$$
$$e = \sqrt{600.25} = 24.5$$

The length of the third side of the larger triangle is: $b + e = 3.5 + 24.5 = 28$ units.
The perimeter is $P = 28 + 9.1 + 25.9 = 63$ units.

18. $12^2 > 5^2 + 9^2$; obtuse

19. $17^2 < 13^2 + 15^2$; acute

20. $25^2 = 7^2 + 24^2$; right

21. $26^2 > 24^2 + 7^2$; obtuse

22. $5^2 = 3^2 + 4^2$; right

23. $30^2 < 25^2 + 25^2$; acute

24. Since the triangle is equilateral, the altitude divides the triangle into two congruent right triangles. The base of each right triangle is 5.
The height of each is: $h^2 + 5^2 = 10^2 \Rightarrow h^2 = 100 - 25 = 75 \Rightarrow h = \sqrt{75}$.
The area of the figure is: $A = \frac{1}{2}(10) \cdot \sqrt{75} = 5\sqrt{75} = 5 \times 5\sqrt{3} = 25\sqrt{3}$ units^2.

25. Let s be the side length of the square. The diagonal of the square divides it into two right triangles with sides s, s and 12. So, $12^2 = s^2 + s^2 \Rightarrow 144 = 2s^2 \Rightarrow 72 = s^2 \Rightarrow s = \sqrt{72}$. The area of the square is: $A = \sqrt{72}^2 = 72$ units^2

26. $c^2 = 5^2 + 5^2 = 50$. Therefore,
diagonal $= \sqrt{50} = 5\sqrt{2}$ cm ≈ 7.07 units

27. Let $s =$ a side of the square.
$s^2 + s^2 = 16^2 \Rightarrow 2s^2 = 256$ or, $s^2 = 128$.
So $s = \sqrt{128} = 8\sqrt{2} \approx 11.31$ units

28. Area shaded region $= x^2$
From the right triangle,
$x^2 = 3.6^2 + 2.4^2 \approx 18.7$ units^2

29. Area shaded region $= \frac{1}{2}\pi\left(\frac{9}{2}\right)^2 = \frac{\pi a^2}{8}$
$a^2 + 21^2 = 29^2$
$a^2 = 841 - 441 = 400$
$A = \frac{\pi \cdot 400}{8} = 50\pi \approx 157.1$ units^2

30. Base of triangle: $b = \sqrt{3^2 + 3^2} = \sqrt{18}$
Height of the rectangle: $h^2 + (\sqrt{18})^2 = 9^2$, so
$h^2 = 81 - 18$
$h = \sqrt{63}$
Area shaded region $= \frac{1}{2}(\sqrt{18})(\sqrt{63}) \approx 16.8$ units^2

31. Sample answer:

m	$\dfrac{m^2 - 1}{2}$	$\dfrac{m^2 + 1}{2}$	triple
3	$\dfrac{9-1}{2} = 4$	$\dfrac{9+1}{2} = 5$	3, 4, 5
5	$\dfrac{25-1}{2} = 12$	$\dfrac{25+1}{2} = 13$	5, 12, 13
7	$\dfrac{49-1}{2} = 24$	$\dfrac{49+1}{2} = 25$	7, 24, 25
9	$\dfrac{81-1}{2} = 40$	$\dfrac{81+1}{2} = 41$	9, 40, 41
11	$\dfrac{121-1}{2} = 60$	$\dfrac{121+1}{2} = 61$	11, 60, 61

Algebraic proof: $m^2 + \left(\dfrac{m^2-1}{2}\right)^2 = m^2 + \left(\dfrac{m^4 - 2m^2 + 1}{4}\right)$
$= \dfrac{4m^2 + m^4 - 2m^2 + 1}{4} = \dfrac{m^4 + 2m^2 + 1}{4} = \dfrac{(m^2+1)^2}{4} = \left(\dfrac{m^2+1}{2}\right)^2$

Since the sum of the squares of the lengths of two sides equals the square of the longest side, any triple generated with this method will represent the sides of a right triangle.

32. Sample answer:

n	$2n$	$n^2 - 1$	$n^2 + 1$	triple
2	4	3	5	3, 4, 5 (since $5^2 = 3^2 + 4^2$)
3	6	8	10	6, 8, 10 (since $10^2 = 6^2 + 8^2$)
4	8	15	17	8, 15, 17 (since $17^2 = 8^2 + 15^2$)
5	10	24	26	10, 24, 26 (since $26^2 = 10^2 + 24^2$)
6	12	35	37	12, 35, 37 (since $37^2 = 12^2 + 35^2$)

Algebraic proof: $(2n)^2 + (n^2 - 1)^2 = 4n^2 + n^4 - 2n^2 + 1 = n^4 + 2n^2 + 1$
$$= (n^2 + 1)^2$$
Since the sum of the squares of the lengths of two sides equals the square of the longest side, any triple generated with this method will represent the sides of a right triangle.

33. Sample answer:

Pythagorean	divisible by 3	divisible by 5
3, 4, 5	3	5
5, 12, 13	12	5
7, 24, 25	24	25
9, 40, 41	9	40
11, 60, 61	60	60
8, 15, 17	15	15
12, 35, 37	12	35
20, 21, 29	21	20
28, 45, 53	45	45
16, 63, 65	63	65

The conjecture is true for all of the triples shown above.

34. If (x, y, z) is a Pythagorean triple, then $x^2 + y^2 = z^2$. Multiply both sides of the equation by a^2.
$$a^2x^2 + a^2y^2 = a^2z^2$$
$$\Rightarrow (ax)^2 + (ay)^2 = (az)^2$$
Thus, (ax, ay, az) is a Pythagorean triple.

35. Let $AB = a$, $AR = EH = b$, and $BR = c$. Then area $ABCD = a^2$, area $EFGH = b^2$, and area $BRST = c^2$. Since $\triangle BAR$ is a right triangle, $a^2 + b^2 = c^2$, showing that $BRST$ is the square whose area is equal to the sum of the areas of the two given squares.

36. Construct the new square so that the diagonal of the given square is the side of the new square, as shown.

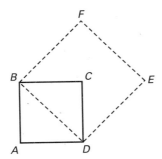

The area of $BDEF$ is the sum of the area of $ABCD$ with itself, or twice the area of $ABCD$.

37. Use the formula for the area of a trapezoid.
$$A = \tfrac{1}{2}h(b_1 + b_2) = \tfrac{1}{2}(a + b)(a + b) = \tfrac{1}{2}(a + b)^2$$
Find the sum of the area of the three right triangles. $A = \tfrac{1}{2}ab + \tfrac{1}{2}ab + \tfrac{1}{2}c^2 = ab + \tfrac{1}{2}c^2$

Set the areas equal to one another and simplify.
$$\tfrac{1}{2}(a + b)^2 = ab + \tfrac{1}{2}c^2$$
$$(a + b)^2 = 2ab + c^2$$
$$a^2 + b^2 + 2ab = 2ab + c^2$$
$$a^2 + b^2 = c^2$$

38. Both squares have the same area, which is $(a + b)^2$. In the first square, after the triangles are removed, a square with side c and area c^2 remains. In the second square, after the triangles are removed, two squares, with sides a and b, and a total area $a^2 + b^2$, remain. Since the areas of the triangles are the same in both squares, $c^2 = a^2 + b^2$

39. a. The area of each inner right triangle $= \frac{1}{2}ab$. By segment subtraction, the side of square $IJKL$ is $(b - a)$. By the Sum of Areas postulate, area $EFGH$ = area $IJKL$ + 4 × (area of one inner right triangle)

$$c^2 = (b - a)^2 + 4\left(\frac{1}{2}ab\right) = (b - a)^2 + 2ab$$

b. The area of each outer right triangle $= \frac{1}{2}ab$. The side of square $ABCD = (a + b)$. By the Sum of Areas postulate, area $ABCD$ = 4 × (area of one outer right triangle) + area $EFGH$.

Thus area $EFGH$ = area $ABCD$ − 4 × (area of one outer right triangle).

Algebraically, $c^2 = (a + b)^2 - 4\left(\frac{1}{2}ab\right) = (a + b)^2 - 2ab$.

Simplifying each equation:
Part **a.**

$$c^2 = (b - a)^2 + 2ab$$
$$= b^2 - 2ab + a^2 + 2ab$$
$$= b^2 + a^2$$

So, $c^2 = a^2 + b^2$.
Part **b.**

$$c^2 = (a + b)^2 - 2ab$$
$$= a^2 + 2ab + b^2 - 2ab$$
$$= a^2 + b^2$$

So, $c^2 = a^2 + b^2$.

40. Let x = the height on the wall from the bottom of the ladder.

$$x^2 + 10^2 = 37^2$$
$$x^2 + 100 = 1369$$
$$x^2 = 1269$$
$$x^2 = \sqrt{1269} \approx 35.6 \text{ ft}$$

Total height $\approx 35.6 + 8.0 = 43.6$ ft

41.

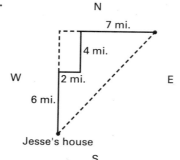

The distance from Jesse's house is the length of the hypotenuse of the right triangle, with sides of length 6 + 4 = 10 miles and 2 + 7 = 9 miles.

$$c^2 = 10^2 + 9^2$$
$$c^2 = 181$$
$$c = \sqrt{181} \approx 13.5 \text{ miles}$$

42. $d^2 = 90^2 + 90^2 = 16200 \Rightarrow d = \sqrt{16200} \approx 127.3$ ft

PAGE 329, LOOK BACK

43. $\dfrac{4}{\sqrt{2}} = \dfrac{4}{\sqrt{2}} \times \dfrac{\sqrt{2}}{\sqrt{2}} = \dfrac{4\sqrt{2}}{\sqrt{2} \times \sqrt{2}} = \dfrac{4\sqrt{2}}{2} = 2\sqrt{2}$

44. $\dfrac{13}{\sqrt{3}} = \dfrac{13}{\sqrt{3}} \times \dfrac{\sqrt{3}}{\sqrt{3}} = \dfrac{13\sqrt{3}}{\sqrt{3} \times \sqrt{3}} = \dfrac{13\sqrt{3}}{3}$

45. $\dfrac{\sqrt{7}}{\sqrt{5}} = \dfrac{\sqrt{7}}{\sqrt{5}} \times \dfrac{\sqrt{5}}{\sqrt{5}} = \dfrac{\sqrt{35}}{\sqrt{5} \times \sqrt{5}} = \dfrac{\sqrt{35}}{5}$

46. False. Sample answer: Any rhombus with an angle other than 90° is not a rectangle.

47. True. Sample answer: Both pairs of opposite sides of a rhombus are parallel, so by definition, the rhombus must be a parallelogram.

48. False. Sample answer: A diagonal of a kite divides the kite, which is a quadrilateral, into two congruent triangles (by SSS), but a kite is not a parallelogram.

49. Hypothesis: In a triangle, the square on one of the sides is equal to the sum of the squares on the remaining two sides of the triangle.
Conclusion: The angle contained by the remaining two sides of the triangle is right.

50.

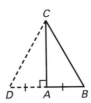

51. No. We cannot be sure than the figure connecting D to B is a single segment. We know only that $\angle DAB$ is a right angle, and have not yet proved that $\angle BAC$ is a right angle.

52. $(DA)^2 = (AB)^2$

53. $(DA)^2 + (AC)^2 = (AB)^2 + (AC)^2$

54. $\angle DAC$ is a right angle, by construction.

55. $(DA)^2 + (AC)^2 = (DC)^2$

56. $(BA)^2 + (AC)^2 = (BC)^2$

57. Substitution Property applied to equations 3, 5, and 6.

58. $DC = BC$

59. By construction

60. Reflexive Property

61. $\triangle DAC \cong \triangle BAC$

62. Corresponding parts of congruent triangles are congruent.

63. 90°

64. Definition of a right triangle

5.5 | **PAGES 335–336, GUIDED SKILLS PRACTICE**

5. $c^2 = 7^2 + 7^2$
$c^2 = 98$
$c = \sqrt{98} = 7\sqrt{2}$

6. Shorter leg = 7. So the hypotenuse is $2 \cdot 7 = 14$ and the longer leg is $7\sqrt{3}$.

7. Shorter leg = 20 ft.
Longer leg = $20\sqrt{3} \approx 34.6$ ft.
The tree is about 34.6 feet tall.

8. The regular hexagon can be divided into 6 equilateral triangles. Their height is the length of the apothem, $h^2 + 6^2 = 12^2$, $h^2 = 108$, $h = \sqrt{108} = 6\sqrt{3}$.
The perimeter is $6 \cdot 12$ in = 72 in.
$A = \frac{1}{2}(6\sqrt{3})(72) = 216\sqrt{3} \approx 374.12$ inches^2.

9. $A = \frac{1}{2}(9.23)(10 \cdot 6) = 276.9$ units^2

PAGES 336–337, PRACTICE AND APPLY

10. $y = 6\sqrt{3}$
$z = 2 \cdot 6 = 12$

11. $y = 6 = x\sqrt{3} \Rightarrow x = \frac{6}{\sqrt{3}} = \frac{6}{\sqrt{3}} \cdot \frac{\sqrt{3}}{\sqrt{3}} = \frac{6\sqrt{3}}{3}$
$= 2\sqrt{3}$
$z = 2x = 2(2\sqrt{3}) = 4\sqrt{3}$

12. $z = 14 = 2x \Rightarrow x = \frac{14}{2} \Rightarrow x = 7$
$y = x\sqrt{3} = 7\sqrt{3}$

13. $y = 4\sqrt{3} = x\sqrt{3} \Rightarrow x = \frac{4\sqrt{3}}{\sqrt{3}} \Rightarrow x = 4$
$z = 2x = 2 \cdot 4 = 8$

14. $q = p = 6$
$r = p\sqrt{2} = 6\sqrt{2}$

15. $r = 6 = p\sqrt{2} \Rightarrow p = \frac{6}{\sqrt{2}} = 3\sqrt{2}$
$q = p = 3\sqrt{2}$

16. $q = 4\sqrt{2} = p$
$r = q\sqrt{2} = 4\sqrt{2} \cdot \sqrt{2} = 8$

17. $r = 10 = p\sqrt{2} \Rightarrow p = \frac{10}{\sqrt{2}} = 5\sqrt{2}$
$q = p = 5\sqrt{2}$

18. $k = 3.4$
$h = k\sqrt{3} = 3.4\sqrt{3}$
$g = 2k = 2(3.4) = 6.8$

19. $k = 6\sqrt{3}$
$h = k\sqrt{3} = (6\sqrt{3}) \cdot \sqrt{3}$
$= 6 \cdot \sqrt{3}^2 = 18$
$g = 2k = 2(6\sqrt{3}) = 12\sqrt{3}$

20. $g = 17 = 2k \Rightarrow k = \frac{17}{2} = 8.5$
$h = k\sqrt{3} = 8.5\sqrt{3}$

21. $h = 2\sqrt{3} = k\sqrt{3} \Rightarrow k = \frac{2\sqrt{3}}{\sqrt{3}} = 2$
$g = 2k = 2(2) = 4$

22. Since two base angles are congruent, the triangle is isosceles.
Thus, $A = \frac{1}{2}(73)(73) = 2664.5$ units^2.

23. The triangle is a 30-60-90 triangle, so the longer leg measures $5.4\sqrt{3}$.
Thus, $A = \frac{1}{2}(5.4)(5.4\sqrt{3}) \approx 25.3$ units^2.

24. The triangle is a 30-60-90 triangle, so the shorter leg is $\frac{5}{2} = 2.5$. The longer leg is then $2.5\sqrt{3}$.
$A = \frac{1}{2}(2.5)(2.5\sqrt{3}) \approx 5.4$ units^2

25. The diagonal of the square forms two 45-45-90 triangles. Let s be the length of a side of the square. The hypotenuse of the 45-45-90 triangle is 15, so
$15 = s\sqrt{2}$ or $s = \frac{15}{\sqrt{2}}$.
$A = \left(\frac{15}{\sqrt{2}}\right)^2 = 112.5$ units^2

26. This is an equilateral triangle. To find the height of the triangle, drop an altitude from a vertex, forming two 30-60-90 triangles. The shortest side is 6, so the height is $6\sqrt{3}$.
$A = \frac{1}{2}(12)(6\sqrt{3}) \approx 62.4$ units^2

27. The altitude forms two 30-60-90 triangles. The side opposite the 60° angle is 16. Let $s =$ the side opposite the shortest angle, so $s\sqrt{3} = 16$ or
$s = \frac{16}{\sqrt{3}}$. The base of the larger triangle is
$2 \cdot \left(\frac{16}{\sqrt{3}}\right)$
$A = \frac{1}{2}\left(2 \cdot \frac{16}{\sqrt{3}}\right)(16) \approx 147.8$ units^2

28. Two sides are equal, and the third side is $\sqrt{2}$ times the other sides. This is a 45-45-90 triangle.

29. Two sides are equal, and the third side is $\sqrt{2}$ times the other sides. This is a 45-45-90 triangle.

30. One side ($2\sqrt{3}$) is twice the shortest side ($\sqrt{3}$) and the third side (3) is $\sqrt{3}$ times the shortest side, so this is a 30-60-90 triangle.

31. One side (2) is twice the shortest side (1) but the third side ($2\sqrt{3}$) is not $\sqrt{3}$ times the shortest side. Also, no two sides are the same length. This triangle is neither 30-60-90 nor 45-45-90.

32. Two sides are equal ($5\sqrt{2}$), but the third side (5) is not $\sqrt{2}$ times the shorter sides. This triangle is neither 30-60-90 nor 45-45-90.

33. One side ($2\sqrt{2}$) is twice the shortest side ($\sqrt{2}$) and one side ($\sqrt{6}$) is $\sqrt{3}$ times the shortest side, so this is a 30-60-90 triangle.

34. The hypotenuse is 18, so the shortest side is $\frac{18}{2} = 9$ and the side opposite the 60° angle is $9\sqrt{3}$.
$P = 18 + 9 + 9\sqrt{3} = 27 + 9\sqrt{3}$ units
$A = \frac{1}{2}(9)(9\sqrt{3}) = \frac{81}{2}\sqrt{3}$ units^2

35. The hypotenuse is 24, so $s\sqrt{2} = 24$ or $s = \frac{24}{\sqrt{2}}$
$= \frac{24}{\sqrt{2}} \cdot \frac{\sqrt{2}}{\sqrt{2}} = \frac{24\sqrt{2}}{2} = 12\sqrt{2}$.
$P = 12\sqrt{2} + 12\sqrt{2} + 24 = 24\sqrt{2} + 24$ units
$A = \frac{1}{2}(12\sqrt{2})(12\sqrt{2}) = 144$ units^2

36. $P = 8 + 8 + 8 = 24$ units
To find the height, divide it into two 30-60-90 triangles. The smallest side is $\frac{8}{2} = 4$ units, so the height is $4\sqrt{3}$ units.
$A = \frac{1}{2}(8)(4\sqrt{3}) = 16\sqrt{3}$ units^2

37. $P = 13 \cdot 6 = 78$ units
The hexagon is divided into 6 equilateral triangles as in Example 4. To find the apothem, look at a 30-60-90 triangle formed by the altitude of one of the triangles. The shorter side is $\frac{13}{2} = 6.5$, so the altitude is $6.5\sqrt{3}$.
$A = \frac{1}{2}(6.5\sqrt{3})(78) = 253.5\sqrt{3}$ units^2

38. The diagonal forms two congruent 45-45-90 triangles. The hypotenuse is 14, so $s\sqrt{2} = 14$ or $s = \frac{14}{\sqrt{2}}$ is the length of a side of the square.

$P = \frac{14}{\sqrt{2}} \cdot 4 = \frac{56}{\sqrt{2}} \cdot \frac{\sqrt{2}}{\sqrt{2}} = \frac{56\sqrt{2}}{2} = 28\sqrt{2}$ units

$A = \left(\frac{14}{\sqrt{2}}\right)^2 = \frac{196}{2} = 98$ units^2

39. The apothem of a regular hexagon is the altitude of an equilateral triangle, or a side of a 30-60-90 triangle (as in Example 4). $a = 5 = s\sqrt{3}$, so $s = \frac{5}{\sqrt{3}}$. The length of a side of the hexagon is $2\left(\frac{5}{\sqrt{3}}\right) = \frac{10}{\sqrt{3}}$.

$P = \frac{10}{\sqrt{3}} \cdot 6 = \frac{60}{\sqrt{3}} = 20\sqrt{3}$ units

$A = \frac{1}{2}(5)(20\sqrt{3}) = 50\sqrt{3}$ units^2

40. The triangles are 45-45-90 triangles, with hypotenuse 10. $10 = x\sqrt{2}$ so

$x = \frac{10}{\sqrt{2}} = \frac{10}{\sqrt{2}} \cdot \frac{\sqrt{2}}{\sqrt{2}} = 5\sqrt{2}$.

41. The apothem is half the side length of the square. The square has side length $x + 10 + x = 5\sqrt{2} + 10 + 5\sqrt{2} = 10\sqrt{2} + 10$. The length of the apothem is $\frac{10\sqrt{2} + 10}{2} = 5\sqrt{2} + 5$ units.

42. $P = 8 \cdot 10 = 80$

$A = \frac{1}{2}(5\sqrt{2} + 5)(80) \approx 482.8$ units^2

43. If the sides have length s, then the value of x will be $\frac{s}{2} \cdot \sqrt{2}$, as in problem 40, and a length of a side of the square will be $\frac{s}{2} \cdot \sqrt{2} + s + \frac{s}{2} \cdot \sqrt{2} = s\sqrt{2} + s$, as in problem 35. The apothem is half the side length of the square, $\frac{s\sqrt{2} + s}{2}$. The perimeter of the octagon is $8s$. So the area is:

$A = \frac{1}{2}\left(\frac{s\sqrt{2} + s}{2}\right) \cdot (8s) = 2s^2\sqrt{2} + 2s^2$ or $2s^2(\sqrt{2} + 1)$

44. Since the tower, guy wire, and the segment to the base of the tower form a 45-45-90 triangle, the guy wire has length $60\sqrt{2} \approx 84.85$ ft.

45. The length of the wire is the hypotenuse of a 30-60-90 triangle. The side opposite the 60° angle is 60, so $s\sqrt{3} = 60$ or $s = \frac{60}{\sqrt{3}} = \frac{60}{\sqrt{3}} \cdot \frac{\sqrt{3}}{\sqrt{3}} = \frac{60\sqrt{3}}{3} = 20\sqrt{3}$. The hypotenuse has length $2s = 2(20\sqrt{3}) = 40\sqrt{3} \approx 69.28$ feet. The guy wire is about 69.28 feet long.

PAGES 337–338, LOOK BACK

46. The lines are parallel because they are each perpendicular to the same line.

47. The lines are parallel because they both form 60° angles with the same line.

48. $P = 2b + 2h$, so $100 = 2b + 2h$ or $50 = b + h$.

Look at values of b and h for which $b + h = 50$.

b	h	$b + h$	$A = bh$
49	1	50	49
48	2	50	96
47	3	50	141
$\vdots$	$\vdots$	$\vdots$	$\vdots$
24	25	50	600
25	25	50	625

The smallest area is found when one side is 49 and another side is 1. The largest is found when the sides are equal length, 25 and 25, or the figure is a square.

49. The outer figure is a trapezoid. Its area is $A = \frac{1}{2}(4 + 7.5)(3) = 17.25$ units^2.

The right triangle has area $A = \frac{1}{2}(4)(3) = 6$ units^2. The area of the other triangle can be found by subtracting the area of the right triangle from the area of the trapezoid: $A = 17.25 - 6 = 11.25$ units^2.

50. $A = \frac{1}{2}(b_1 + b_2)h$, so
$$103.5 = \frac{1}{2}(17.5 + 5.5)h$$
$$103.5 = \frac{1}{2}(23)h$$
$$207 = 23h$$
$$h = \frac{207}{23} = 9 \text{ units}$$

51. The height of the isosceles triangle is: $h^2 + 4^2 = 6^2$
$$h^2 = 20$$
$$h = \sqrt{20}$$
$$A = \frac{1}{2}(8)\sqrt{20} = 4\sqrt{20} = 8\sqrt{5} \text{ units}^2$$
$$P = 8 + 6 + 6 = 20 \text{ units}$$

52. $(PR)^2 + 100^2 = 240^2$
$$(PR)^2 = 240^2 - 100^2$$
$$(PR)^2 = 47600$$
so $PR = \sqrt{47600} \approx 218.17$ units

53. $A = \frac{1}{2}ab$

The third side is: $c^2 = a^2 + b^2$
$$c = a^2 + b^2$$
so, $P = a + b + \sqrt{a^2 + b^2}$.

PAGE 338, LOOK BEYOND

54. tangent of $45° = 1$ on the calcualtor. In a 45-45-90 triangle $x = y$, so the ratio is $\frac{y}{x} = \frac{y}{y} = 1$. The answers agree.

55. tangent of $30° \approx 0.58$. In a 30-60-90 triangle, $x = y\sqrt{3}$, so the ratio is $\frac{y}{x} = \frac{y}{y\sqrt{3}} = \frac{1}{\sqrt{3}}$
$$= \frac{1}{\sqrt{3}} \cdot \frac{\sqrt{3}}{\sqrt{3}} = \frac{\sqrt{3}}{3} \approx 0.58. \text{ The answers agree.}$$

56. The angle measures in a regular pentagon are:
$$m = 180 - \frac{360}{5} = 108°.$$
The segment from A to a vertex of the pentagon divides the vertex angle in half, $\frac{108}{2} = 54°$.
So $m\angle A = 180 - 54 - 90 = 36°$.

57. tangent of $\angle A \approx 0.73$

58. $y = \frac{12}{2} = 6$

59. tangent of $\angle A = 0.73 = \frac{y}{x}$ and $y = 6$, so $0.73 = \frac{6}{x}$ or $x = \frac{6}{0.73} \approx 8.22$.

60. length of apothem $= x = 8.22$

$P = 5 \cdot 12 = 60$

$A = \frac{1}{2}(8.22)(60) = 246.6$ units^2

61. The angle measures in a regular 9-gon are:

$m = 180 - \frac{360}{9} = 140°$.

Using the same procedure as in problem 50,

$\angle A = 180 - 90 - \frac{140}{2} = 20°$.

tangent of $\angle A \approx 0.36 = \frac{y}{x}$ and $y = \frac{4}{2} = 2$

so, $0.36 = \frac{2}{x}$ or $x = \frac{2}{0.36} \approx 5.56$, the length of the apothem.

$P = 9 \cdot 4 = 36$

$A \approx \frac{1}{2}(5.56)36 = 100.08$ units^2

5.6 **PAGES 342–343, GUIDED SKILLS PRACTICE**

---

5. $AB = \sqrt{(2-0)^2 + (3-0)^2} = \sqrt{2^2 + 3^2} = \sqrt{4+9} = \sqrt{13}$

6. $CD = \sqrt{(-1-5)^2 + (3-1)^2} = \sqrt{(-6)^2 + 2^2} = \sqrt{36+4} = \sqrt{40} = 2\sqrt{10}$

7. y-values:

$A: y = \sqrt{4^2 - 0^2} = \sqrt{16} = 4$

$B: y = \sqrt{4^2 - 1^2} = \sqrt{15}$

$C: y = \sqrt{4^2 - 2^2} = \sqrt{12} = 2\sqrt{3}$

$D: y = \sqrt{4^2 - 3^2} = \sqrt{7}$

Area $= 1 \cdot 4 + 1 \cdot \sqrt{15} + 1 \cdot 2\sqrt{3} + 1 \cdot \sqrt{7}$

≈ 13.98 units^2

8. y-values:

$B: y = \sqrt{4^2 - 1^2} = \sqrt{15}$

$C: y = \sqrt{4^2 - 2^2} = \sqrt{12} = 2\sqrt{3}$

$D: y = \sqrt{4^2 - 3^2} = \sqrt{7}$

$E: y = \sqrt{4^2 - 4^2} = \sqrt{0} = 0$

Area $= 1 \cdot \sqrt{15} + 1 \cdot 2\sqrt{3} + 1 \cdot \sqrt{7} + 1 \cdot 0$

$= 9.98$ units^2

PAGES 343–345, PRACTICE AND APPLY

---

9. $d = \sqrt{(5-0)^2 + (8-0)^2} = \sqrt{25+64} = \sqrt{89}$

≈ 9.43

10. $d = \sqrt{(4-1)^2 + (6-2)^2}$

$= \sqrt{3^2 + 4^2}$

$= \sqrt{9+16}$

$= \sqrt{25} = 5$

11. $d = \sqrt{(3-1)^2 + (9-4)^2} = \sqrt{2^2 + 5^2} = \sqrt{29} \approx 5.39$

12. $d = \sqrt{(-3-6)^2 + (-3-12)^2} = \sqrt{(-9)^2 + (-15)^2} = \sqrt{306} \approx 17.49$

13. $d = \sqrt{(-6-(-1))^2 + (16-4)^2} = \sqrt{(-5)^2 + 12^2} = \sqrt{25+144} = \sqrt{169} = 13$

14. $d = \sqrt{(-2-(-6))^2 + (-3-(-12))^2} = \sqrt{4^2 + 9^2} = \sqrt{97} \approx 9.85$

15. The coordinates of A are $(-5, 3)$ and the coordinates of B are $(1, 4)$.

$AB = \sqrt{(-5-1)^2 + (3-4)^2} = \sqrt{(-6)^2 + (-1)^2} = \sqrt{37} \approx 6.08$

16. The coordinates of B are $(1, 4)$ and the coordinates of C are $(5, 5)$.

$BC = \sqrt{(5-1)^2 + (5-4)^2} = \sqrt{4^2 + 1^2} = \sqrt{17} \approx 4.12$

17. The coordinates of G are $(4, -1)$ and the coordinates of J are $(2, 2)$.

$GJ = \sqrt{(2-4)^2 + (2-(-1))^2} = \sqrt{(-2)^2 + 3^2} = \sqrt{4+9} = \sqrt{13} \approx 3.61$

18. The coordinates are: $E(-5, -3), J(2, 2), H(5, -4)$.

$EJ = \sqrt{(-5-2)^2 + (-3-2)^2} = \sqrt{(-7)^2 + (-5)^2} = \sqrt{74}$

$JH = \sqrt{(5-2)^2 + (-4-2)^2} = \sqrt{3^2 + (-6)^2} = \sqrt{9+36} = \sqrt{45}$

$EH = \sqrt{(5-(-5))^2 + (-4-(-3))^2} = \sqrt{10^2 + (-1)^2} = \sqrt{101}$

$P = \sqrt{74} + \sqrt{45} + \sqrt{101} \approx 25.36$ units

19. The coordinates are: $A(-5, 3)$, $B(1, 4)$, $F(2, -3)$.

$AB = \sqrt{(-5-1)^2 + (3-4)^2} = \sqrt{(-6)^2 + (-1)^2} = \sqrt{36+1} = \sqrt{37}$

$BF = \sqrt{(1-2)^2 + (4-(-3))^2} = \sqrt{(-1)^2 + 7^2} = \sqrt{50}$

$AF = \sqrt{(-5-2)^2 + (3-(-3))^2} = \sqrt{(-7)^2 + 6^2} = \sqrt{49+36} = \sqrt{85}$

$P = \sqrt{37} + \sqrt{50} + \sqrt{85} \approx 22.37$ units

20. The coordinates are :$F(2, -3)$, $G(4, -1)$, $H(5, -4)$, $I(3, -5)$.

$FG = \sqrt{(4-2)^2 + (-1-(-3))^2} = \sqrt{2^2 + 2^2} = \sqrt{8}$

$GH = \sqrt{(5-4)^2 + (-4-(-1))^2} = \sqrt{1^2 + (-3)^2} = \sqrt{10}$

$HI = \sqrt{(3-5)^2 + (-5-(-4))^2} = \sqrt{(-2)^2 + (-1)^2} = \sqrt{5}$

$IF = \sqrt{(3-2)^2 + (-5-(-3))^2} = \sqrt{1^2 + (-2)^2} = \sqrt{5}$

$P = \sqrt{8} + \sqrt{10} + \sqrt{5} + \sqrt{5} \approx 10.46$ units

21. The coordinates are: $B(1, 4)$, $C(5, 5)$, $D(5, 3)$.

$BC = \sqrt{(5-1)^2 + (5-4)^2} = \sqrt{16+1} = \sqrt{17}$

$BD = \sqrt{(5-1)^2 + (3-4)^2} = \sqrt{16+1} = \sqrt{17}$

$CD = |5-3| = 2$

Since two sides are equal in length, the triangle is isosceles.

22. The coordinates are: $A(-5, 3)$, $J(2, 2)$, $E(-5, -3)$.

$AJ = \sqrt{(2-(-5))^2 + (2-3)^2} = \sqrt{7^2 + (-1)^2} = \sqrt{50}$

$JE = \sqrt{(-5-2)^2 + (-3-2)^2} = \sqrt{(-7)^2 + (-5)^2} = \sqrt{74}$

$EA = \sqrt{(-5-(-5))^2 + (-3-3)^2} = \sqrt{0^2 + (-6)^2} = \sqrt{36} = 6$

The triangle is not isosceles, since no two sides are equal in length.

23. The coordinates are: $B(1, 4)$, $H(5, -4)$, $E(-5, -3)$

$BH = \sqrt{(5-1)^2 + (-4-4)^2} = \sqrt{4^2 + (-8)^2} = \sqrt{16+64} = \sqrt{80}$

$HE = \sqrt{(-5-5)^2 + (-3-(-4))^2} = \sqrt{(-10)^2 + (1)^2} = \sqrt{101}$

$EB = \sqrt{(-5-1)^2 + (-3-4)^2} = \sqrt{(-6)^2 + (-7)^2} = \sqrt{36+49} = \sqrt{85}$

The triangle is not equilateral, since the three sides do not have equal length.

24. $a = \sqrt{(-4-2)^2 + (9-1)^2} = \sqrt{(-6)^2 + 8^2} = \sqrt{36+64} = \sqrt{100} = 10$

$b = \sqrt{(6-2)^2 + (4-1)^2} = \sqrt{4^2 + 3^2} = \sqrt{16+9} = \sqrt{25} = 5$

$c = \sqrt{(-4-6)^2 + (9-4)^2} = \sqrt{(-10)^2 + 5^2} = \sqrt{125}$

$a^2 + b^2 = 10^2 + 5^2 = 125$, $c^2 = \sqrt{125}^2 = 125$.

Since $a^2 + b^2 = c^2$, the triangle is a right triangle.

25. $a = \sqrt{(1-(-2))^2 + (5-2)^2} = \sqrt{3^2 + 3^2} = \sqrt{18}$

$b = \sqrt{(6-1)^2 + (0-5)^2} = \sqrt{5^2 + (-5)^2} = \sqrt{50}$

$c = \sqrt{(6-(-2))^2 + (0-2)^2} = \sqrt{8^2 + (-2)^2} = \sqrt{68}$

$a^2 + b^2 = (\sqrt{18})^2 + (\sqrt{50})^2 = 18 + 50 = 68$, $c^2 = (\sqrt{68})^2 = 68$.

Since $a^2 + b^2 = c^2$, the triangle is a right triangle.

26. $a = \sqrt{(4-1)^2 + (2-4)^2} = \sqrt{3^2 + (-2)^2} = \sqrt{13}$

$b = \sqrt{(4-6)^2 + (2-6)^2} = \sqrt{(-2)^2 + (-4)^2} = \sqrt{20}$

$c = \sqrt{(6-1)^2 + (6-4)^2} = \sqrt{5^2 + 2^2} = \sqrt{29}$

$a^2 + b^2 = (\sqrt{13})^2 + (\sqrt{20})^2 = 13 + 20 = 33$, $c^2 = (\sqrt{29})^2 = 29$.

Since $a^2 + b^2 \neq c^2$, the triangle is not a right triangle.

27. The hypotenuse has length $\sqrt{(9-3)^2 + (2-3)^2} = \sqrt{6^2 + (-1)^2} = \sqrt{37}$.

Since the hypotenuse of a 45-45-90 triangle with legs of length s is $s\sqrt{2}$,

$$\sqrt{37} = s\sqrt{2} \text{ or } s = \frac{\sqrt{37}}{\sqrt{2}} = \sqrt{\frac{37}{2}} = \frac{\sqrt{74}}{2}$$

The legs are of length $\dfrac{\sqrt{74}}{2}$ units.

28. $s = \sqrt{(4-0)^2 + (-1-3)^2} = \sqrt{4^2 + (-4)^2} = \sqrt{32}$.

The hypotenuse is $s\sqrt{2} = \sqrt{32} \cdot \sqrt{2} = \sqrt{64} = 8$ units.

29. The hypotenuse has length $\sqrt{(7-4)^2 + (2-(-2))^2} = \sqrt{3^2 + 4^2} = \sqrt{25} = 5$.

So $5 = 2s$ or $s = \frac{5}{2}$. The shorter leg is $\frac{5}{2}$ units. The longer leg is $\frac{5}{2} \cdot \sqrt{3}$ units.

30. The longer leg has length $\sqrt{(2-(-3))^2 + (1-5)^2} = \sqrt{5^2 + (-4)^2} = \sqrt{25+16}$

$= \sqrt{41}$. So $s\sqrt{3} = \sqrt{41}$ or $s = \frac{\sqrt{41}}{\sqrt{3}} = \sqrt{\frac{41}{3}} = \frac{\sqrt{123}}{3}$.

The length of the shorter leg is $\frac{\sqrt{123}}{3}$.

The length of the hypotenuse is $2s = \frac{2\sqrt{123}}{3}$.

31. The hypotenuse is the segment between $(6, 0)$ and $(0, 8)$. The midpoint of the

hypotenuse is $M = \left(\frac{6+0}{2}, \frac{0+8}{2}\right) = (3, 4)$.

Let $A = (0, 0)$, $B = (6, 0)$, $C = (0, 8)$,

$AM = \sqrt{(3-0)^2 + (4-0)^2} = \sqrt{3^2 + 4^2} = \sqrt{25} = 5$

$BM = \sqrt{(6-3)^2 + (0-4)^2} = \sqrt{3^2 + (-4)^2} = \sqrt{25} = 5$

$CM = \sqrt{(3-0)^2 + (4-8)^2} = \sqrt{3^2 + (-4)^2} = \sqrt{25} = 5$

The distances are the same, 5 units.

32. The hypotenuse is the segment between $(4, 0)$ and $(0, 7)$.

The midpoint of the hypotenuse is: $M = \left(\frac{4+0}{2}, \frac{0+7}{2}\right) = \left(2, \frac{7}{2}\right)$

Let $A = (0, 0)$, $B = (4, 0)$, $C = (0, 7)$.

$AM = \sqrt{(2-0)^2 + \left(\frac{7}{2} - 0\right)^2} = \sqrt{2^2 + \left(\frac{7}{2}\right)^2} = \sqrt{4 + \frac{49}{4}} = \sqrt{\frac{65}{4}} = \frac{\sqrt{65}}{2}$

$BM = \sqrt{(2-4)^2 + \left(\frac{7}{2} - 0\right)^2} = \sqrt{(-2)^2 + \left(\frac{7}{2}\right)^2} = \sqrt{4 + \frac{49}{4}} = \sqrt{\frac{65}{4}} = \frac{\sqrt{65}}{2}$

$CM = \sqrt{(2-0)^2 + \left(\frac{7}{2} - 7\right)^2} = \sqrt{2^2 + \left(-\frac{7}{2}\right)^2} = \sqrt{4 + \frac{49}{4}} = \sqrt{\frac{65}{4}} = \frac{\sqrt{65}}{2}$

The distances are the same, $\frac{\sqrt{65}}{2}$ units.

33. Conjecture: The distances from the midpoint of the hypotenuse to all vertices of a right triangle are equal.

34. The hypotenuse is the segment between $(x, 0)$ and $(0, y)$.

The midpoint of the hypotenuse is: $M = \left(\frac{x+0}{2}, \frac{0+y}{2}\right) = \left(\frac{x}{2}, \frac{y}{2}\right)$.

Let $A = (0, 0)$, $B = (x, 0)$, $C = (0, y)$.

$AM = \sqrt{\left(\frac{x}{2} - 0\right)^2 + \left(\frac{y}{2} - 0\right)^2} = \sqrt{\left(\frac{x}{2}\right)^2 + \left(\frac{y}{2}\right)^2} = \sqrt{\frac{x^2}{4} + \frac{y^2}{4}}$

$BM = \sqrt{\left(\frac{x}{2} - x\right)^2 + \left(\frac{y}{2} - 0\right)^2} = \sqrt{\left(\frac{-x}{2}\right)^2 + \left(\frac{y}{2}\right)^2} = \sqrt{\frac{x^2}{4} + \frac{y^2}{4}}$

$CM = \sqrt{\left(\frac{x}{2} - 0\right)^2 + \left(\frac{y}{2} - y\right)^2} = \sqrt{\left(\frac{x}{2}\right)^2 + \left(\frac{-y}{2}\right)^2} = \sqrt{\frac{x^2}{4} + \frac{y^2}{4}}$

Thus, $AM = BM = CM$.

35. Sample answer: Use the Left-Hand Rule with rectangles of width $\frac{1}{2}$.

$A = \frac{1}{2} \cdot 3 + \frac{1}{2} \cdot 3 + \frac{1}{2} \cdot 2 + \frac{1}{2} \cdot 2.25 + \frac{1}{2} \cdot 3 + \frac{1}{2} \cdot 4 + \frac{1}{2} \cdot 5 + \frac{1}{2} \cdot (4.5) + \frac{1}{2} \cdot 3 + \frac{1}{2} \cdot (1.75)$

$= 1.5 + 1.5 + 1 + 1.125 + 1.5 + 2 + 2.5 + 2.25 + 1.5 + 0.875 = 15.75$ units^2

Use the Right-Hand Rule with rectangles of width $\frac{1}{2}$.

$A = \frac{1}{2} \cdot 3 + \frac{1}{2} \cdot 2 + \frac{1}{2} \cdot 2.25 + \frac{1}{2} \cdot 3 + \frac{1}{2} \cdot 4 + \frac{1}{2} \cdot 5 + \frac{1}{2} \cdot (4.5) + \frac{1}{2} \cdot 3 + \frac{1}{2}(1.75) + \frac{1}{2}(5.5)$

$= 1.5 + 1 + 1.125 + 1.5 + 2 + 2.5 + 2.25 + 1.5 + 0.875 + 2.75 = 17$ units^2

Average of the two areas $= \frac{15.75 + 17}{2} = 16.375$ units^2

36. Sample answer: Using rectangles of width 0.5, the height of each rectangle can be found by substituting the value of x into the equation $y = x^2$.

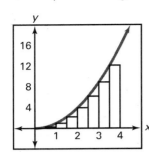 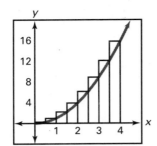

From left: $(0.5)(0) + (0.5)(0.25) + (0.5)(1) + (0.5)(2.25) + (0.5)(4)$
$+ (0.5)(6.25) + (0.5)(9) + (0.5)(12.25) = 17.5$ units$^2 \rightarrow$ low estimate

From right: $(0.5)(0.25) + (0.5)(1) + (0.5)(2.25) + (0.5)(4) + (0.5)(6.25)$
$+ (0.5)(9) + (0.5)(12.5) + (0.5)(16) = 25.5$ units$^2 \rightarrow$ high estimate

The average of the areas $= \dfrac{17.5 + 25.5}{2} = 21.5$ units^2

37. Sample answer: Using rectangles of width 0.5, the height of each rectangle can be found by substituting the value of x into the equation $y = x^2 + 2$.

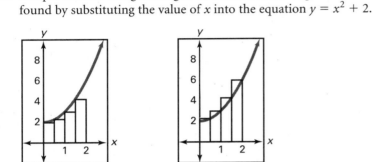

From left: $A = (0.5)(2) + (0.5)(2.25) + (0.5)(3) + (0.5)(4.25) = (0.5)(11.5)$
$= 5.75$ units$^2 \rightarrow$ low estimate

From right: $A = (0.5)(2.25) + (0.5)(3) + (0.5)(4.25) + (0.5)(6) = (0.5)(15.5)$
$= 7.75$ units$^2 \rightarrow$ high estimate

The average of the areas $= \dfrac{7.75 + 5.75}{2} = 6.75$ units^2

If rectangles of width 1 are used, the estimate is about 7.

38. Sample answer: Using rectangles of width 0.5, the height of each rectangle can be found by substituting the value of x into the equation $y = -x^2 + 4$.

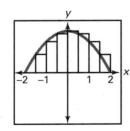 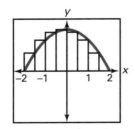

From left: $(0.5)(0) + (0.5)(1.75) + (0.5)(3) + (0.5)(3.75) + (0.5)(4)$
$+ (0.5)(3.75) + (0.5)(3) + (0.5)(1.75) = 10.5$ units^2

From right: $(0.5)(1.75) + (0.5)(3) + (0.5)(3.75) + (0.5)(4) + (0.5)(3.75)$
$+ (0.5)(3) + (0.5)(1.75) + (0.5)(0) = 10.5$ units^2

The two methods give the same result: 10.5 units^2. A better estimate can be obtained by using a smaller rectangle width.

39. $d = \sqrt{(11 - 5)^2 + (7 - 1)^2} = \sqrt{6^2 + 6^2}$
$= \sqrt{72}$
≈ 8.5 miles

40. Sample answer: Using rectangles of width 1:

From left: $A = 1 \cdot 0 + 1 \cdot 1.5 + 1 \cdot 2.2 + 1 \cdot 2.4 + 1 \cdot 1.4$
$= 7.5$

From right $A = 1 \cdot 1.5 + 1 \cdot 2.2 + 1 \cdot 2.4 + 1 \cdot 1.4 + 1 \cdot 0$
$= 7.5$

The two methods give the same result: $7 \cdot 5(10{,}000 \text{ ft}^2) = 75{,}000 \text{ ft}^2$

PAGE 345, LOOK BACK

41. slope $\overline{AB} = \dfrac{-5 - 1}{7 - (-1)} = \dfrac{-6}{8} = -\dfrac{3}{4}$

slope $\overline{BC} = \dfrac{-5 - (-1)}{7 - 10} = \dfrac{-4}{-3} = \dfrac{4}{3}$

slope $\overline{CD} = \dfrac{-1 - 5}{10 - 2} = \dfrac{-6}{8} = -\dfrac{3}{4}$

slope $\overline{AD} = \dfrac{1 - 5}{-1 - 2} = \dfrac{-4}{-3} = \dfrac{4}{3}$

Since slope $\overline{AB}$ = slope $\overline{CD}$ and slope $\overline{BC}$ = slope $\overline{AD}$, $\overline{AB} \parallel \overline{CD}$ and $\overline{BC} \parallel \overline{AD}$, and $ABCD$ is a parallelogram by definition.

Since (slope $\overline{AB}$)(slope $\overline{BC}$) $= \left(-\dfrac{3}{4}\right)\left(\dfrac{4}{3}\right) = -1$, $\overline{AB} \perp \overline{BC}$ and $ABCD$ is a rectangle since it has a right angle (Theorem 4.6.4).

42. Using a square, each side would be $100 \div 4 = 25$ and the area would be $25^2 = 625 \text{ m}^2$. Using a circle, since $100 = 2\pi r$, $r = \dfrac{50}{\pi}$ and the area would be $\pi\left(\dfrac{50}{\pi}\right)^2 = \dfrac{2500}{\pi} \approx 795.77 \text{ m}^2$. The circular area is larger.

43. For the circle, since $\pi r^2 = 225$, $r = \dfrac{15}{\sqrt{\pi}}$. So the circumference would be $2\pi\left(\dfrac{15}{\sqrt{\pi}}\right) \approx 53.17$ cm. For the square, $s^2 = 225$, so $s = 15$ and $P = 4(15) = 60$ cm. The square has the greater perimeter.

44. $4.5^2 + 8^2 = c^2$; $c = \sqrt{4.5^2 + 8^2} = \sqrt{84.25} \approx 9.18$ cm

45. $a = \sqrt{5^2 - 4^2} = \sqrt{25 - 16} = 3$ cm. Students may also recognize that the triangle is 3-4-5.

46. The altitude divides the triangle into two 30-60-90 triangles. Since the side opposite the 30° angle is $\dfrac{8}{2} = 4$, the altitude $h = 4\sqrt{3}$
$A = \dfrac{1}{2}(8)(4\sqrt{3}) = 16\sqrt{3} \text{ in}^2$
$P = 3(8) = 24$ in.

47. If the length of a side is s, the altitude is $\dfrac{1}{2}\sqrt{3}s$.
$A = \dfrac{1}{2} \cdot s \cdot \left(\dfrac{1}{2}s\sqrt{3}\right) = \dfrac{s^2}{4}\sqrt{3} \text{ units}^2$
$P = 3s$ units

PAGE 346, LOOK BEYOND

48. Area $\approx \dfrac{1}{2}(5 + \sqrt{24}) \cdot 1 + \dfrac{1}{2}(\sqrt{24} + \sqrt{21}) \cdot 1 + \dfrac{1}{2}(\sqrt{21} + 4) \cdot 1$
$+ \dfrac{1}{2}(4 + 3) \cdot 1 + \dfrac{1}{2}(3 + 0) \cdot 1$
$= 18.98 \text{ units}^2$

49. The answer is an underestimate since the curve is above the top leg of each trapezoid.

50. Sample answer: An estimate for the area of the full circle is $4 \times 18.98 = 75.92 \text{ units}^2$, which is the answer we obtained by averaging the Left-Hand and Right-Hand Rules.

51. The rectangle with width h and height y_1 has area $y_1 h$. The rectangle with width h and height y_2 has area $y_2 h$. The average of the two is: $\dfrac{y_1 h + y_2 h}{2}$
$= \dfrac{(y_1 + y_2)h}{2} = \dfrac{1}{2}(y_1 + y_2)h$

The area of the trapezoid is $\dfrac{1}{2}(y_1 + y_2)h$

The areas are the same using both methods.

5. $\sqrt{(p-0)^2 + (q-0)^2} = \sqrt{p^2 + q^2}$

6. $\sqrt{(r-p)^2 + (s-q)^2}$

7. $\left(\dfrac{0+2p}{2}, \dfrac{0+2q}{2}\right) = (p, q)$

8. $\left(\dfrac{2p+2r}{2}, \dfrac{2q+2s}{2}\right) = (p+r, q+s)$

9. $\text{slope} = \dfrac{q-0}{p-0} = \dfrac{q}{p}$

10. $\text{slope} = \dfrac{s-q}{r-p}$

PAGES 350–353, PRACTICE AND APPLY

11. C will have the same x-coordinate as D and the same y-coordinate as B.
$C(q, p)$

12. $F(2p, 0)$

13. I has same y-coordinate as H, so $I(x, q)$.
$HI = GJ = r$
Thus, the x-coordinate of I is $r + x$-coordinate of H, or $r + p$.
$I(r + p, q)$

14. M has same x-coordinate as N and L has same y-coordinate as M. So if $M(a, b)$ and $L(c, d)$, one can conclude that $c = 0$ (L on y-axis), so $a = p$ and $b = d$.
$M(p, b)$ $L(0, b)$
But all sides of a square are equal so $b = p$.
Thus $M(p, p)$ and $L(0, p)$.

15. R has same y-coordinate as Q. The y-coordinate of S is zero. The x-coordinate of R and the x-coordinate of S cannot be determined by the given vertices. $R(a, q)$; $S(b, 0)$

16. Since $TUVW$ is a rhombus,
$TU = UV = VW = TW$
$TU = \sqrt{p^2 + q^2}$
$V(p + \sqrt{p^2 + q^2}, q)$
$W(\sqrt{p^2 + q^2}, 0)$

17. $C(2r, 2q)$ and $D(2s, 0)$; C has to be on the same horizontal line that B is on so it must have the same y-coordinate as B. D is on the x-axis.

18. $M(p, q)$; $N(r + s, q)$

19. $AD = \sqrt{(2s-0)^2 + (0-0)^2} = 2s$
$BC = \sqrt{(2r-2p)^2 + (2q-2q)^2} = 2r - 2p$
$MN = \sqrt{(r+s-p)^2 + (q-q)^2}$
$\quad = s + r - p$
Now find the average of the two bases:
$\dfrac{(AD + BC)}{2} = \dfrac{(2s + 2r - 2p)}{2} = s + r - p = MN$

20. slope of $\overline{AD} = 0$
slope of $\overline{BC} = 0$
slope of $\overline{MN} = 0$

All three segments have the same slope. Therefore the midsegment is parallel to the bases.

21. Hypotenuse of $\triangle JKL$ is KL.
The midpoint of KL is $M\left(\dfrac{q}{2}, \dfrac{p}{2}\right)$.

22. Find the lengths of $\overline{JM}, \overline{KM}, \overline{LM}$. If they are equal then the midpoint of the hypotenuse is equidistant from the three vertices.

$JM = \sqrt{\left(\dfrac{q}{2} - 0\right)^2 + \left(\dfrac{p}{2} - 0\right)^2} = \sqrt{\left(\dfrac{q}{2}\right)^2 + \left(\dfrac{p}{2}\right)^2}$

$KM = \sqrt{\left(\dfrac{q}{2} - 0\right)^2 + \left(\dfrac{p}{2} - p\right)^2} = \sqrt{\left(\dfrac{q}{2}\right)^2 + \left(\dfrac{p}{2}\right)^2}$

$LM = \sqrt{\left(\dfrac{q}{2} - q\right)^2 + \left(\dfrac{p}{2} - 0\right)^2} = \sqrt{\left(\dfrac{q}{2}\right)^2 + \left(\dfrac{p}{2}\right)^2}$

23. M is equidistant from the three vertices. Therefore, a circle can be drawn by joining J, K, and L and its radius is $JM = KM = LM$.

24. Midpoint of $\overline{EF} = M(p, q)$
Midpoint of $\overline{FG} = N(p + r, q + s)$
Midpoint of $\overline{GH} = O(r + t, s)$
Midpoint of $\overline{EH} = P(t, 0)$

25.

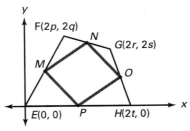

F(2p, 2q)
N
G(2r, 2s)
M
O
E(0, 0)   P   H(2t, 0)

27. Let $ABCD$ be a parallelogram formed by the following vertices:
$A(0, 0)$, $B(p, q)$, $C(r, q)$, and $D(s, 0)$ where AB and CD are opposite sides and BC and AD are opposite sides. Slopes of opposite sides are equal:

$$\frac{q}{p} = \frac{q}{r - s}$$

$$p = r - s$$

$$AB = \sqrt{p^2 + q^2}$$

$$CD = \sqrt{(s - r)^2 + q^2} = \sqrt{p^2 + q^2}$$

$$AB = CD$$

$$BC = \sqrt{(r - p)^2} = \sqrt{[r - (r - s)]^2} = \sqrt{s^2}$$

$$AD = \sqrt{s^2}$$

$$BC = AD$$

26. slope of $\overline{NO} = \dfrac{s - (q + s)}{r + t - (p + r)} = \dfrac{-q}{t - p}$

slope of $\overline{PM} = \dfrac{0 - q}{t - p} = \dfrac{-q}{t - p}$

slope of $\overline{MN} = \dfrac{q + s - q}{p + r - p} = \dfrac{s}{r}$

slope of $\overline{PO} = \dfrac{0 - s}{t - (r + t)} = \dfrac{-s}{-r} = \dfrac{s}{r}$

Slopes of opposite sides are equal, thus $MNOP$ is a parallelogram.

28. Let $ABCD$ be a square formed by:
$A(0, 0)$, $B(0, p)$, $C(p, p)$, and $D(p, 0)$

slope of diagonal $AC = \dfrac{p - 0}{p - 0} = 1$

slope of diagonal $BD = \dfrac{0 - p}{p - 0} = -1$

The two diagonals have slopes that are negative reciprocals of each other. Therefore they are perpendicular to each other.

29. Let $ABCD$ be a rectangle formed by:
$A(0, 0)$, $B(0, p)$, $C(p, q)$, and $D(q, 0)$

diagonal $AC = \sqrt{(p - 0)^2 + (q - 0)^2} = \sqrt{p^2 + q^2}$

diagonal $BD = \sqrt{(q - 0)^2 + (0 - p)^2} = \sqrt{p^2 + q^2}$

so the diagonals of a rectangle are congruent.

30. Let $ABCD$ be a parallelogram formed by
$A(0, 0)$, $B(p, q)$, $C(s, q)$, $D(r, 0)$
Since opposite sides of a parallelogram are congruent, $s = p + r$, so we have:

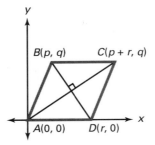

Since the diagonals of the parallelograms are perpendicular:

slope $\overline{AC} \cdot$ slope $\overline{BD} = -1$

$$\frac{q}{p + r} \cdot \frac{q}{p - r} = -1$$

$$q^2 = -1(p + r)(p - r)$$

$$q^2 = -(p^2 - r^2)$$

$$q^2 = r^2 - p^2$$

$$p^2 + q^2 = r^2$$

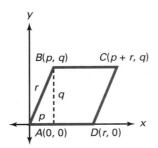

This implies that the length of side $\overline{AB}$ is r, by the Pythagorean Theorem. Since one pair of adjacent sides of the parallelogram are congruent, it is a rhombus.

31. Let $\triangle ABC$ be a triangle formed by $A(0, 0)$, $B(2p, 2q)$, and $C(2r, 0)$. Then the midpoints of the sides of the triangle are $M(p, q)$, $N(p + r, q)$, and $P(r, 0)$.

The slope of median $\overleftrightarrow{AN}$ is $m = \dfrac{q - 0}{p + r - 0} = \dfrac{q}{p + r}$.

The equation of $\overleftrightarrow{AN}$ is $y = \dfrac{q}{p + r}x$

The slope of median $\overleftrightarrow{CM}$ is $m = \dfrac{q - 0}{p + r - 0} = \dfrac{q}{p + r}$.

The equation of $\overleftrightarrow{CM}$ is $y = \dfrac{q}{p - 2r}(x - 2r)$

The slope of median $\overleftrightarrow{BP}$ is $m = \dfrac{2q - 0}{2p - r} = \dfrac{2q}{2p - r}$.

The equation of $\overleftrightarrow{BP}$ is $y = \dfrac{2q}{2p - r}(x - r)$

Set the first two equations equal and solve for x:

$$\frac{q}{p + r}x = \frac{q}{p - 2r}(x - 2r)$$

$$qx(p - 2r) = q(x - 2r)(p + r)$$

$$xp - 2rx = xp + xr - 2rp - 2r^2$$

$$2r + 2r^2p = 3rx$$

$$x = \frac{2}{3}(p + r)$$

Substitute into the first equation to find y:

$$y = \frac{q}{p+r}\frac{2}{3}(p+r) = \frac{2}{3}q$$

So the first two equations intersect at $\left(\frac{2}{3}(p+r), \frac{2}{3}q\right)$.

Substitute these values into the third equation to verify that the three lines intersect at a single pint.

$$\frac{2}{3}q = \frac{2q}{2p-r}\left(\frac{2}{3}(p+r) - r\right)$$
$$= \frac{2q}{2p-r}\frac{2p + 2r - 3r}{3}$$
$$= \frac{2q}{2p-r}\frac{2p-r}{3}$$
$$= \frac{2}{3}q$$

Since this results in a true statement, the three medians intersect at a single point.

32. a.
$$d_L = \sqrt{(100-x)^2 + (0-0)^2}$$
$$= \sqrt{(100-x)^2}$$
$$= |100 - x| \text{ meters}$$
$$d_w = \sqrt{(x-0)^2 + (0-30)^2}$$
$$= \sqrt{x^2 + 900} \text{ meters}$$

b. $C(x) = |100 - x| + 2\sqrt{x^2 + 900}$ (cost in thousands of dollars)

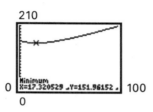

For the lowest cost road, the coordinates of X should be approximately (17.32, 151.962). This will make a road that costs about $151,962.

PAGE 352, LOOK BACK

33. $BC = 5$, since $AB = 5$ and $ABCD$ is a square.

34. $CD = 5$, since $AB = 5$ and $ABCD$ is a square.

35. $AD = 5$, since $AB = 5$ and $ABCD$ is a square.

36. $AC = \sqrt{5^2 + 5^2} = \sqrt{50} = 5\sqrt{2}$

37. $BD = 5\sqrt{2}$, since the diagonals of a square are congruent and $AC = 5\sqrt{2}$.

38. $AE = \frac{5\sqrt{2}}{2}$, since the diagonals of a square bisect each other and $AC = 5\sqrt{2}$.

39. $EC = \frac{5\sqrt{2}}{2}$, since the diagonals of a square bisect each other and $AC = 5\sqrt{2}$.

40. $BE = \frac{5\sqrt{2}}{2}$, since the diagonals of a square bisect each other and $BD = 5\sqrt{2}$.

41. $ED = \frac{5\sqrt{2}}{2}$, since the diagonals of a square bisect each other and $BD = 5\sqrt{2}$.

42. $m\angle BEA = 90°$, since the diagonals of a square are perpendicular.

43. $m\angle BEC = 90°$; since the diagonals of a square are perpendicular.

44. $MN = \frac{5.7 + 8.5}{2} = 7.1$

45. $A = \frac{1}{2}(b_1 + b_2)h = \frac{1}{2}(5.7 + 8.5)(3.5) = 24.85$ square units

46.

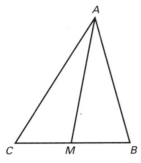

Given: scalene $\triangle ABC$ with median $\overline{AM}$
Prove: area of $\triangle ABM$ = area of $\triangle ACM$

Proof: M is the midpoint of $\overline{BC}$, so $CM = MB$ by the definition of midpoint. A is the vertex opposite $\overline{CM}$ in $\triangle ACM$ and also the vertex opposite $\overline{MB}$ in $\triangle ABM$. Therefore, the two triangles have the same altitude and thus the same height. Since the bases and heights of the two triangles are the same, their areas must be equal.

47.

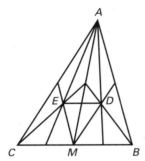

Sample answer: $\overline{AM}$ bisects $\overline{DE}$; $\overline{DE}$ is parallel to $\overline{BC}$.

5.8   **PAGE 356, GUIDED SKILLS PRACTICE**

---

6. Blue area = $7 \times 4 = 28$. Total area = $10 \times 5 = 50$ Probability dart will land in the blue area = $\frac{28}{50} = \frac{14}{25} = 0.56$

7. Unshaded area = $4 - \pi$
Probability penny will not land touching an intersection point = $\frac{4 - \pi}{4} \approx 0.21$

8. The width of the penny is considered 1 unit, so the shaded area is π, and the square has area 4. The probability that the penny will cover a vertex is $\frac{\pi}{4}$, theoretically. Repeated tosses should produce a number close to $\frac{\pi}{4}$. The more tosses, the more accurate the estimate should be. Multiply by 4 to give an estimate for π.

PAGES 356–358, PRACTICE AND APPLY

---

9. $P = \dfrac{\text{Shaded area}}{\text{Total area}} = \dfrac{8}{9}$

10. $P = \dfrac{\text{Shaded area}}{\text{Total area}} = \dfrac{4.5}{9} = \dfrac{1}{2}$

11. $P = \dfrac{\text{Shaded area}}{\text{Total area}} = \dfrac{\pi(1.5)^2 - \pi(0.5)^2}{\pi(1.5)^2}$
$= \dfrac{2.25\pi - 0.25\pi}{2.25\pi} = \dfrac{2\pi}{2.25\pi} = \dfrac{2}{2.25} = \dfrac{8}{9}$

12. $P = \dfrac{1}{4}$

13. $P = \dfrac{1}{3}$

14. $P = \dfrac{2}{3}$

15. $0.75 = \frac{75}{100}$, or 75%

16. $\frac{1}{4} = \frac{25}{100}$, or 25%

17. $\frac{2}{3} = 0.66$ or $66\frac{2}{3}$%

18. 60% means $\frac{60}{100} = 0.6$

19. 50% means $\frac{500}{100} = 0.5$

20. $33\frac{1}{3}$% means $\frac{33\frac{1}{3}}{100} = \frac{\frac{100}{3}}{100} = \frac{100}{300} = \frac{1}{3} = 0.3\overline{33}$

21. 45% means $\frac{45}{100} = \frac{9}{20}$

22. 80% means $\frac{80}{100} = \frac{4}{5}$

23. $66\frac{2}{3}\% = \left(66 + \frac{2}{3}\right)\% = \left(\frac{198}{3} + \frac{2}{3}\right)\% = \left(\frac{200}{3}\right)\%$, which means $\frac{\frac{200}{3}}{100} = \frac{2}{3}$

24. $P = \frac{1}{5}$

25. $P = \frac{3}{5}$

26. $P = \frac{2}{5}$

27. $\frac{3}{5} + \frac{2}{5} = 1$. One of these events must occur.

28. Sample answer:

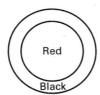

The dart board must be constructed so that the ratio of the red area to the total area $= 0.5$. So $P = \frac{\text{red area}}{\text{total area}} = 0.5$. Let r_1 be the radius of the red circle and let r_2 be the radius of the outer circle. Then $P = \frac{\pi r_1^2}{\pi r_2^2} = 0.5$ or $\frac{r_1^2}{r_2^2} = 0.5$. So $r_1^2 = 0.5 \cdot r_2^2$ or $r_1 = \sqrt{0.5} \cdot r_2 \approx 0.71 r_2$. The radius of the red circle must be approximately 0.71 or 71% of the outer circle. For example, if the radius of the outer circle is 12 inches, the radius of the red circle should be about $0.71 \times 12 = 8.52$ inches.

29. Sample answer:

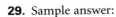
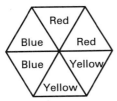

The ratio of the red area to total area is: $P = \frac{2}{6} = \frac{1}{3}$.

30. Let r = radius of inner circle. Then the radius of the outer circle is $2r$.

Area of red regions $= \frac{1}{2}$(area of outer ring)
$$= \frac{\pi(2r)^2 - \pi r^2}{2} = \frac{3\pi r^2}{2}$$

$$PR = \frac{\text{Area of red regions}}{\text{Area of dart board}} = \frac{\frac{3\pi r^2}{2}}{\pi(2r)^2} = \frac{\frac{3\pi r^2}{2}}{4\pi r^2} = \frac{3}{8}$$

31. Area of black regions $= \frac{1}{2}$(area of inner circle)
$$= \frac{\pi r^2}{2}$$

$$PB = \frac{\text{Area of black regions}}{\text{Area of dart board}} = \frac{\frac{\pi r^2}{2}}{\pi(2r)^2} = \frac{\frac{\pi r^2}{2}}{4\pi r^2} = \frac{1}{8}$$

32. $Pw = 1 - PR - PB = 1 - \frac{3}{8} - \frac{1}{8} = \frac{1}{2}$

33. The probability that the center of a penny will land within any square is 1 minus the probability calculated in Example 2. The grid consists of one half white squares and one half black squares, so

$$P = \frac{1 - \frac{\pi}{4}}{2} = \frac{\frac{4-\pi}{4}}{2} = \frac{4-\pi}{8} \approx 0.11$$

34. $80\% = \frac{80}{100} = \frac{8}{10} = \frac{4}{5} = 0.8$

35. $P = \dfrac{\text{Area of Oklahoma City}}{\text{Area of state of Oklahoma}} = \dfrac{625}{69{,}956} \approx 0.0089$ or 0.89%, a less than 1% chance.

36. $PA = \dfrac{\text{area of target } A}{\text{area of field}} = \dfrac{\frac{1}{2}(10)(20)}{70 \cdot 100} = \dfrac{1}{70} \approx 0.014$

37. $PB = \dfrac{\text{area of target } B}{\text{area of field}} = \dfrac{25 \cdot 25}{70 \cdot 100} = \dfrac{625}{7000} = \dfrac{5}{56}$
≈ 0.089

38. $PC = \dfrac{\text{area of target } C}{\text{area of field}} = \dfrac{\pi(20)^2}{70 \cdot 100} = \dfrac{400\pi}{7000} = \dfrac{4\pi}{70}$
≈ 0.180

39. $P = 1 - PA - PB - PC$
$= 1 - 0.014 - 0.089 - 0.180 = 0.717$

40. The probability that the train is there is $\frac{1}{5}$. The probability that it will arrive within 2 minutes is $\frac{1}{5} + \frac{1}{5} = \frac{2}{5}$. The probability that you'll have to wait at most 2 minutes is $\frac{1}{5} + \frac{2}{5} = \frac{3}{5}$.

PAGE 358, LOOK BACK

41. $16^2 + 30^2 \ \underline{?}\ 34^2$
$1156 = 1156$
The triangle is right.

42. $27^2 + 36^2 \ \underline{?}\ 50^2$
$2025 < 2500$
The triangle is obtuse.

43. The figure is not a triangle since $16 + 63 < 82$.

44. $11^2 + 60^2 \ \underline{\quad}\ 61^2$
$3721 = 3721$
The triangle is right.

45. $10^2 + 24^2 \ \underline{\quad}\ 25^2$
$676 > 625$
The triangle is acute.

46. The figure is not a triangle since $8 + 15 = 23$.

47. $A = \frac{1}{2}(4)(7.5) = 15 \text{ units}^2$

48. $A = 4(7.5) = 30 \text{ units}^2$

49. $A = \frac{1}{2}(12.6)(20 + 30) = 315 \text{ units}^2$

50. $A = \frac{22}{7}(16)^2 \approx 804.57 \text{ units}^2$

PAGE 359, LOOK BEYOND

51. The area of the board is $\pi(170)^2 = 28{,}900\pi \text{ mm}^2$
The area of the double ring $= \pi(170)^2 - \pi(170-8)^2$
$\qquad\qquad\qquad\qquad = 28{,}900\pi - 26{,}244\pi = 2656\pi \text{ mm}^2$

$P = \dfrac{2656\pi}{28{,}900\pi} \approx 0.09$

52. The area of the triple ring $= \pi(117)^2 - \pi(117-8)^2$
$\qquad\qquad\qquad\qquad = 13{,}689\pi - 11{,}881\pi = 1808\pi \text{ mm}^2$

$P = \dfrac{1808\pi}{28{,}900\pi} \approx 0.06$

53. Each sector is $\frac{1}{20}$th of the board, so
$P = \dfrac{0.06}{20} = 0.003$

54. Area of inner bull's eye $= \pi\left(\dfrac{12.7}{2}\right)^2 \approx 40.32\pi \text{ mm}^2$
$P = \dfrac{40.32\pi}{28{,}900\pi} \approx 0.001$

55. Area of outer bull's eye $= \pi\left(\frac{31}{2}\right)^2 - \pi\left(\frac{12.7}{2}\right)^2 \approx 199.93\pi$ mm^2

$P = \frac{199.93\pi}{28,900\pi} \approx 0.007$

56. When playing darts, you are aiming at a certain section instead of throwing at random.

CHAPTER REVIEW AND ASSESSMENT

1. $P = 23 + 30 + 37 = 90$ units.

2. $P = 2.0 + 1.3 + 3.8 + 2.8 + 2.4 + 2.5 = 14.8$ units

3. $A = 28 \times 40 = 1120$ units^2

4. $A = 15^2 = 225$ units^2

5. $A = \frac{1}{2}(4)(12) = 24$ units^2

6. $A = \frac{1}{2}(7)(5) = 17.5$ units^2

7. $A = 4(9) = 36$ units^2

8. $A = \frac{1}{2}(3 + 5)(6) = 24$ units^2

9. $C = \pi d = 11\pi$ units

10. $C = 2\pi(2.5) = 5\pi$ units

11. $A = \pi \cdot (10)^2 = 100\pi$ units^2

12. $d = 1, r = \frac{1}{2}$
$A = \pi\left(\frac{1}{2}\right)^2 = \frac{1}{4}\pi$ units^2

13. $c^2 = 20^2 + 21^2$
$c^2 = 400 + 441$
$c^2 = 841$
$c = \sqrt{841} = 29$

14. $a^2 + 99^2 = 101^2$
$a^2 = 10201 - 9801$
$a^2 = 400$
$a = \sqrt{400} = 20$

15. $48^2 + 63^2 \underline{\ ?\ } 65^2$
$6273 > 4225$
The triangle is acute.

16. $48^2 + 55^2 \underline{\ ?\ } 73^2$
$5329 = 5329$
The triangle is right.

17. $s\sqrt{2} = 100$, so $s = \frac{100}{\sqrt{2}} = \frac{100}{\sqrt{2}} \cdot \frac{\sqrt{2}}{\sqrt{2}} = \frac{100\sqrt{2}}{2}$
$= 50\sqrt{2}$

18. Length of hypotenuse is $34\sqrt{2}$.

19. $s\sqrt{3} = 27$, so the shorter leg is $s = \frac{27}{\sqrt{3}} = \frac{27}{\sqrt{3}} \cdot \frac{\sqrt{3}}{\sqrt{3}} = \frac{27\sqrt{3}}{3} = 9\sqrt{3}$ units.
The hypotenuse is $2s = 2(9\sqrt{3}) = 18\sqrt{3}$ units.

20. $2s = 16$, so $s = 8$ units, the length of the shorter leg.
The longer leg has length $8\sqrt{3}$ units.

21. $d = \sqrt{(6 - 0)^2 + (8 - 0)^2} = \sqrt{36 + 64} = \sqrt{100} = 10$

22. $d = \sqrt{(7 - 3)^2 + (2 - 3)^2} = \sqrt{4^2 + (-1)^2} = \sqrt{17}$

23. $d = \sqrt{(3 - (-2))^2 + (-1 - 4)^2} = \sqrt{5^2 + (-5)^2} = \sqrt{50} = 5\sqrt{2}$

24. Sample answer: The y values are:
$x = 1, y = 1^3 = 1$
$x = 2, y = 2^3 = 8$
$x = 3, y = 3^3 = 27$
$x = 4, y = 4^3 = 64$

Left-Hand Rule:
$A = 1 \cdot 1 + 1 \cdot 8 + 1 \cdot 27 = 36$ units^2
Right-Hand Rule:
$A = 1 \cdot 8 + 1 \cdot 27 + 1 \cdot 64 = 99$ units^2

The average of the two methods is: $\frac{36 + 99}{2} = 67.5$ units^2

Answers will vary due to the differences in approximation.
The exact answer is 63.75 units^2.

25.

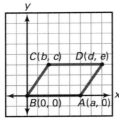

Let $D(d, e)$ be the fourth vertex.

D should have the same y-coordinate as C so $D(d, c)$.

Since opposite sides of a parallelogram are parallel which means they have the same slope, then

slope of $\overline{BC}$ = slope of $\overline{AD}$

$$\frac{c}{b} = \frac{c}{d - a}$$
$$b = d - a$$
$$d = a + b$$

Thus, $D(a + b, c)$.

26. $BD = \sqrt{(a + b - 0)^2 + (c - 0)^2} = \sqrt{(a + b)^2 + c^2}$

$AC = \sqrt{(b - a)^2 + (c - 0)^2} = \sqrt{(b - a)^2 + c^2}$

27. $BD = AC$

$$\sqrt{(a + b)^2 + c^2} = \sqrt{(b - a)^2 + c^2}$$
$$(a + b)^2 + c^2 = (b - a)^2 + c^2$$
$$(a + b)^2 = (b - a)^2$$
$$a^2 + 2ab + b^2 = b^2 - 2ab + a^2$$
$$4ab = 0$$

$a = 0$ or $b = 0$

28. If $a = 0$, then A and B are the same point and C and D are the same point. $ABCD$ becomes a segment, not a parallelogram. Thus, $a \neq 0$.

If $b = 0$, then C is $(0, c)$

so C is on the y-axis and therefore $m\angle A = 90°$. Thus $ABCD$ is a rectangle.

29. $P = \frac{1}{4} = 0.25$

30. $P = \frac{3}{8} = 0.375$

31. The large triangle is composed of 4 medium size triangles, and 3 of them are composed of 4 small triangles. If the middle medium size triangle were composed of 4 small triangles as the others, there would be 16 small triangles, 3 of which are shaded. So, $P = \frac{3}{16} = 0.1875$

32. $P = \frac{2}{3} \approx 0.67$

33. The Pythagorean theorem gives: $x^2 = \left(\frac{1}{4}x\right)^2 + 16^2$

$$x^2 = \frac{1}{16}x^2 + 256$$
$$x^2 - \frac{1}{16}x^2 = 256$$
$$\frac{15}{16}x^2 = 256$$
$$x^2 = \frac{4096}{15}$$
$$x = \sqrt{\frac{4096}{15}} \approx 16.52$$

The ladder is long enough since $16.52 + 3 < 20$ feet.

The ladder should be approximately $\frac{16.52}{4} = 4.13$ feet from the wall.

34. Perimeter of outside edge $= 160 + 160 + 2\pi \cdot 80 \approx 822.65$ m long.
Perimeter of inside edge $= 160 + 160 + 2\pi \cdot 40 \approx 571.33$ m long.
Area of running surface = Area of outside circle − Area of inside circle + Area of 2 rectangles

Since the outside radius of the circle is 80 and the inside radius is 40, the width of the track is 40 m.

$$\text{Area} = \pi \cdot 80^2 - \pi \cdot 40^2 + 2(160 \cdot 40)$$
$$= 6400\pi - 1600\pi + 12{,}800$$
$$= 4800\pi + 12{,}800$$
$$= 27{,}879.64 \text{ m}^2$$

CHAPTER TEST

For Exercises 1–4, the following information is needed: $\overline{AB} \perp \overline{AE}$ and $\overline{CG} \perp \overline{BF}$, so opposite sides of $ABFG$ are congruent. Because $AB = 4$ cm, $FG = 4$ cm. $\overline{CG} \perp \overline{AE}$ and $\overline{AE} \perp \overline{DE}$, so opposite sides of $CDEG$ are congruent.

Because $GE = 6$ cm, $CD = 6$ cm, and because $CG = 5 + 4 = 9$ cm, $DE = 9$ cm.
$BF = \sqrt{13^2 - 5^2} = \sqrt{144} = 12$ cm, so $AG = 12$ cm and $AE = 12 + 6 = 18$ cm.

1. $AB + BC + CD + DE + AE =$
$4 + 13 + 6 + 9 + 18 = 50$ cm

2. $AB + BF + FC + CD + DE + AE =$
$4 + 12 + 5 + 6 + 9 + 18 = 54$ cm

3. $(GE)(DE) = (6)(9) = 54$ cm^2

4. $(AB)(AG) + (GE)(DE) = (4)(12) + 54 = 102$ cm^2

5. $\frac{1}{2}(9)(15) = 67.5$ units^2

6. $(8)(12) = 96$ units^2

7. $\frac{1}{2}(7)(5 + 9) = 49$ units^2

8. $C = \pi(2)(8) = 16\pi \approx 50.24$ units
$A = \pi(8)^2 = 64\pi \approx 200.96$ units^2

9. $C = \pi(2)(26) = 52\pi \approx 163.28$ units
$A = \pi(26)^2 = 676\pi \approx 2122.64$ units^2

10. $C = \pi(30) = 30\pi \approx 94.2$ units
$A = \left(\frac{30}{2}\right)^2 = 225\pi \approx 706.5$ units^2

11. $C = \pi(42) = 42\pi \approx 131.88$ units
$A = \pi\left(\frac{42}{2}\right)^2 = 441\pi \approx 1384.74$ units^2

12. Area of round pizza: $A = \pi\left(\frac{12}{2}\right)^2 = 36\pi \approx 113.1$ sq in.
Area of 12-inch square pizza: $12^2 = 144$ sq in.
For the same price of \$8, the square pizza offers more; therefore, the square pizza is a better buy.

13. $(24)(12) - (3.14)(6)^2 = 174.96$ units^2

14. $7^2 + 24^2 = c^2$; $c = \sqrt{7^2 + 24^2} = 25$ units

15. $10^2 + b^2 = 26^2$; $b = \sqrt{26^2 - 10^2} = 24$ units

16. $a^2 + 12^2 = 16^2$; $a = \sqrt{16^2 - 12^2} = 4\sqrt{7}$

17. $6^2 + 6^2 = c^2$; $c = \sqrt{6^2 + 6^2} = 6\sqrt{2}$

18. $2x^2 = 80^2$; $x = \sqrt{\frac{80^2}{2}} = \sqrt{3200} = 40\sqrt{2}$

19. $2 \cdot 28^2 = c^2$; $x = \sqrt{1568} = 28\sqrt{2}$

20. $x\sqrt{3} = 15$; $x = \frac{15}{\sqrt{3}} = \frac{15\sqrt{3}}{3} = 5\sqrt{3}$
The shorter leg is $5\sqrt{3}$ units long and the hypotenuse is $2(5\sqrt{3}) = 10\sqrt{3}$ units long.

21. $2x = 24$; $x = 12$
The shorter leg is 12 units long, and the longer leg is $12\sqrt{3}$ units long.

22. $d = \sqrt{(5 + 3)^2 + (3 + 3)^2} = \sqrt{100} = 10$ units

23. $d = \sqrt{(5-2)^2 + (7-3)^2} = \sqrt{25} = 5$ units

24. $d = \sqrt{(-1-0)^2 + (-3-0)^2} = \sqrt{10}$ units

25. $d = \sqrt{(4+6)^2 + (9-3)^2} = 2\sqrt{34}$ units

26. $2q; 0$

27. $C(p, q); D\left(2r + \frac{2s - 2r}{2}, \frac{2q - 0}{2}\right) = D(r + s, q)$

28. $JM = \sqrt{(2s - 0)^2 + (0 - 0)^2} = 2s$

$KL = \sqrt{(2r - 2p)^2 + (2q - 2q)^2} = 2(r - p)$

$CD = \sqrt{(r + s - p)^2 + (q - q)^2} = r + s - p$

29. $\frac{KL + JM}{2} = \frac{2(r - p) + 2s}{2} = r - p + s = CD$

30. slope $\overline{JM} = \frac{0 - 0}{2s - 0} = 0$, slope $\overline{KL} = \frac{2q - 2q}{2r - 2p} = 0$,

slope $\overline{CD} = \frac{q - q}{r + s - p} = 0$; $\overline{JM}, \overline{KL},$ and $\overline{CD}$ are all horizontal and parallel.

31. $\frac{5}{8} = 0.625$

32. Area of figure $= (8)(6) = 48$ units^2
Area of shaded region $= (6)(4) = 24$ units^2
$P(\text{shaded}) = \frac{24}{48} = \frac{1}{2} = 0.5$

CHAPTERS 1–5 CUMMULATIVE ASSESSMENT

1. $a + b$ must be greater than c, so the answer is A.

2. RS must be equal to TU if the figure is a parallelogram, so the answer is C.

3. Area of circle $= \pi(5)^2 = 25\pi \approx 78.54$
Area of rectangle $= 6 \cdot 13 = 78$
$78.74 > 78$, so the answer is A.

4. Parallelogram

5. Hexagon

6. Trapezoid

7. Circle

8. c

9. b

10. a

11. c

12. d

13. slope of $AB = \frac{3 - 8}{4 - (-1)} = \frac{-5}{5} = -1$

slope of $BC = \frac{2 - 3}{1 - 4} = \frac{-1}{-3} = \frac{1}{3}$

slope of $AC = \frac{2 - 8}{1 - (-1)} = \frac{-6}{2} = -3$

14. Since the slope of $\overline{BC}$ and the slope of $\overline{AC}$ are negative reciprocals, or $\frac{1}{3} \cdot (-3) = -1$, sides $\overline{BC}$ and $\overline{AC}$ are perpendicular, so $\triangle ABC$ is a right triangle.

15. midpoint of $AB = \left(\frac{-1 + 4}{2}, \frac{8 + 3}{2}\right) = \left(\frac{3}{2}, \frac{11}{2}\right)$

midpoint of $BC = \left(\frac{4 + 1}{2}, \frac{3 + 2}{2}\right) = \left(\frac{5}{2}, \frac{5}{2}\right)$

midpoint of $AC = \left(\frac{-1 + 1}{2}, \frac{8 + 2}{2}\right) = (0, 5)$

16. length of $AB = \sqrt{(4 - (-1))^2 + (3 - 8)^2} = \sqrt{5^2 + (-5)^2} = \sqrt{50} \approx 7.07$

length of $BC = \sqrt{(1 - 4)^2 + (2 - 3)^2} = \sqrt{(-3)^2 + (-1)^2} = \sqrt{10} \approx 3.16$

length of $AC = \sqrt{(1 - (-1))^2 + (2 - 8)^2} = \sqrt{2^2 + (-6)^2} = \sqrt{40} \approx 6.32$

17. $(XZ)^2 + 21^2 = 29^2$
$(XZ)^2 = 841 - 441$
$(XZ)^2 = 400$
$XZ = \sqrt{400} = 20$

18. $(WZ) = 15^2 + 20^2$
$(WZ)^2 = 225 + 400$
$(WZ)^2 = 625$
$WZ = \sqrt{625} = 25$

19. $A = \frac{1}{2}(21 + 15) \cdot 20 = 360$ units^2

20. Area of a circle of radius 10: $A = \pi \cdot 10^2 = 100\pi \approx 314.16$ units^2. The area of a square is s^2, so $s^2 = 314.16$ or $s \approx 17.7$ units.

CHAPTER 6

Shapes in Space

6.1 **PAGE 375, GUIDED SKILLS PRACTICE**

5.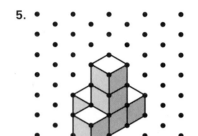

6.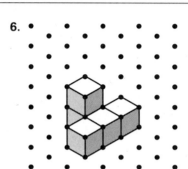

7. Answers may vary. Sample answer:

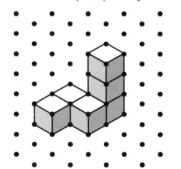

8.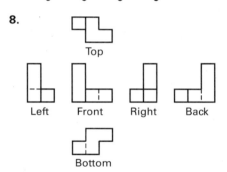

9. $v = 6$ cubic units

10. $s = 26$ square units

PAGES 375–377, PRACTICE AND APPLY

11. a

12. e

13. c

14. b

15. f

16. d

17. To indicate the hidden faces, add 3 dashed segments as shown.

18.
Top

Left Front Right Back

Bottom

19. Five cubes are used thus the volume is 5 cubic units.

20. Surface area = area of left face + area of front face + area of right face
+ area of back face + area of top face + area of bottom face
= 5 + 3 + 5 + 3 + 2 + 2
= 20 square units

21.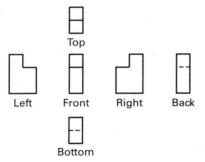
Top

Left Front Right Back

Bottom

22. Answers may vary. Sample answer:

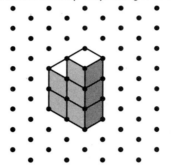

23.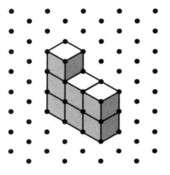

24. 10 faces

25.

26. No

27. Yes. The top view has 2 distinct faces which are congruent because they have the same size and shape. They are rectangles that are 1 unit by 2 units. There is also a face on the back side that has the same size and shape.

28. a. 18 square units

b. 16 square units

c. 18 square units

Shape b has the least surface area.

29. a.

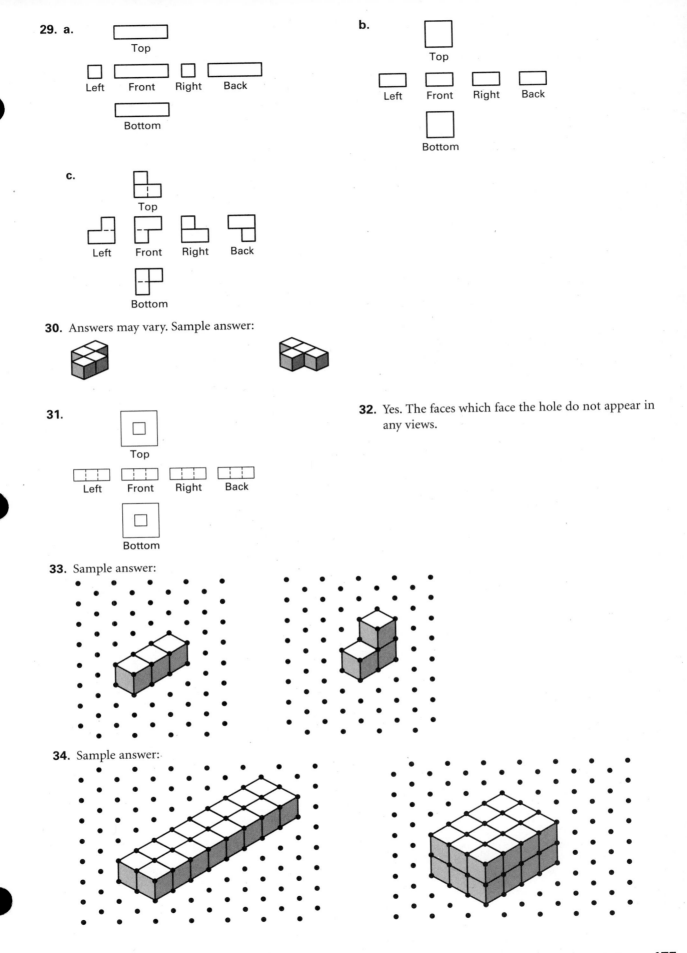

Top

Left Front Right Back

Bottom

b.

Top

Left Front Right Back

Bottom

c.

Top

Left Front Right Back

Bottom

30. Answers may vary. Sample answer:

31.

Top

Left Front Right Back

Bottom

32. Yes. The faces which face the hole do not appear in any views.

33. Sample answer:

34. Sample answer:

35. Answers may vary. Sample answer:

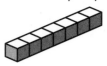

36. Answers may vary. Sample answer:

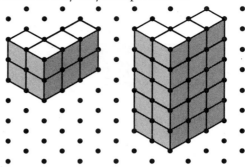

37. Answers may vary. Sample answer:

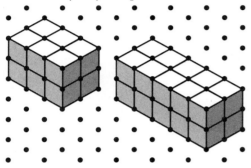

38. $2^3 = 8$
$3^3 = 27$
$4^3 = 64$
n^3

39. $4 \times 6 = 24$ square units
$9 \times 6 = 54$ square units
$16 \times 6 = 96$ square units
$n^2 \times 6 = 6n^2$ square units

40. 8 unit cubes will have 3 red faces.
12 unit cubes will have 2 red faces.
6 unit cubes will have 1 red face.
1 unit cube will have no red face.

41. 8 unit cubes will have 3 red faces.
24 unit cubes will have 2 red faces
24 unit cubes will have 1 red face.
8 unit cubes will have no red face.

42.

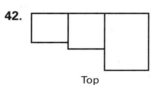

Top

Left

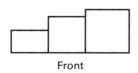

Front

43.

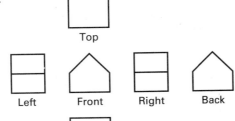

These views might be used when writing instructions on how to assemble the tent.

44. $\frac{1}{2}(6)(8) = 24$ units^2

45. $(10.7)(2.3) = 24.61$ m^2

46. $\frac{1}{2}(5 + 8)(3.7) = 24.05$ ft^2

47. $\pi(5)^2 \approx 78.54$ m^2

48. $\sqrt{6^2 + 3^2} \approx 6.71$ cm

49. $\sqrt{315^2 - 265^2} \approx 170.29$ yd

50.

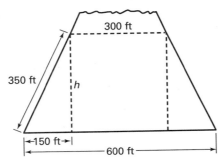

$h = \sqrt{350^2 - 150^2} \approx 316.23$

Since the labeled segment near the top is one-half the base, that segment is a midsegment of the triangular cross section of the original pyramid. The portion above the labeled segment is thus the same height as the portion below it.

$2h \approx 2(316.23) \approx 632.46 \approx 632$ ft

51. Answers may vary. Sample answer:

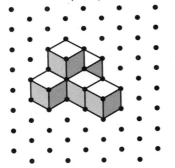 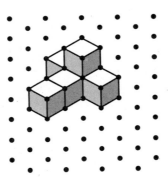

52. Yes. The object uses 6 cubes but only 5 are seen in the isometric drawings in answer to Exercise 51.

53. 3 left + 3 right + 4 front + 1 back + 4 top + 1 bottom = 16 faces

54. The solid has eleven congruent faces of one square and three congruent faces of two squares.

7. Sample answer:
$\overline{HI}$ and $\overline{KG}$; $\overline{LM}$ and $\overline{ON}$

8. Sample answer:
$\overline{HI}$ and $\overline{MN}$; $\overline{HL}$ and $\overline{IG}$

9. Sample answer:
HIML and *KGNO*; *HKOL* and *IGNM*

10. Sample answer:
$\overline{HI}$ and *IGNM*; $\overline{HL}$ and *HIGK*

11. Sample answer:
$\overline{HI}$ and $\overline{ON}$; $\overline{LO}$ and $\overline{IG}$

12. Line *p* and line *q* are perpendicular.
Line *p* and line *r* are perpendicular.

13. There is a line in S that is parallel to line *m*.

14. The measure of the dihedral angle is m∠*EFG*.

PAGES 384–385, PRACTICE AND APPLY

15. Line *p* and plane M are parallel.

16. No, we would only be able to draw a conclusion if we knew that $m \parallel n$.

17. Line *p* is not perpendicular to plane R.

18. True. If $r \parallel s$ and $t \parallel s$, then $r \parallel t$.

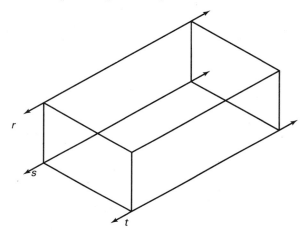

19. True. If $R \parallel S$ and $T \parallel S$, then $R \parallel T$.

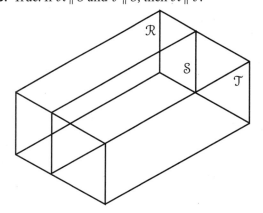

20. False. $R \perp T$ and $S \perp T$, but R is not perpendicular to S.

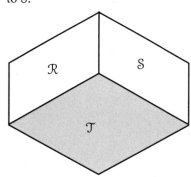

21. True. If $S \perp \ell$ and $R \perp \ell$, then $S \parallel R$.

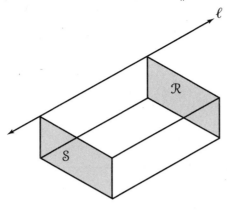

22. True. If $\ell \perp R$ and $m \perp R$, then $\ell \parallel m$. There is a line in the plane that contains both points where the two perpendicular lines intersect the plane. So both lines will be perpendicular to that line and therefore parallel.

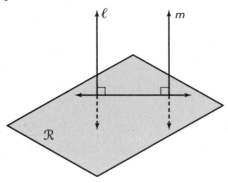

23. Check student constructions.

24. The insert in $\overline{CD}$.

25. The angle formed by the insert in $\overline{AB}$.

26. The angle formed by the insert in $\overline{AB}$ is larger than the angle formed by the insert in $\overline{CD}$.

27. intersect

28. skew

29. plane

30. Acute; the angle the airplane makes with the ground is less than 90°.

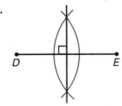

PAGE 386, LOOK BACK

31.

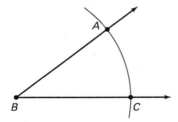

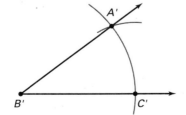

32.

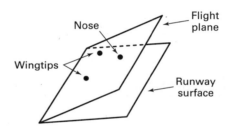

33.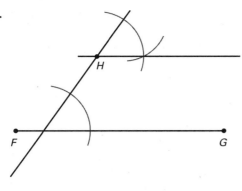

34.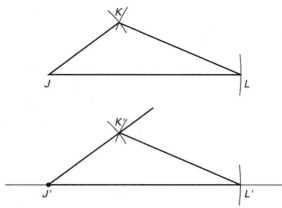

35. Area of a square = $s^2 = 14^2 = 196$ units^2

36. Area of a trapezoid = $\frac{1}{2}(b_1 + b_2)h$
$$= \frac{1}{2}(22 + 32)16$$
$$= 432 \text{ units}^2$$

37. Area of a parallelogram = $l \cdot h$
$$= 28 \cdot 18$$
$$= 504 \text{ units}^2$$

38. Area of a triangle = $\frac{1}{2}bh$
$$= \frac{1}{2}(45)(28)$$
$$= 630 \text{ units}^2$$

39. $h = \sqrt{(15)^2 - (7.5)^2} \approx 12.99$

Area of equilateral triangle = $\frac{1}{2}(15)(12.99)$
$$\approx 97.4 \text{ units}^2$$

40. 6 triangles with
$$h = \sqrt{(9)^2 - (4.5)^2} \approx 7.79$$
$$A \approx \frac{1}{2}(9)(7.79) \approx 35.07$$

Area of regular hexagon $\approx 6(35.07) \approx 210.4$ units^2

PAGE 386, LOOK BEYOND

41.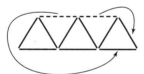

42.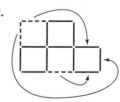

6.3 | PAGE 391, GUIDED SKILLS PRACTICE

5. Rectangle (because it is the base).

7. Rectangle.

6. Parallelogram.

8. First, find the length of the diagonal of the unit cell. Each edge is 430 pm, so:
$$d = \sqrt{l^2 + w^2 + h^2}$$
$$d = \sqrt{430^2 + 430^2 + 430^2} = \sqrt{3(430)^2} = 430\sqrt{3}$$
$$d \approx 744.78$$

The diagonal is the length of four radii, so divide by 4.
$$r \approx \frac{744.78}{4} \approx 186$$

The radius of a sodium atom is approximately 186 pm.

9. Hexagonal prism.

10. Pentagonal prism.

11. Not a prism; prisms do not have curved edges.

12. Triangular prism.

13. Not a prism; there are no parallel bases.

14. Pentagonal prism.

15. $\triangle ABC \cong \triangle DEF$; In a prism, the bases are translated images of each other.

16. $\overline{AD}$ and $\overline{CF}$

17. Rectangle.

18. $\overline{BC}$, $\overline{AB}$, and $\overline{DE}$

19. $ABED$ and $BCFE$

20. $\angle ADE$, $\angle ADF$, $\angle DAB$, $\angle DAC$, $\angle BED$, $\angle BEF$, $\angle CFD$, $\angle CFE$, $\angle FCA$, $\angle FCB$, $\angle EBA$, $\angle EBC$

21. $\overline{HM}$

22. Parallelogram.

23. Parallelogram.

24. $GKLH \cong JNMI$; $GKNJ \cong HLMI$; $KLMN \cong GHIJ$

25. Give that $m\angle GKN = 60°$ and $m\angle GKL = 80°$, we know that angles supplementary to these angles are obtuse. These angles are part of parallelograms, so the two angles adjacent to these in their parallelogram will be obtuse. Also, angles corresponding to these obtuse angles in the congruent opposite face will also be obtuse. Thus, $\angle KNJ$, $\angle JGK$, $\angle LMI$, $\angle IHL$, $\angle NJI$, $\angle IMN$, $\angle KGH$, and $\angle HLK$ are obtuse.

26. $\sqrt{4^2 + 12^2 + 3^2} = 13$

27. $\sqrt{10^2 + 5^2 + 12^2} \approx 16.40$

28. $\sqrt{7.5^2 + 8^2 + 8.5^2} \approx 13.87$

29. $\sqrt{a^2 + a^2 + a^2} = a\sqrt{3}$

30. $\sqrt{\ell^2 + 8^2 + 24^2} = 26$
$\ell^2 + 8^2 + 24^2 = 26^2$
$\ell^2 = 26^2 - 8^2 - 24$
$\ell = \sqrt{26^2 - 8^2 - 24^2} = 6$

31. $\sqrt{7^2 + w^2 + 10^2} = 15$
$7^2 + w^2 + 10^2 = 15^2$
$w^2 = 15^2 - 7^2 - 10^2$
$w = \sqrt{15^2 - 7^2 - 10^2} \approx 8.72$

32. $d = \sqrt{a^2 + a^2 + (2a)^2} = \sqrt{6a^2} = a\sqrt{6}$

33. $\sqrt{x^2 + x^2 + x^2} = 10$
$\sqrt{3x^2} = 10$
$x\sqrt{3} = 10$
$x = \dfrac{10}{\sqrt{3}} \approx 5.77$

	Number of faces	Number of vertices	Number of edges
34.	5	6	9
35.	6	8	12
36.	7	10	15
37.	8	12	18
38.	$n + 2$	$2n$	$3n$

39. faces: $20 + 2 = 22$; vertices: $2(20) = 40$; edges: $3(20) = 60$

40. Triangular prism: $6 - 9 + 5 = 2$
Rectangular prism: $8 - 12 + 6 = 2$
Pentagonal prism: $10 - 15 + 7 = 2$
Hexagonal prism: $12 - 18 + 8 = 2$
$V - E + F = 2$

41. The distance between opposite vertices on a regular hexagon with side length a is $2a$. d is the hypotenuse of a right triangle with legs of length $2a$ and h.
$d = \sqrt{(2a)^2 + h^2}$
$= \sqrt{4a^2 + h^2}$

42. Fold the left and right rectangle up to meet above the center rectangle. Then fold the triangles up to meet the edges of these rectangles.

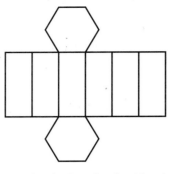

43. Let x be the length of a side of a unit-cell cube.

$$d = 4a = \sqrt{x^2 + x^2 + x^2} = x\sqrt{3}$$

$$x = \frac{4a}{\sqrt{3}} = \frac{4a\sqrt{3}}{3}$$

44. The length, f, of a diagonal of a face of the cube is 4 radii.

$$f = 4(107) = 788$$

If e is the length of the edge of the cube, then

$$f = \sqrt{e^2 + e^2} = \sqrt{2e^2} = e\sqrt{2}.$$

Solve for e:

$$e = \frac{f}{\sqrt{2}} = \frac{788}{\sqrt{2}} \approx 557.2 \text{ pm}$$

The length, d, of a diagonal of the cube is

$$d = \sqrt{e^2 + e^2 + e^2} = \sqrt{3e^2} = e\sqrt{3}$$
$$= 557.2\sqrt{3} \approx 965.1 \text{ pm}.$$

PAGE 394, LOOK BACK

45. True, because the diagonals of a rectangle bisect each other.

46. Sample answer: False. In a rectangle made by joining two congruent 30-60-90° triangles, the diagonal divides the corner angles into 60° and 30° angles.

47. False.

48. False. The diagonals of a rectangle divide the rectangle into two pairs of congruent triangles.

49. $\sqrt{(3 - 3)^2 + (-4 - 4)^2} = 8$

$\left(\frac{3 + 3}{2}, \frac{4 + (-4)}{2}\right) = (3, 0)$

50. $\sqrt{(-2 - 4)^2 + (3 - (-2))^2} = \sqrt{36 + 25} \approx 7.81$

$\left(\frac{4 + (-2)}{2}, \frac{-2 + 3}{2}\right) = \left(1, \frac{1}{2}\right)$

51. $\sqrt{(5 - 2)^2 + [6 - (-2)]^2} = \sqrt{9 + 64} \approx 8.54$

$\left(\frac{2 + 5}{2}, \frac{-2 + 6}{2}\right) = \left(\frac{7}{2}, 2\right)$

52. $180° - (136° + 29°) = 15°$

PAGE 395, LOOK BEYOND

53. $AG = \sqrt{10^2 + 7^2 + 8^2} \approx 14.59 \text{ ft}$

54. ≈ 18.03 ft. The correct answer to this problem depends on drawing an appropriate net for the room, such as net B below. Notice that net A gives a larger value for the distance *AG*.

A.

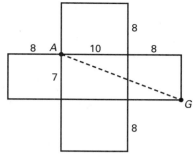

$$AG = \sqrt{(10 + 8)^2 + 7^2}$$
$$\approx 19.31 \text{ ft.}$$

B.

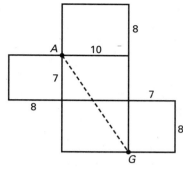

$$AG = \sqrt{(8 + 7)^2 + 10^2}$$
$$\approx 18.03$$

6.4 **PAGE 399, GUIDED SKILLS PRACTICE**

6.

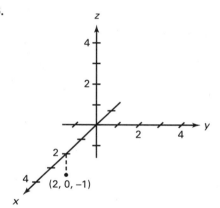

7.

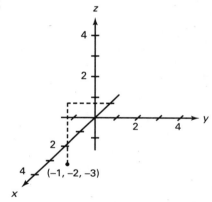

8.

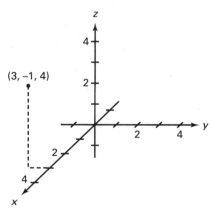

9. (0, 0, 0); (2, 0, 0); (0, 7, 0); (2, 7, 0); (0, 0, 6); (2, 0, 6); (0, 7, 6); (2, 7, 6)

10. $\sqrt{(3 - 0)^2 + (6 - 0)^2 + (1 - 0)^2} = \sqrt{46} \approx 6.78$ **11.** $\sqrt{(5 - 3)^2 + (-3 - 1)^2 + (1 - 0)^2} = \sqrt{21} \approx 4.58$

12. $\sqrt{[2 - (-4)]^2 + (3 - 7)^2 + [-5 - (-2)]^2} = \sqrt{61} \approx 7.81$

13.

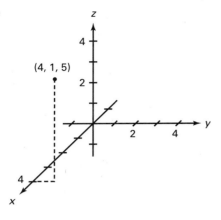

14.

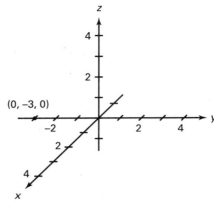

15.

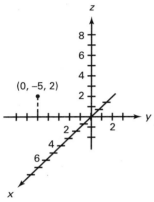

16.

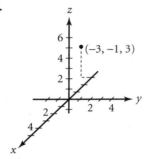

17. top-front-right

18. bottom-back-right

19. x-z plane

20. y-axis

21. top-back-right

22. x-y plane

23. x-axis

24. bottom-back-left

25. Signs of numbers are: $(+, +, +)$.
Sample answer: $(1, 2, 3)$

26. Signs of numbers are: $(+, +, -)$.
Sample eanswer: $(1, 2, -3)$

27. Signs of numbers are: $(-, -, +)$.
Sample answer: $(-1, -2, 3)$

28. Signs of numbers are: $(-, -, -)$.
Sample answer: $(-1, -2, -3)$

29. $\sqrt{(-1 - 1)^2 + (-1 - 1)^2 + (-1 - 1)^2} = \sqrt{12}$
≈ 3.46

30. $\sqrt{(5 - 2)^2 + (-2 - 1)^2 + (7 - 3)^2} = \sqrt{34}$
≈ 5.83

31. $\sqrt{(-5 - 2)^2 + (-6 - 0)^2 + (-5 - 1)^2} = \sqrt{121}$
$= 11$

32. $\sqrt{(6 - 7)^2 + [(4 - (-6)]^2 + (3 - 5)^2} = \sqrt{105}$
≈ 10.25

33. $(0, 10, 7)$

34. $(6, 10, 7)$

35. $(6, 10, 0)$

36. $(0, 0, 7)$

37. $A(0, 0, 0)$ and $D(6, 0, 0)$
$AD = \sqrt{(6 - 0)^2 + (0 - 0)^2 + (0 - 0)^2}$
$= \sqrt{36}$
$= 6$

38. $A(0, 0, 0)$ and $G(6, 0, 7)$
$AG = \sqrt{(6 - 0)^2 + (0 - 0)^2 + (7 - 0)^2}$
$= \sqrt{85}$
≈ 9.22

39. $A(0, 0, 0)$ and $H(6, 10, 7)$
$AH = \sqrt{(6 - 0)^2 + (10 - 0)^2 + (7 - 0)^2}$
$= \sqrt{185}$
≈ 13.6

40. $D(6, 0, 0)$ and $F(0, 10, 7)$
$DF = \sqrt{(0 - 6)^2 + (10 - 0)^2 + (7 - 0)^2}$
$= \sqrt{185}$
≈ 13.6

41. $7 \times 6 = 42$ units^2

42. $10 \times 6 = 60$ units^2

43. $x_1 = 5, y_1 = -2, z_1 = 3$
$x_2 = 6, y_2 = -7, z_2 = 4$
$$\left(\frac{x_1 + x_2}{2}, \frac{y_1 + y_2}{2}, \frac{z_1 + z_2}{2}\right)$$
$$\left(\frac{5 + 6}{2}, \frac{-2 + (-7)}{2}, \frac{3 + 4}{2}\right)$$
$$\left(\frac{11}{2}, \frac{-9}{2}, \frac{7}{2}\right)$$

44. $x_1 = 3, y_1 = 2, z_1 = 1$
$x_2 = 1, y_2 = 2, z_2 = 3$
$$\left(\frac{x_1 + x_2}{2}, \frac{y_1 + y_2}{2}, \frac{z_1 + z_2}{2}\right)$$
$$\left(\frac{3 + 1}{2}, \frac{2 + 2}{2}, \frac{1 + 3}{2}\right)$$
$$\left(\frac{4}{2}, \frac{4}{2}, \frac{4}{2}\right)$$
$$(2, 2, 2)$$

45. $x_1 = -1, y_1 = -4, z_1 = -5$
$x_2 = 6, y_2 = 1, z_2 = 0$
$$\left(\frac{x_1 + x_2}{2}, \frac{y_1 + y_2}{2}, \frac{z_1 + z_2}{2}\right)$$
$$\left(\frac{-1 + 6}{2}, \frac{-4 + 1}{2}, \frac{-5 + 0}{2}\right)$$
$$\left(\frac{5}{2}, \frac{-3}{2}, \frac{-5}{2}\right)$$

46. $x_1 = 2, y_1 = -1, z_1 = 0$
$x_2 = 0, y_2 = 0, z_2 = 1$
$$\left(\frac{x_1 + x_2}{2}, \frac{y_1 + y_2}{2}, \frac{z_1 + z_2}{2}\right)$$
$$\left(\frac{2 + 0}{2}, \frac{-1 + 0}{2}, \frac{0 + 1}{2}\right)$$
$$\left(\frac{2}{2}, \frac{-1}{2}, \frac{1}{2}\right)$$
$$\left(1, \frac{-1}{2}, \frac{1}{2}\right)$$

47. $x_1 = 1, y_1 = 1, z_1 = 1$
$x_2 = -1, y_2 = -1, z_2 = -1$
$$\left(\frac{x_1 + x_2}{2}, \frac{y_1 + y_2}{2}, \frac{z_1 + z_2}{2}\right)$$
$$\left(\frac{1 + (-1)}{2}, \frac{1 + (-1)}{2}, \frac{1 + (-1)}{2}\right)$$
$$\left(\frac{0}{2}, \frac{0}{2}, \frac{0}{2}\right)$$
$$(0, 0, 0)$$

48. $x_1 = a, y_1 = b, z_1 = c$
$x_2 = -a, y_2 = -b, z_2 = -c$
$$\left(\frac{x_1 + x_2}{2}, \frac{y_1 + y_2}{2}, \frac{z_1 + z_2}{2}\right)$$
$$\left(\frac{a + (-a)}{2}, \frac{b + (-b)}{2}, \frac{c + (-c)}{2}\right)$$
$$\left(\frac{0}{2}, \frac{0}{2}, \frac{0}{2}\right)$$
$$(0, 0, 0)$$

49. Sample answer:
$$1 = \sqrt{x^2 + y^2 + z^2}$$
If $x = y = z$, then
$$1 = \sqrt{x^2 + x^2 + x^2}$$
$$1 = \sqrt{3x^2}$$
$$1^2 = 3x^2$$
$$\frac{1}{3} = x^2$$
$$x = \sqrt{\frac{1}{3}} = \frac{\sqrt{3}}{3}$$
$$\left(\frac{\sqrt{3}}{3}, \frac{\sqrt{3}}{3}, \frac{\sqrt{3}}{3}\right)$$

50. The red light source is located above the positive y-axis and the blue light source is located above the negative y-axis.

PAGE 401, LOOK BACK

51. $\dfrac{4 - 0}{6 - 0} = \dfrac{2}{3}$

52. $\dfrac{3 - 5}{3 - 1} = \dfrac{-2}{2} = -1$

53. $\dfrac{-5 - 1}{2 - 3} = \dfrac{-6}{-1} = 6$

54. $\dfrac{-4 - 1}{6 - (-3)} = \dfrac{-5}{9}$

55.

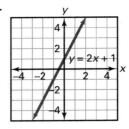

x-intercept:
Let $y = 0$.
$$y = 2x + 1$$
$$0 = 2x + 1$$
$$-2x = 1$$
$$x = -\frac{1}{2}$$
$$\left(-\frac{1}{2}, 0\right)$$
y-intercept:
Let $x = 0$.
$$y = 2x + 1$$
$$y = 2(0) + 1$$
$$y = 1$$
$$(0, 1)$$

56.

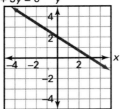

x-intercept:
Let $y = 0$.
$$2x + 3y = 6$$
$$2x + 3(0) = 6$$
$$2x = 6$$
$$x = 3$$
$$(3, 0)$$
y-intercept:
Let $x = 0$.
$$2x + 3y = 6$$
$$2(0) + 3y = 6$$
$$3y = 6$$
$$y = 2$$
$$(0, 2)$$

57.

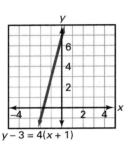

x-intercept:
Let $y = 0$.
$$y - 3 = 4(x + 1)$$
$$0 - 3 = 4(x + 1)$$
$$-3 = 4x + 4$$
$$-7 = 4x$$
$$-\frac{7}{4} = x$$
$$\left(-\frac{7}{4}, 0\right)$$
y-intercept:
Let $x = 0$.
$$y - 3 = 4(x + 1)$$
$$y - 3 = 4(0 + 1)$$
$$y - 3 = 4(1)$$
$$y - 3 = 4$$
$$y = 7$$
$$(0, 7)$$

58.

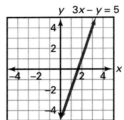

x-intercept:
Let $y = 0$.
$$3x - y = 5$$
$$3x - 0 = 5$$
$$3x = 5$$
$$x = \frac{5}{3}$$
$$\left(\frac{5}{3}, 0\right)$$
y-intercept:
Let $x = 0$.
$$3x - y = 5$$
$$3(0) - y = 5$$
$$-y = 5$$
$$y = -5$$
$$(0, -5)$$

59. slope $= \frac{212 - 32}{100 - 0} = 1.8$

Let b be the *y*-intercept.
$$32 = 1.8(0) + b$$
$$b = 32$$
The equation is $y = 1.8x + 32$.

60. $x_1 = 0.819, y_1 = -0.546, z_1 = -0.237$
$x_2 = -2.238, y_2 = 4.328, z_2 = 1.910$
$d = \sqrt{(x_2 - x_1)^2 + (y_2 - y_1)^2 + (z_2 - z_1)^2}$
$\quad = \sqrt{(-2.238 - 0.819)^2 + [4.328 - (-0.546)]^2 + [1.910 - (-0.237)]^2}$
$\quad \approx 6.141$
$6.141 \text{ au}\left(\dfrac{92{,}960{,}000 \text{ miles}}{1 \text{ au}}\right) = 570{,}867{,}360 \text{ miles}$

61. $x_1 = 1.389, y_1 = 0.145, z_1 = 0.029$
$x_2 = 3.935, y_2 = -8.443, z_2 = -3.656$
$d = \sqrt{(x_2 - x_1)^2 + (y_2 - y_1)^2 + (z_2 - z_1)^2}$
$\quad = \sqrt{(3.935 - 1.389)^2 + (-8.443 - 0.145)^2 + (-3.656 - 0.029)^2}$
$\quad \approx 9.686$
$9.686 \text{ au}\left(\dfrac{149{,}600{,}000}{1 \text{ au}}\right) = 1{,}449{,}025{,}600 \text{ km}$

62. The coordinates of the Sun are $(0, 0, 0)$. Distance from the Sun to Mars:
$d = \sqrt{(1.389 - 0)^2 + (0.145 - 0)^2 + (0.029 - 0)^2}$
$\quad = \sqrt{1.389^2 + 0.145^2 + 0.029^2}$
$\quad \approx 1.397$
Mars is about 1.397 au from the Sun.
$227{,}940{,}000 \text{ km}\left(\dfrac{1 \text{ au}}{149{,}600{,}000 \text{ km}}\right) \approx 1.524 \text{ au}$
The mean distance between Mars and the Sun is 1.524 au.
$1.524 - 1.397 = 0.127$
The estimate is off by 0.127 au because the orbit of Mars is elliptical.

63. The coordinates of the Sun are $(0, 0, 0)$.
Earth's distance from the Sun:
$d = \sqrt{(0.819 - 0)^2 + (-0.546 - 0)^2 + (-0.237 - 0)^2}$
$\quad = \sqrt{(0.819)^2 + (-0.546)^2 + (-0.237)^2}$
$\quad \approx 1.012 \text{ au}$
Mars' distance from the Sun:
$d = \sqrt{(1.389 - 0)^2 + (0.145 - 0)^2 + (0.029 - 0)^2}$
$\quad = \sqrt{1.389^2 + 0.145^2 + 0.029^2}$
$\quad \approx 1.397 \text{ au}$
Jupiter's distance from the Sun:
$d = \sqrt{(-2.238 - 0)^2 + (4.328 - 0)^2 + (1.910 - 0)^2}$
$\quad = \sqrt{(-2.238)^2 + 4.328^2 + 1.910^2}$
$\quad \approx 5.233 \text{ au}$
Saturn's distance from the Sun:
$d = \sqrt{(3.935 - 0)^2 + (-8.443 - 0)^2 + (-3.656 - 0)^2}$
$\quad = \sqrt{3.935^2 + (-8.443)^2 + (-3.656)^2}$
$\quad \approx 10.007 \text{ au}$
Earth is the closest planet in the table to the Sun.

6. *x*-intercept:
Let $y = 0$ and $z = 0$.
$$3x + 6y + 4z = 12$$
$$3x + 6(0) + 4(0) = 12$$
$$3x = 12$$
$$x = 4$$
$(4, 0, 0)$

y-intercept:
Let $x = 0$ and $z = 0$.
$$3x + 6y + 4z = 12$$
$$3(0) + 6y + 4(0) = 12$$
$$6y = 12$$
$$y = 2$$
$(0, 2, 0)$

z-intercept:
Let $x = 0$ and $y = 0$.
$$3x + 6y + 4z = 12$$
$$3(0) + 6(0) + 4z = 12$$
$$4z = 12$$
$$z = 3$$
$(0, 0, 3)$

7. *x*-intercept:
Let $y = 0$ and $z = 0$.
$$2x + 5y - z = 2$$
$$2x + 5(0) - 0 = 2$$
$$2x = 2$$
$$x = 1$$
$(1, 0, 0)$

y-intercept:
Let $x = 0$ and $z = 0$.
$$2x + 5y - z = 2$$
$$2(0) + 5y - 0 = 2$$
$$5y = 2$$
$$y = \frac{2}{5}$$
$\left(0, \frac{2}{5}, 0\right)$

z-intercept:
Let $x = 0$ and $y = 0$.
$$2x + 5y - z = 2$$
$$2(0) + 5(0) - z = 2$$
$$-z = 2$$
$$z = -2$$
$(0, 0, -2)$

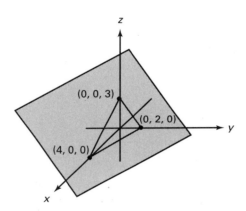

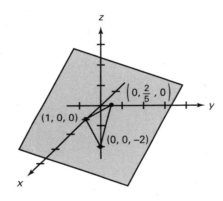

8.

t	x	y
1	3	4
2	4	8
3	5	12

9.

t	x	y
1	0	5
2	1	7
3	2	9

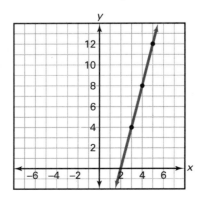

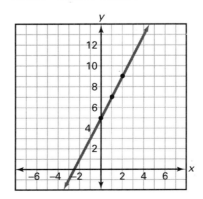

10.

t	x	y	z
1	1	2	2
2	2	4	3
3	3	6	4

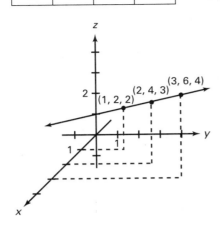

11.

t	x	y	z
1	1	7	−1
2	4	8	−2
3	7	9	−3

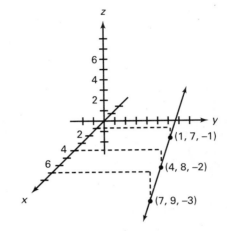

PAGE 399, PRACTICE AND APPLY

12. $3x + 2y + 7z = 4$

x-intercept:
Let $y = 0$ and $z = 0$.
$$3x + 2(0) + 7(0) = 4$$
$$3x = 4$$
$$x = \frac{4}{3}$$
$\left(\frac{4}{3}, 0, 0\right)$

y-intercept:
Let $x = 0$ and $z = 0$.
$$3(0) + 2y + 7(0) = 4$$
$$2y = 4$$
$$y = 2$$
$(0, 2, 0)$

z-intercept:
Let $x = 0$ and $y = 0$.
$$3(0) + 2(0) + 7z = 4$$
$$7z = 4$$
$$z = \frac{4}{7}$$
$\left(0, 0, \frac{4}{7}\right)$

13. $2x - 4y + z = -2$

x-intercept:
Let $y = 0$ and $z = 0$
$2x - 4(0) + 0 = -2$
$2x = -2$
$x = -1$
$(-1, 0, 0)$

y-intercept:
Let $x = 0$ and $z = 0$.
$2(0) - 4y + 0 = -2$
$-4y = -2$
$y = \frac{1}{2}$
$\left(0, \frac{1}{2}, 0\right)$

z-intercept:
Let $x = 0$ and $y = 0$.
$2(0) - 4(0) + z = -2$
$z = -2$
$(0, 0, -2)$

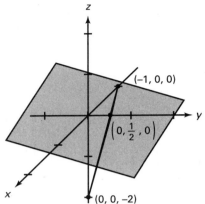

14. $x - 2y - 2z = -4$

x-intercept:
Let $y = 0$ and $z = 0$.
$x - 2(0) - 2(0) = -4$
$x = -4$
$(-4, 0, 0)$

y-intercept:
Let $x = 0$ and $z = 0$.
$0 - 2y - 2(0) = -4$
$-2y = -4$
$y = 2$
$(0, 2, 0)$

z-intercept:
Let $x = 0$ and $y = 0$.
$0 - 2(0) - 2z = -4$
$-2z = -4$
$z = 2$
$(0, 0, 2)$

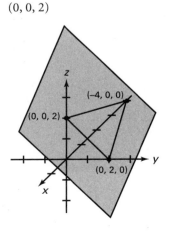

15. $-3x + y = 4$

x-intercept:
Let $y = 0$ and $z = 0$.
$-3x + 0 = 4$
$-3x = 4$
$x = -\frac{4}{3}$
$\left(-\frac{4}{3}, 0, 0\right)$

y-intercept:
Let $x = 0$ and $z = 0$.
$-3(0) + y = 4$
$y = 4$
$(0, 4, 0)$

There is no z-intercept.
The plane is parallel to the z-axis.

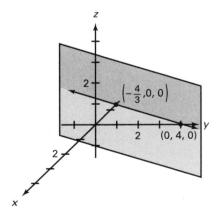

16. $x - 2y = 2$
x-intercept:
Let $y = 0$ and $z = 0$.
$x - 2(0) = 2$
$\qquad x = 2$
$(2, 0, 0)$

y-intercept:
Let $x = 0$ and $z = 0$.
$0 - 2y = 2$
$\quad -2y = 2$
$\qquad y = -1$
$(0, -1, 0)$

There is no z-intercept.
The plane is parallel to the z-axis.

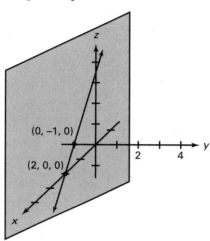

17. $x = -4$
x-intercept:
Let $y = 0$ and $z = 0$.
$x = -4$
$(-4, 0, 0)$
There is no y-intercept.
There is no z-intercept.
The plane is parallel to the y-axis and z-axis.

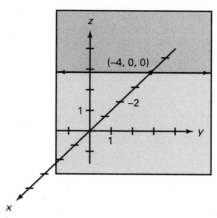

18.

t	x	y
1	4	0
2	5	-1
3	6	-2

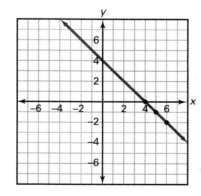

19.

t	x	y
1	1	-1
2	3	-2
3	5	-3

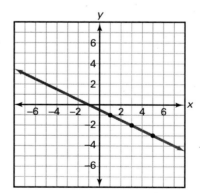

20.

t	x	y
1	3	5
2	6	5
3	9	5

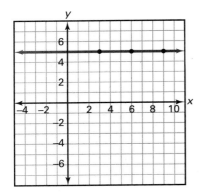

21.

t	x	y
1	2	2
2	2	−2
3	2	−6

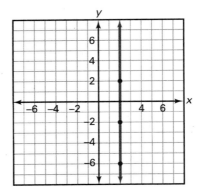

22.

t	x	y	z
0	0	0	0
1	1	2	3
2	2	4	6

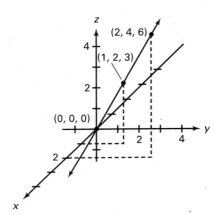

23.

t	x	y	z
0	1	1	0
1	2	3	0
2	3	5	0

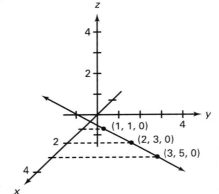

24.

t	x	y	z
0	0	0	1
1	1	1	0
2	2	2	−1

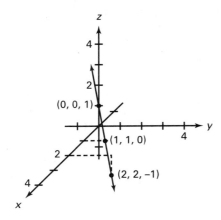

25.

t	x	y	z
0	0	4	−7
1	3	4	−3
2	6	4	1

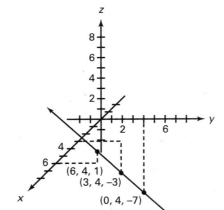

26. $x + 3y - z = 7$
To find trace, let $z = 0$.
$x + 3y - 0 = 7$
$\qquad x + 3y = 7$
trace: $x + 3y = 7$
Sketch the plane by using the intercepts:
x-intercept: $(7, 0, 0)$
y-intercept: $\left(0, \frac{7}{3}, 0\right)$
z-intercept: $(0, 0, -7)$

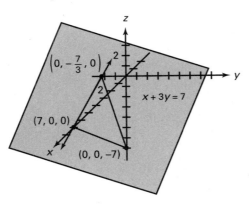

27. $5x - 2y + z = 2$
To find trace, let $z = 0$.
$5x - 2y + 0 = 2$
$\qquad 5x - 2y = 2$
trace: $5x - 2y = 2$
Sketch the plane by using the intercepts.
x-intercept: $\left(\frac{2}{5}, 0, 0\right)$
y-intercept: $(0, -1, 0)$
z-intercept: $(0, 0, 2)$

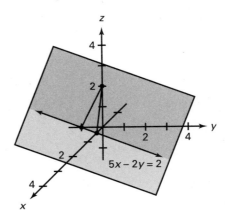

28. $2x + 7y + 3z = 2$
To find trace, let $z = 0$.
$2x + 7y + 3(0) = 2$
$\qquad 2x + 7y = 2$
trace: $2x + 7y = 2$
Sketch the plane by using the intercepts.
x-intercept: $(1, 0, 0)$
y-intercept: $\left(0, \frac{2}{7}, 0\right)$
z-intercept: $\left(0, 0, \frac{2}{3}\right)$

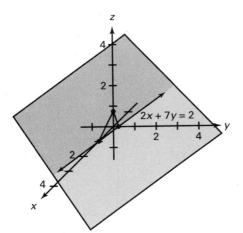

29. $-4x - 2y + 2z = 1$
To find trace, let $z = 0$.
$-4x - 2y + 2(0) = 1$
$\qquad -4x - 2y = 1$
trace: $-4x - 2y = 1$
Sketch the plane by using the intercepts:
x-intercept: $\left(-\frac{1}{4}, 0, 0\right)$
y-intercept: $\left(0, -\frac{1}{2}, 0\right)$
z-intercept: $\left(0, 0, \frac{1}{2}\right)$

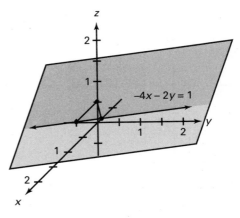

30. Sample answer:

If $x = t$, then

$t = 0 \Rightarrow (0, 0, 0)$ and

$t = 2 \Rightarrow (2, -4, -1)$.

$y = at + b$ for specific values for a and b.

$0 = a(0) + b$

$0 = b$

$4 = a(2) + 0$

$4 = 2a$

$2 = a$

$y = 2t$

$z = ct + d$ for specific values for c and d.

$0 = c(0) + d$

$0 = d$

$-1 = c(2) + 0$

$-1 = 2c$

$-\frac{1}{2} = c$

$z = -\frac{1}{2}t$

The parametric questions are

$x = t$

$y = 2t$

$z = -\frac{1}{2}t.$

31. Sample answer:

If $x = t$, then

$t = -2 \Rightarrow (-2, 1, 5)$ and

$t = 6 \Rightarrow (6, -4, 7)$.

$y = at + b$ for specific values for a and b.

$1 = a(-2) + b$

$-4 = a(6) + b$ so

$1 = -2a + b$

$\underline{4 = -6a - b}$

$5 = -8a$

$-\frac{5}{8} = a$

$1 = -2\left(-\frac{5}{8}\right) + b$

$1 = \frac{10}{8} + b$

$-\frac{1}{4} = b$

$y = -\frac{5}{8}t - \frac{1}{4}$

$z = ct + d$ for specific values for c and d.

$5 = c(-2) + d$

$7 = c(6) + d$ so

$15 = -6c + 3d$

$\underline{7 = 6c + d}$

$22 = \phantom{-6c + {}}4d$

$\frac{11}{2} = d$

$7 = 6c + \frac{11}{2}$

$\frac{3}{2} = 6c$

$\frac{1}{4} = c$

$z = \frac{1}{4}t + \frac{11}{2}$

The parametric questions are:

$x = t$

$y = -\frac{5}{8}t - \frac{1}{4}$

$z = \frac{1}{4}t + \frac{11}{2}.$

32.

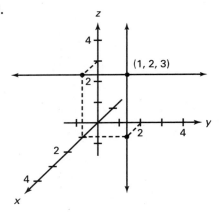

The lines intersect when $t = 3$ and $s = 2$.

33.

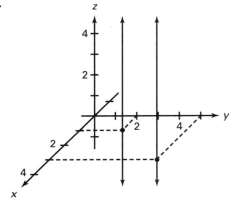

The lines are parallel. They are both vertical lines, perpendicular to the xy-plane at $(1, 2, 0)$ and $(3, 5, 0)$.

34.

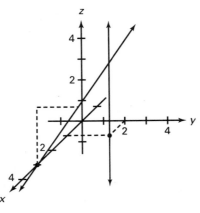

The lines are skew.
Line 1 always has x-coordinate 1 and line 2 always has x-coordinate 3. Because the x-coordinates are always different, the lines do not intersect.

36. $z = 5x + 25y + 200$ in standard form
$5x + 25y - z = -200$
Sketch the plane using the intercepts.
x-intercept: $(-40, 0, 0)$
y-intercept: $(0, -8, 0)$
z-intercept: $(0, 0, 200)$

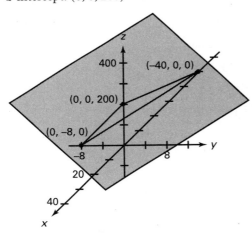

35. By setting the x-coordinates equal to each other, we get
$t - 1 = s - 3$
$s = t + 2$

By setting the y-coordinates equal to each other, we get
$2t + 1 = s$
$2t + 1 = t + 2$
$t = 1$
$s = t + 2 = 3$

By setting the z-coordinates equal to each other, we get
$-t = 2s - 7$
$-1 = 2(3) - 7$
$-1 = -1$
Since a true statement results, the two lines intersect.
The point of intersection is $(0, 3, -1)$.

37. $z = 7x + 4y - 50$ in standard form is
$7x + 4y - z = 50$
Sketch the plane using the intercepts.
x-intercept: $\left(\frac{50}{7}, 0, 0\right)$
y-intercept: $\left(0, \frac{25}{2}, 0\right)$
z-intercept: $(0, 0, -50)$

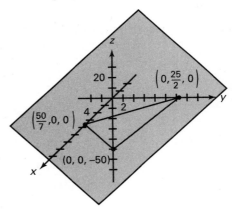

Find the trace by letting $z = 0$.
The trace of the plane is $7x + 4y = 50$.
$z = 0$ for points in the trace thus the profit is $0.

38.

t	x	y	z
0	0	0	0
1	-1	1	$\frac{1}{2}$
2	-2	2	1

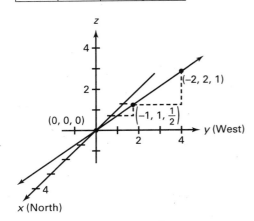

The plane took off in a southwest direction.

PAGE 407, LOOK BACK

39. Parallel, because they have the same slope, 3, and different y-intercepts.

40. Perpendicular, because their slopes, 1 and -1 are negative reciprocals of each other.

41. Neither, because their slopes, 2 and -2, are not equal and their product is not -1.

42. Perpendicular.

$2x + 3y = 6$ or $y = -\frac{2}{3}x + 2$

$3x - 2y = 6$ or $y = \frac{3}{2} - 3$

The product of their slopes is $\left(-\frac{2}{3}\right)\left(\frac{3}{2}\right) = -1$.

43. Given: $\angle 1 \cong \angle 2$; $\overline{EH} \cong \overline{FG}$

Prove: $\overline{EF} \cong \overline{HG}$

Statements	Reasons
$\angle 1 \cong \angle 2$; $\overline{EH} \cong \overline{FG}$	Given
$\overline{EH} \parallel \overline{FG}$	Converse of the Corresponding Angles Postulate
$EFGH$ is a parallelogram.	Theorem 4.6.2
$\overline{EF} \cong \overline{HG}$	Corollary 4.5.2

44.

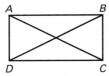

Statements	Reasons
Rectangle $ABCD$ with diagonals $\overline{AC}$ and $\overline{BD}$	Given
$ABCD$ is a parallelogram.	Theorem 4.5.7
$\overline{AD} \cong \overline{BC}$	Theorem 4.5.2
$\overline{DC} \cong \overline{DC}$	Reflexive Property of Congruence
$\overline{AC} \cong \overline{BD}$	Theorem 4.5.9
$\triangle ADC \cong \triangle BCD$	SSS

45.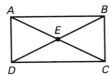

Statements	Reasons
$ABCD$ is a rectangle.	Given
$ABCD$ is a parallelogram.	Theorem 4.5.6
$\overline{AD} \cong \overline{BC}$	Theorem 4.5.2
$\overline{AE} \cong \overline{CE}$ $\overline{BE} \cong \overline{DE}$	Theorem 4.5.4
$\triangle AED \cong \triangle CEB$	SSS

46.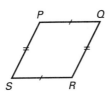

Statements	Reasons
Quadrilateral $PQRS$ $\overline{PQ} \cong \overline{RS}$ $\overline{PS} \cong \overline{QR}$	Given
$PQRS$ is a parallelogram.	Theorem 4.6.1

PAGE 408, LOOK BEYOND

47. $\dfrac{x-2}{3} = t$

$x - 2 = 3t$

$\quad x = 3t + 2$

$\dfrac{y-5}{-1} = t$

$y - 5 = -t$

$\quad y = -t + 5$

$\dfrac{z+1}{4} = t$

$z + 1 = 4t$

$\quad z = 4t - 1$

Parametric form:

$x = 3t + 2$

$y = -t + 5$

$z = 4t - 1$

48. $x - 2 = t$

$\quad x = t + 2$

$\dfrac{y+6}{2} = t$

$y + 6 = 2t$

$\quad y = 2t - 6$

$\dfrac{z-3}{-5} = t$

$z - 3 = -5t$

$\quad z = -5t + 3$

Parametric form:

$x = t + 2$

$y = 2t - 6$

$z = -5t + 3$

49.
$$x = 2t + 1$$
$$x - 1 = 2t$$
$$\frac{x-1}{2} = t$$
$$y = -3t + 6$$
$$y - 6 = -3t$$
$$\frac{y-6}{-3} = t$$
$$z = 4t$$
$$\frac{z}{4} = t$$
Symmetric form:
$$\frac{x-1}{2} = \frac{y-6}{-3} = \frac{z}{4}$$

50.
$$x = t - 4$$
$$x + 4 = t$$
$$y = 2t$$
$$\frac{y}{2} = t$$
$$z = 6t + 2$$
$$z - 2 = 6t$$
$$\frac{z-2}{6} = t$$
Symmetric form:
$$x + 4 = \frac{y}{2} = \frac{z-2}{6}$$

6.6 | **PAGE 413, GUIDED SKILLS PRACTICE**

8. Parallel to each other but not to the picture plane. Parallel to each other but not to the picture plane.

9. The lines containing the vertical segments will not meet because they are parallel to each other and parallel to the picture plane.

10. The lines containing the nonvertical sides of the building meet at the horizon of the drawings.

PAGES 414–417, PRACTICE AND APPLY

11. Students follow the steps to produce a one-point perspective drawing of a cube.

12.

13.

14. It will remain a two-dimensional drawing of the square.

15–16.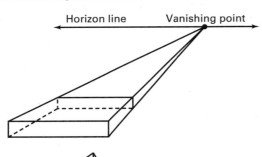

Horizon line   Vanishing point

17. Students follow the steps to produce a two-point perspective drawing of a cube.

18.

19.

20. When the points are moved closer together, the cube becomes narrower. When they are moved farther apart, the view becomes broader. For a realistic drawing the points must be an appropriate distance apart.

21. If both vanishing points are moved to the same side of the original vertical line, the cube can't be drawn properly.

If one vanishing point moves toward the position directly above the vertical line, the other vanishing point must move farther away. When one vanishing point is directly above the vertical line, the second vanishing point is infinitely far away. One of the faces of the cube becomes invisible, and the front and back faces are parallel to the picture plane.

22–23.

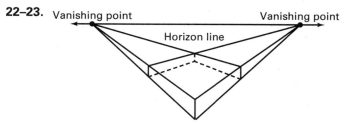

24. Students follow the steps to produce the indicated drawing.

25. Check student drawings.

26. Segments radiating from point P divide $\overleftrightarrow{AB}$ into a number of congruent segments. $\overleftrightarrow{CD}$ is drawn parallel to $\overleftrightarrow{AB}$. Diagonals $\overline{AW}$ and $\overline{BY}$ are drawn through points C and D, respectively. The intersections of the diagonals with the segments from point P determine the vertical placement of the parallel lines.

27. Simplest method: Make a perspective drawing of a square with horizontal and vertical diagonals. Divide the horizontal diagonal equally. Pass lines from the vanishing points through the division points.

28. Check student drawings.

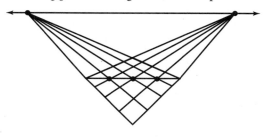

PAGE 417, LOOK BACK

29. Sample answer:

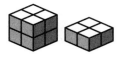

30. The base of the front is 2 units with a hidden edge. The base of the right is 2 units with a hidden edge. From the top we see an 'L' shape. One of the unit squares is 2 units high so from the top we see an edge.

31.

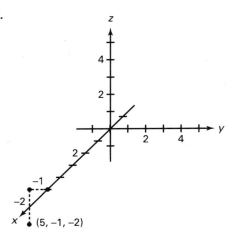

32.

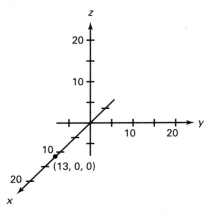

33.

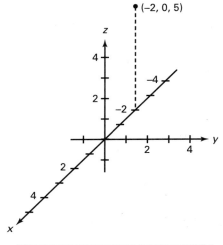

34. $\sqrt{(-5-4)^2 + (2-3)^2 + (-1-2)^2} \approx 9.54$

35. $\sqrt{[15-(-1)]^2 + (5-0)^2 + (-2-1)^2} \approx 17.03$

36. $\left(\dfrac{5-3}{2}, \dfrac{5-3}{2}, \dfrac{5-3}{2}\right) = \left(\dfrac{2}{2}, \dfrac{2}{2}, \dfrac{2}{2}\right) = (1, 1, 1)$

37. $\left(\dfrac{0-1}{2}, \dfrac{0+10}{2}, \dfrac{0+9}{2}\right) = \left(-\dfrac{1}{2}, \dfrac{10}{2}, \dfrac{9}{2}\right) = \left(-\dfrac{1}{2}, 5, \dfrac{9}{2}\right)$

PAGE 418, LOOK BEYOND

38.

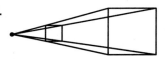

39. Answers may vary. Sample answer:
The projection process of the slide projector translates an image onto a plane, like a perspective drawing. The projection starts with a 2-dimensional object, while a perspective drawing starts with a 3-dimensional drawing.

1. Sample answer:

2.

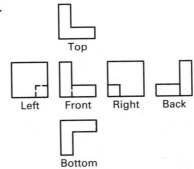

3. $V = 3(3)(1) + 2(1)(1) = 11$ units^3

4. $S = 2(3)(3) + 2[3(1) + 2(1)] + 2[3(1) + 2(1)]$
$= 18 + 10 + 10 = 38$ units^2

5. Sample answer:
OPQRST and *UVWXYZ*
PQWV and *TSYZ*

6. Sample answer:
PQWV and *UVWXYZ*

7. Sample answer: $\overline{PO}$ and $\overline{WX}$

8. Sample answer: $\overline{PV}$ and *UVWXYZ*

9. Sample answer: *DEF*

10. Sample answer: *CBEF*

11. Sample answer: $\overline{CF}$

12. $\sqrt{24^2 + 8^2 + 6^2} = 26$

13.

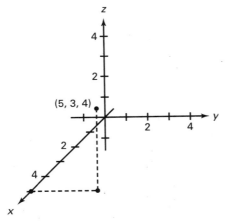

top-front-right

14.

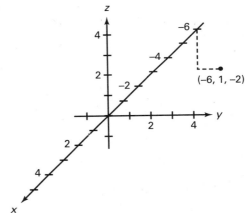

bottom-back-right

15. $\sqrt{(0-2)^2 + (7-4)^2 + (-1-5)^2} = 7$

16. $\sqrt{[-5-(-1)]^2 + (2-6)^2 + [4-(-2)]^2} \approx 8.25$

17. Sketch the plane using intercepts.
x-intercept: $(5, 0, 0)$
y-intercept: $(0, 10, 0)$
z-intercept: $(0, 0, -10)$

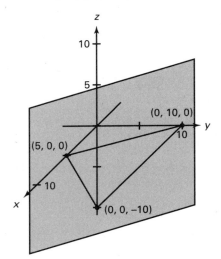

18. Sketch the plane using intercepts.
x-intercept: $(-8, 0, 0)$
y-intercept: $(0, 2, 0)$
z-intercept: $(0, 0, 8)$

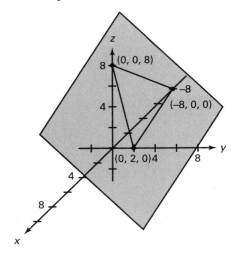

19.

t	x	y	z
0	3	-1	0
1	4	0	2
2	5	1	4
3	6	2	6

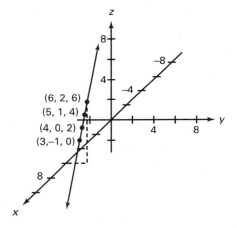

20.

t	x	y	z
0	2	-4	1
1	1	-3	-2
2	0	-2	-5
3	-1	-1	-8

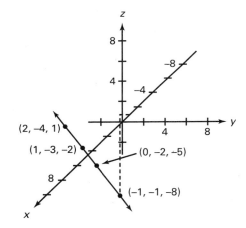

21.

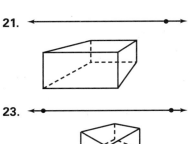

22.

23.

24.

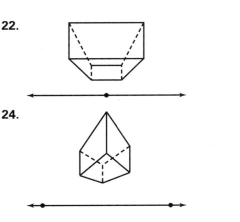

25.

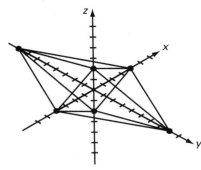

26. $3x + 5y - z = 20$

Sketch the plane using intercepts:

x-intrercept: $\left(\dfrac{20}{3}, 0, 0\right)$

y-intercept: $(0, 4, 0)$

z-intercept: $(0, 0, -20)$

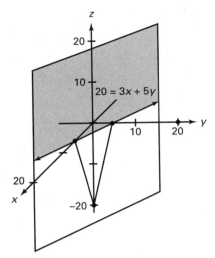

The trace represents number of washes that will neither raise nor lose money (break even).

27.

CHAPTER TEST

1. Sample answer:

3. There are 12 cubes, so the volume is 12 units³.

5. Sample answer: *ONKJ* and *PQLM, OPMN* and *JQLK*

7. Sample answer: $\overline{JK}$ and *KLMN*

9. $\triangle PKJ$

11. Sample answer: $\overline{KL}$

13. $d = \sqrt{12^2 + 9^2 + 5^2} = \sqrt{250} = 5\sqrt{10} \approx 15.8$ feet

15. The point $(-1, 5, -3)$ has a negative x-coordinate (back), a positive y-coordinate (right), and a negative z-coordinate (bottom), so it lies in the bottom-back-right octant.

2.

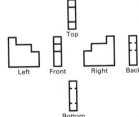

4. There are 40 square faces around the outside, so the surface area is 40 units².

6. Sample answer: $\overline{OP}$ and $\overline{ML}$

8. plane

10. Sample answer: *JMQP*

12. Sample answer: $\angle LKJ$ and $\angle MLK$

14. Because the sign of each coordinate in $(2, 5, 9)$ is positive, the point lies in the first octant.

16. The point $(1, 0, -3)$ has a positive x-coordinate (front), a zero y-coordinate (xz-plane), and a negative z-coordinate (lower), so it lies in the front-lower quadrant of the xz-plane.

17. The point $(0, -4, 0)$ has a zero x-coordinate (yz-plane), and negative y-coordinate (left), and a zero z-coordinate (xy-plane), so it lies on the left side of the intersection of the xy-plane and the yz-plane; that is, the left side of the y-axis.

19. The point $(2, 0, 0)$ has a positive x-coordinate (front), a zero y-coordinate (xz-plane), and a zero z-coordinate (xy-plane), so it lies in the front half of the intersection of the xz-plane and xy-plane; that is, the front half of the x-axis.

21. $d = \sqrt{(3 + 4)^2 + (-2 - 0)^2 + (1 - 1)^2}$
$= \sqrt{53} \approx 7.28$

23. x-intercept: $3x + 3(0) + 2(0) = 18$; $x = 6$; $(6, 0, 0)$
y-intercept: $3(0) + 3y + 2(0) = 18$; $y = 6$, $(0, 6, 0)$
z-intercept: $3(0) + 3(0) + 2z = 18$; $z = 9$; $(0, 0, 9)$
Check students' graphs.

25. For $t = 0$:
$x = 0, y = -0.5(0) = 0, z = 2(0)$; $(0, 0, 0)$
For $t = 2$:
$x = 2, y = -0.5(2) = -1, z = 2(2) = 4$; $(2, -1, 4)$
Check students' graphs.
The line goes between the positive x-axis (North) and negative y-axis (East), so the plane took off in a northeast direction.

27.

29.

18. The point $(-6, 6, 0)$ has a negative x-coordinate (back), a positive y-coordinate (right), and a zero z-coordinate (xy-plane), so it lies in the back-right quadrant of the xy-plane, or Quadrant II.

20. $d = \sqrt{(0 - 1)^2 + (1 - 0)^2 + (1 - 0)^2} = \sqrt{3} \approx 1.73$

22. x-intercept: $x + 2(0) + 4(0) = 12$; $x = 12$; $(12, 0, 0)$
y-intercept: $0 + 2y + 4(0) = 12$; $y = 6$; $(0, 6, 0)$
z-intercept: $0 + 2(0) + 4z = 12$; $z = 3$; $(0, 0, 3)$
Check students' graphs.

24. For $t = 0$:
$x = 0 + 1 = 1, y = 2(0) = 0, z = 0 - 3 = -3$;
$(1, 0, -3)$
For $t = 1$:
$x = 1 + 1 = 2, y = 2(1) = 2, z = 1 - 3 = -2$;
$(2, 2, -2)$
Check students' graphs.

26.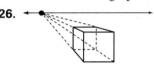

28.

1. $AB = \sqrt{4^2 - 2^2}$

$= \sqrt{16 - 4}$

$= \sqrt{12}$

$= 2\sqrt{3}$

$XZ = \sqrt{2^2 + 2^2}$

$= \sqrt{4 + 4}$

$= \sqrt{8}$

$= 2\sqrt{2}$

$AB > XZ$

The answer is A.

2. $d_A = \sqrt{(9 - 0)^2 + (9 - 0)^2 + (9 - 0)^2}$

$= \sqrt{9^2 + 9^2 + 9^2}$

$= \sqrt{3 \cdot 9^2}$

$= 9\sqrt{3}$

$d_B = \sqrt{(15 - 6)^2 + (0 - 9)^2 + (19 - 10)^2}$

$= \sqrt{9^2 + (-9)^2 + 9^2}$

$= \sqrt{3 \cdot 9^2}$

$= 9\sqrt{3}$

$d_A = d_B$

The answer is C.

3. $\text{Area}_A = \frac{1}{2}(14 + 25)6$

$= 117$

$\text{Area}_B = 6 \cdot 10$

$= 60$

$\text{Area}_A > \text{Area}_B$

The answer is A.

4. $\text{Area}_A = \pi(1.5)^2 \approx 7.1$

$\text{Area}_B = \frac{3^2}{2} = \frac{9}{2} = 4.5$

$\text{Area}_A > \text{Area}_B$

The area is A.

5. $\frac{180°(9 - 2)}{9} = \frac{180°(7)}{9}$

$= \frac{1260°}{9}$

$= 140°$

The answer is c.

6. $A = \pi r^2$

$154 = \pi r^2$

$\frac{154}{\pi} = r^2$

$\sqrt{\frac{154}{\pi}} = r$

$C = 2\pi r$

$= 2\pi\sqrt{\frac{154}{\pi}}$

≈ 44

The answer is b.

7. $\left(\frac{x_1 + x_2}{2}, \frac{y_1 + y_2}{2}\right)$

$\left(\frac{9 + (-12)}{2}, \frac{12 + 9}{2}\right)$

$\left(\frac{-3}{2}, \frac{21}{2}\right)$

$(-1.5, 10.5)$

The answer is a.

8. $\left(\frac{x_1 + x_2}{2}, \frac{y_1 + y_2}{2}, \frac{z_1 + z_2}{2}\right)$

$\left(\frac{2 + 11}{2}, \frac{10 + 10}{2}, \frac{0 + (-6)}{2}\right)$

$\left(\frac{13}{2}, \frac{20}{2}, \frac{-6}{2}\right)$

$(6.5, 10, -3)$

The answer is c.

9. $x° + 80° = 180°$

$x + 80 = 180$

$x = 100$

The answer is c.

10. $s = 2(6) + 2(5) + 2(6)$

$= 12 + 10 + 12$

$= 34$

The solid is 34 square units.

The answer is a.

11. Statement: If lines are skew, then they are not parallel and do not intersect.

Converse: If lines are not parallel and do not intersect, then they are skew.

Yes, the statement and its converse are true.

12. Let $x =$ the length of a side of the square
$x^2 = 441$ square inches
$x = 21$ inches
perimeter of the square $= 4(21) = 84$ inches
Let $r =$ the radius of the circle
$\pi r^2 = 441$ square inches
$r = \sqrt{\dfrac{441}{\pi}}$ inches
circumference of the circle $= 2\pi\left(\sqrt{\dfrac{441}{\pi}}\right) \approx 74.44$ inches

13.

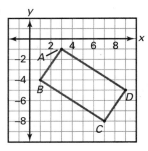

Let A be the vertex at $(3, -1)$, B be the vertex $(1, -4)$, C be the vertex $(7, -8)$ and D be vertex $(9, -5)$.

Slope of $\overline{AB} = \dfrac{-1 - (-4)}{3 - 1} = \dfrac{3}{2}$
Slope of $\overline{BC} = \dfrac{-4 - (-8)}{1 - 7} = \dfrac{4}{-6} = -\dfrac{2}{3}$
Slope of $\overline{DC} = \dfrac{-8 - (-5)}{7 - 9} = \dfrac{-3}{-2} = \dfrac{3}{2}$
Slope of $\overline{AD} = \dfrac{-5 - (-1)}{9 - 3} = \dfrac{-4}{6} = -\dfrac{2}{3}$
The product of the slopes of $\overline{AD}$ and $\overline{AB}$ is $\left(-\dfrac{2}{3}\right)\left(\dfrac{3}{2}\right) = -1$.

Since the slopes of $\overline{AD}$ and $\overline{BC}$ and the slopes of $\overline{AB}$ and $\overline{DC}$ are equal, the opposite sides are parallel and $ABCD$ is a parallelogram. Since the product of the slopes of $\overline{AD}$ and $\overline{AB}$ is -1, $\overline{AD} \perp \overline{AB}$. Thus, $ABCD$ is a rectangle.

14. $p = 8(6) = 48$
$A = \frac{1}{2}ap = \frac{1}{2}(4\sqrt{3})(48) \approx 166.3 \text{ m}^2$

15. Sample answer:

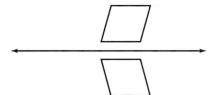

16. Area of $\triangle PEG = \frac{1}{2}(12)(5) = 30 \text{ feet}^2$

17. Area $= 6(6) + (7)10 + 4(4) = 122 \text{ units}^2$

18. A 7-gon has interior angles which sum to $180°(7 - 2) = 900°$
If the exterior angle is $35°$, the interior angle is $180° - 35° = 145°$.
If the exterior angle is $50°$, the interior angle is $180° - 50° = 130°$.
$x° = 900° - 150° - 130° - 130° - 130° - 145° - 110°$
$x° = 105°$
$x = 105$

19. $\dfrac{2.5 - 2}{8 - 2} = \dfrac{0.5}{6} \approx 0.08$ or 8%

CHAPTER 7

Surface Area and Volume

7.1 **PAGE 433, GUIDED SKILLS PRACTICE**

6. $S = 2(5)(1) + 2(5)(1) + 2(1)(1) = 22$
$V = 5(1)(1) = 5$
$\frac{S}{V} = \frac{22}{5} = 4.4$

7. $S = 2(10)(1) + 2(10)(1) + 2(1)(1) = 42$
$V = 10(1)(1) = 10$
$\frac{S}{V} = \frac{42}{10} = 4.2$

8. $S = 2(5)(5) + 2(5)(5) + 2(5)(5) = 150$
$V = 5(5)(5) = 5^3 = 125$
$\frac{S}{V} = \frac{150}{125} = 1.2$

9. $S = 2(10)(10) + 2(10)(10) + 2(10)(10) = 600$
$V = 10(10)(10) = 10^3 = 1000$
$\frac{S}{V} = \frac{600}{1000} = 0.6$

10. $V = (48 - 2x)(42 - 2x)x$

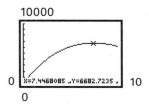

Maximum volume ≈ 6682.72 in^3
squares should be about 7.45 in. $\times$ 7.45 in.

11. Surface area of box A:
$2(4)(2) + 2(4)(6) + 2(2)(6) = 16 + 48 + 24 = 88$
Surface area of box B:
$2(3)(2) + 2(3)(8) + 2(2)(8) = 12 + 48 + 32 = 92$
Box A has the smaller surface area.

PAGES 434–435, PRACTICE AND APPLY

Length	Width	Height	Surface Area	Volume	$\dfrac{\text{Surface area}}{\text{Volume}}$
2	2	1	**12.** 16	**13.** 4	**14.** 4
4	4	1	**15.** 48	**16.** 16	**17.** 3
7	3	5	**18.** 142	**19.** 105	**20.** 1.35
4	**21.** 7	3	**22.** 122	84	**23.** 1.45
2	5	**24.** 6	104	**25.** 60	**26.** 1.73

27. $V = s^3 = 64 \Rightarrow s = 4$
$S = 6s^2 = 6(4)^2 = 96$ in^2
$\frac{S}{V} = \frac{96}{64} = 1.5$

28. $V = s^3 = 1000 \Rightarrow s = 10$ cm
$S = 6s^2 = 6(10)^2 = 600$ cm^2
$\frac{S}{V} = \frac{600}{1000} = 0.6$

29. $S = 2(n)(n) + 2(n)(1) + 2(n)(1) = 2n^2 + 4n$
$V = n(n)(1) = n^2$
$\dfrac{S}{V} = \dfrac{2n^2 + 4n}{n^2} = \dfrac{2n + 4}{n} = 2 + \dfrac{4}{n}$

30. $S = 2(2s)(s) + 2(2s)(s) + 2(s)(s)$
$\quad = 4s^2 + 4s^2 + 2s^2 = 10s^2$
$V = (2s)(s)(s) = 2s^3$
$\dfrac{S}{V} = \dfrac{10s^2}{2s^3} = \dfrac{5}{s}$

31. Sample answer: Maximize the volume. Assuming you use all available lumber, the surface area is fixed. You want to create the maximum amount of storage space.

32. Sample answer; Minimize the surface area. The volume is constant, so you may want to minimize the surface area to save on packaging costs.

33. Sample answer: Minimize the surface area. Assuming the item(s) to be mailed have a fixed volume, you would want to minimize the amount of packaging to reduce the weight of the package.

34. Sample answer: Minimize the surface area. The volume is constant, so you may want to minimize the surface area to save on construction materials.

35. For a prism with dimensions $n \times n \times 1$, the ratio is $2 + \dfrac{4}{n}$, which approaches 2 as n increases. For a prism with dimensions $n \times n \times n$, the ratio is $\dfrac{6n^2}{n^3} = \dfrac{6}{n}$, which approaches 0 as n increases.

36. The prism with dimensions $n \times n \times n$, or a cube, has the smallest surface area for a given volume.

37. $S = 6s^2$
$V = s^3$
$\dfrac{S}{V} = \dfrac{6s^2}{s^3} = \dfrac{6}{s} = 1$
$\qquad\qquad s = 6$

38. Let x be the length of a base edge.
$S = 2(x)(x) + 2(x)(2x) + 2(2x)(x) = 10x^2$
$V = x(x)(2x) = 2x^3$
$\dfrac{S}{V} = \dfrac{10x^2}{2x^3} = \dfrac{5}{x} = 1$
$\qquad\qquad x = 5$

39. Sample answer: The flat shape of the worm maximizes the surface area of its body exposed to the air which maximizes the amount of oxygen it can take in through its skin.

40. Sample answer: The thousands of air sacs make the surfacearea of the lungs large, thereby maximizing the ability of the lungs to absorb oxygen. This surface area is much greater than that of the skin, so the human lungs can absorb enough oxygen to support a much larger body than the skin could absorb.

41. Sample answer: Tall trees with large, broad leaves have a large surface area exposed. The large surface area is important if there is competition for light, but it also increases water loss. In desert conditions there is plenty of light but little water, so conservation of water is important; broadleaved trees would soon die.

42. Sample answer: In a large roast, there is more meat to cook, so the surface-area-to-volume ratio is less than for a smaller roast. There is relatively less surface through which heat can enter to cook a large amount of meat.

43. Sample answer: When broken into smaller pieces, a block of ice will have a larger surface-area-to-volume ratio which maximizes heat exchange.

44. Sample answer: Completely chewing food "flattens" it out, thus increasing the surface area on which digestive enzymes and juices can act.

45. Sample answer: The short, squat form maximizes the volume of the cactus and therefore the amount of water it can hold. Also the surface area is kept to a minimum which minimizes water loss through transpiration and evaporation.

PAGES 435–436, LOOK BACK

46. Sample counterexamples: Any two isosceles right triangles, one with legs of length a and one with legs of length b, with $a \neq b$; or any two equilateral triangles, one with a side of length a and one with a side of length b with $a \neq b$. These will satisfy AAA but will not be congruent.

47. $A = 29(15) = 435$ units^2

48. $A = \frac{1}{2}(45 \times 28) = 630$ units^2

49. $A = 12.2 \times 5.5 = 67.1$ units^2

50. $A = \frac{1}{2}(46 + 33)(16) = 632$ units^2

51. No. Sample answer:

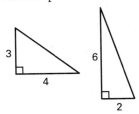

These two triangles have the same area, but are not congruent.

52. $2(\ell + w) = 20$
$\ell + w = 10$

$\ell = 9$	$w = 1$	$A = 9$
$\ell = 8$	$w = 2$	$A = 16$
$\ell = 7$	$w = 3$	$A = 21$
$\ell = 6$	$w = 4$	$A = 24$
$\ell = 5$	$w = 5$	$A = 25$

53. $C = 2\pi r$
$10\pi = 2\pi r$
$r = 5$
$A = \pi r^2 = \pi(5)^2 = 25\pi$ units^2

54.

$h = \sqrt{15^2 - 8^2} = \sqrt{161} \approx 12.69$ ft
$A = \frac{1}{2}(16)(12.69) \approx 101.51$ ft^2

55. $6\sqrt{2} \approx 8.49$ cm

PAGE 436, LOOK BEYOND

56. Sample answer: 40 in.^{2}

57. Sample answer: $\frac{120 \text{ lb}}{40 \text{ in}^2} = 3$ lb/in^2

58. The area of the feet would increase by a factor of 4.
Area $= 4 \cdot 40$ in$^2 = 160$ in^2. The weight would increase by a factor of 8.
Weight $= 8 \cdot 120$ lb $= 960$ lb.

59. Sample answer: $\frac{W}{A} = \frac{960 \text{ lb}}{160 \text{ in}^2} = 6$ lb/in^2
As the size increases, the pressure on the feet increases. Because there is a limit to the pressure that the feet can withstand, the size of the human body cannot exceed certain limits.

7.2   PAGE 441, GUIDED SKILLS PRACTICE

5. The area of the base is
$B = \frac{1}{2}(6)(16) = 48$.

The perimeter of the base is
$p = 10 + 16 + 10 = 36$.
The lateral area is
$L = hp = 22(36) = 792$.
Thus the surface area is
$S = L + 2B = 792 + 2(48) = 888$ units^2.

6. The area of the base is
$B = 4^2 = 16$.
The perimeter of the base is
$p = 4 + 4 + 4 + 4 = 16$.
The lateral area is
$L = hp = 9(16) = 144$.
Thus the surface area is
$S = L + 2B = 144 + 2(16) = 176$ units^2.

7. The volume of the right rectangular prism is
$V = Bh = lwh = (10)(7)(12) = 840$ units^3.

8. The base of the right regular octagonal prism has a perimeter of $(12)(8) = 96$ and an apothem of $6 + 6\sqrt{2}$. So the area is
$$B = \tfrac{1}{2}ap = \tfrac{1}{2}(96)(6 + 6\sqrt{2})$$
$$= 288 + 288\sqrt{2} \approx 695.29$$
The volume is $V = Bh = (288 + 288\sqrt{2})(40)$
$\approx 27{,}811.74$ units^3

PAGES 442–444, PRACTICE AND APPLY

9. Sample answer:

10. Sample answer:

11. Sample answer:

12. Sample answer:

13. $V = (7)(5) = 35$ cm^3

14. $V = (9)(6) = 54$ m^3

15. $V = (17)(23) = 391$ in^3

16. $V = (32)(17) = 544$ ft^3

17. $B = (5)(7) = 35$ units^2
$p = 2(5 + 7) = 24$ units
$L = hp = (2)(24) = 48$ units^2
$S = L + 2B = 48 + 2(35) = 118$ units^2
$V = (5)(7)(2) = 70$ units^3

18. $B = (16)(9) = 144$ units^2
$p = 2(16 + 9) = 50$ units
$L = hp = (10)(50) = 500$ units^2
$S = L + 2B = 500 + 2(144) = 788$ units^2
$V = (16)(9)(10) = 1440$ units^3

19. $B = \left(\tfrac{1}{2}\right)\left(\tfrac{2}{3}\right) = \tfrac{1}{3}$ units^2
$p = 2\left(\tfrac{1}{2} + \tfrac{2}{3}\right) = \tfrac{7}{3}$ units
$L = hp = (1)\left(\tfrac{7}{3}\right) = \tfrac{7}{3}$ units^2
$S = L + 2B = \tfrac{7}{3} + 2\left(\tfrac{1}{3}\right) = 3$ units^3
$V = \left(\tfrac{1}{2}\right)\left(\tfrac{2}{3}\right)(1) = \tfrac{1}{3}$ units^3

20. $B = (1.3)(4) = 5.2$ units^2
$p = 2(1.3 + 4) = 10.6$ units
$L = hp = (0.5)(10.6) = 5.3$ units^2
$S = L + 2B = 5.3 + 2(5.2) = 15.7$ units^2
$V = (1.3)(4)(0.5) = 2.6$ units^3

21. $S = L + 2B = hp + 2B$
$286 = 32h + 2(7)(9)$
$286 = 32h + 126$
$160 = 32h$
$h = 5$ in.

22. The apothem measures $9\sqrt{3}$ cm.
$$S = (25)(108) + 2\left(\tfrac{1}{2}\right)(9\sqrt{3})(108)$$
$$= 2700 + 972\sqrt{3}$$
$$\approx 4383.55 \text{ cm}^2$$

23. A side of the hexagon measures 8 cm and the perimeter is $(6)(8) = 48$ cm.
$$S = (20)(48) + 2\left(\tfrac{1}{2}\right)(4\sqrt{3})(48) = 960 + 192\sqrt{3}$$
$$\approx 1292.55 \text{ cm}^2$$

24. $343 = s^3$
$s = 7$
$S = 6(7^2) = 294$ yd^2

25. $B = (3)(2) = 6$ units^2
$V = (2)(3)(2.8) = 16.8$ units^3
$S = 2(2 \cdot 3) + 2(2 \cdot 2.8) + 2(3 \cdot 3) = 41.2$ units^2

26. $B = \tfrac{1}{2}(2.5)(1.9) = 2.375$ units^2
$V = Bh = (2.375)(3.6) = 8.55$ units^3
$S = 2\left(\tfrac{1}{2} \cdot 2.5 \cdot 1.9\right) + 2.1 \cdot 3.6 + 2.5 \cdot 4 + 2.5 \cdot 3.6$
$= 31.31$ units^2

27. $V = \left(\dfrac{7(6 + 8)}{2}\right)18 = 882$ m^3

28. The legs of the base triangle measure $5\sqrt{2}$.
$V = \tfrac{1}{2}(5\sqrt{2})(5\sqrt{2})(23) = 575$ cm^3

29. $B = \frac{1}{2}(6)(\sqrt{4^2 - 3^2}) = 3(\sqrt{7})$ in^2

$S = (13)(4 + 4 + 6) + (2)(3\sqrt{7})$

$\quad = 182 + 6\sqrt{7} \approx 197.87$ in^2

$V = 13(3\sqrt{7}) = 39\sqrt{7} \approx 103.18$ in^3

30. Since the volume of a prism $= Bh$, doubling the height of a prism while the base is unchanged gives $B(2h) = 2Bh$, which will double the volume.

31. Since the surface area of a cube $= 6s^2$ and the volume of a cube $= s^3$, doubling the length of an edge gives $S = 6(2s)^2 = 24s^2$ and $V = (2s)^3 = 8s^3$, which multiplies the surface area by $2^2 = 4$, and multiplies the volume by $2^3 = 8$.

32. Tripling the edge multiplies the surface area by $3^2 = 9$ and multiplies the volume by $3^3 = 27$.

33. Let x be the side length of the base; x is also equal to the height. Since the volume of a right hexagonal prism $V = Bh = \frac{1}{2}(6x)\left(\frac{x}{2}\sqrt{3}\right) = \frac{3}{2}x^3\sqrt{3}$ and the surface area $S = L + 2B = x(6x) + 2\left(\frac{3}{2}x^2\sqrt{3}\right)$ $= 6x^2 + 3x^2\sqrt{3} = x^2(6 + 3\sqrt{3})$, doubling all the edges multiplies the volume by $2^3 = 8$, and multiplies the surface area by $2^2 = 4$.

34. $V = s^3$

$2V = 2s^3$

$2V = (\sqrt[3]{2}s)^3$

When the volume of a cube is doubled, the length of its side is multiplied $\sqrt[3]{2}$. Since $S = 6s^2$, the surface area is multiplied by $(\sqrt[3]{2})^2 = 2^{2/3}$.

35. Feline Feast: surface area $= 198$ in^2

volume $= 162$ in^3

Kitty Krunchies:

surface area $= 190$ in^2

volume $= 126$ in^3

36. The prism can be divided into 2 rectangular prisms and one trapezoidal prism.

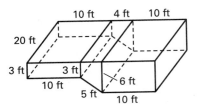

$V = (20)(10)(3) + \frac{1}{2}(6 + 3)(4)(20) + (20)(10)(6)$

$\quad = 600 + 360 + 1200 = 2160$ ft$^3 \approx 16{,}119.4$ gal.

37. The surface area of the swimming pool is

$[(10)(20) + (10)(3) + (20)(3) + (10)(3)] + \left[2 \times \frac{1}{2}(6 + 3)(4) + (20)(5)\right]$

$+ [(10)(20) + (10)(6) + (20)(6) + (10)(6)] = 896$ ft.^{2}

It will take $\frac{896}{400} = 2.24$ gallons of paint to paint the inside of the pool.

38. $S = (4.5)(7.0) + 2\left(\frac{1}{2}\right)(4.5)(3.5) + 2\sqrt{3.5^2 + 2.25^2}(7.0) \approx 105.5$ ft^2

39. There are 10 stories in the building.

$L = (10)(12)(4)(48) = 23{,}040$

$23{,}040$ ft^2 of glass was used.

PAGE 444, LOOK BACK

40. $\sqrt{20} = \sqrt{(5)(4)} = 2\sqrt{5}$

41. $(\sqrt{18})(\sqrt{2}) = \sqrt{36} = 6$

42. $(5\sqrt{7})^2 = 5^2(\sqrt{7})^2 = (25)(7) = 175$

43. $(2\sqrt{2})(3\sqrt{8}) = 6\sqrt{16} = (6)(4) = 24$

44. $\frac{2}{\sqrt{2}} = \frac{2}{\sqrt{2}} \cdot \frac{\sqrt{2}}{\sqrt{2}} = \frac{2\sqrt{2}}{2} = \sqrt{2}$

45. $\sqrt{27} + \sqrt{12} = \sqrt{(9)(3)} + \sqrt{(4)(3)}$

$\quad\quad\quad\quad\quad = 3\sqrt{3} + 2\sqrt{3} = 5\sqrt{3}$

46. Let s be the length of the shorter leg.

$s = \frac{1}{2}(10) = 5$ in.

47. Let ℓ be the length of the longer leg.

$\ell = \frac{1}{2}(10)\sqrt{3} \approx 8.66$ in.

48.

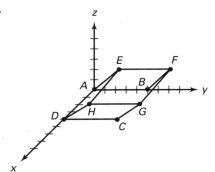

It is a rectangular prism.

49. $V = Bh = 5 \times 4 \times 2 = 40$ units^3

50. Area of $ABCD$ = Area of $EFGH = 4 \cdot 5 = 20$

Area of $ADHE$ = Area of $BCGF = 4 \cdot$ (distance from E to the x-axis)
$$= 4 \cdot \sqrt{2^2 + 2^2}$$
$$= 4 \cdot \sqrt{8}$$
$$= 4 \cdot 2\sqrt{2}$$
$$= 8\sqrt{2}$$

Area of $ABFE$ = Area of $DCGH = 5 \cdot$ (distance from E to the y-axis)
$$= 5 \cdot \sqrt{2^2 + 2^2}$$
$$= 5 \cdot \sqrt{8}$$
$$= 5 \cdot 2\sqrt{2}$$
$$= 10\sqrt{2}$$

Then $S = 20 + 20 + 8\sqrt{2} + 8\sqrt{2} + 10\sqrt{2} + 10\sqrt{2}$
$$= 40 + 36\sqrt{2} \approx 90.9 \text{ units}^2$$

7.3 | **PAGE 449, GUIDED SKILLS PRACTICE**

---

6. $S = \frac{1}{2}\ell p + B = \frac{1}{2}(10)(21) + \frac{1}{2}(7)\left(\frac{7}{2}\sqrt{3}\right)$
$$= 105 + \frac{49}{4}\sqrt{3} \approx 126.22 \text{ units}^2$$

7. $S = \frac{1}{2}\ell p + B = \frac{1}{2}(8)(4)(6) + (6)^2 = 132$ units^2

8. $L = \frac{1}{2}\ell p = \frac{1}{2}(7)(10)(4) = 140$ ft^2

9. The volume of the pyramid is $V = \frac{1}{3}Bh = \frac{1}{3}(708^2)(471)$
$$= 78{,}698{,}448 \text{ ft}^3$$
The weight in pounds is 78,698,448 cubic feet $\times$ 167 pounds per cubic foot
$\approx 13{,}142{,}640{,}826$ pounds

PAGES 449–451, PRACTICE AND APPLY

10.

11.

12.

13.

14. $S = \frac{1}{2}\ell p + B = \left(\frac{1}{2}\right)(9)(3)(8) + \frac{1}{2}(8)(4\sqrt{3}) \approx 135.71$ units^2

15. $S = \frac{1}{2}\ell p + B = \left(\frac{1}{2}\right)(7)(4)(6) + 6^2 = 120$ units^2

16. $S = \frac{1}{2}\ell p + B = \frac{1}{2}\ell p + \frac{1}{2}ap$

$a = 5\sqrt{3} \qquad p = 6(10) = 60$

$S = \frac{1}{2}(12)(60) + \frac{1}{2}(5\sqrt{3})(60) \approx 619.81$ units^2

17. $V = \frac{1}{3}Bh = \frac{1}{3}(4)(3)(7.6) = 30.4$ m^3

18. $V = \frac{1}{3}Bh = \frac{1}{3}(7)(9)(8) = 168$ m^3

19. $V = \frac{1}{3}Bh = \frac{1}{3}(12)(13)(9) = 468$ in^3

20.

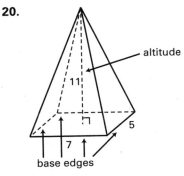

$V = \frac{1}{3}Bh = \frac{1}{3} \times 5 \times 7 \times 11 = \frac{385}{3}$ units^3

21.

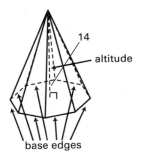

$V = \frac{1}{3}Bh = \frac{1}{3}(16)(14)$
$= \frac{224}{3}$ units^3

22.

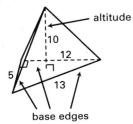

$V = \frac{1}{3}Bh = \frac{1}{3} \times \frac{(5 \times 12)}{2} \times 10 = 100$ units^3

23.

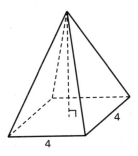

height $= \sqrt{4^2 + 4^2} = \sqrt{32} = 4\sqrt{2}$ units

$V = \frac{1}{3}Bh = \frac{1}{3}(4 \times 4)(4\sqrt{2}) = \frac{64\sqrt{2}}{3}$ units^3

24. $A = (3 + 4)(2 + 5) = (7)(7) = 49$ units^2

25. The slant height ℓ is the hypotenuse of a right triangle with legs 3 and 4.
$\ell = \sqrt{3^2 + 4^2} = \sqrt{25} = 5$ units.
Area of $\triangle ABC = \frac{1}{2}(2 + 5) \cdot 5 = 17.5$ units^2

26. The slant leight ℓ is the hypotenuse of a right triangle with legs 3 and 5.
$\ell = \sqrt{3^2 + 5^2} = \sqrt{34} \approx 5.83$ units.
Area of $\triangle ACD = \frac{1}{2}(3 + 4)\sqrt{34} \approx 20.41$ units^2

27. The slant height ℓ is the hypotenuse of an isosceles right triangle with legs of length 3.
$\ell = \sqrt{3^2 + 3^2} = \sqrt{18} \approx 4.24$ units.
Area of $\triangle ADE = \frac{1}{2}(2 + 5)\sqrt{18} \approx 14.85$ units^2

28. The slant height ℓ is the hypotenuse of a right triangle with legs of length 2 and 3.
$\ell = \sqrt{2^2 + 3^2} = \sqrt{13} \approx 3.61$ units.
Area of $\triangle AEB = \frac{1}{2}(3 + 4)\sqrt{13} \approx 12.62$ units^2

29. $S \approx \frac{1}{2}(5)(7) + \frac{1}{2}(5.83)(7) + \frac{1}{2}(4.24)(7) + \frac{1}{2}(3.61)(7) + 49$

$= 114.38 \text{ units}^2$

30. $V = \frac{1}{3}Bh$

$= \frac{1}{3}(49)(3)$

$= 49 \text{ units}^3$

31. $V = \frac{1}{3}Bh = \frac{1}{3}(24h) = 104 \text{ cubic units}$

$8h = 104$

$h = \frac{104}{8} = 13 \text{ units}$

32. $V = \frac{1}{3}Bh = \frac{1}{3}(10)^2 h = 500 \text{ cubic units}$

$\frac{100}{3}h = 500$

$h = 500\left(\frac{3}{100}\right) = 15 \text{ units}$

33. The base of the pyramid is an equilateral triangle which has a perimeter of 12 units, so its side length is 4 units and its area is $\frac{1}{2} \cdot 4 \cdot 2\sqrt{3} = 4\sqrt{3} \text{ units}^2$. The volume of the pyramid is 8 units^3, so

$8 = \frac{1}{3}Bh = \frac{1}{3} \cdot 4\sqrt{3} \cdot h$. Thus, $h = \frac{3 \cdot 8}{4\sqrt{3}} = 2\sqrt{3} \approx 3.46 \text{ units}$

34. The apothem $a = \sqrt{3}$, so $h = \sqrt{2^2 - (\sqrt{3})^2} = \sqrt{1} = 1 \text{ unit}$

Number of sides of base, n	Number of vertices, V	Number of edges, E	Number of faces, F
3	4	6	4
4	**35.** 5	**36.** 8	**37.** 5
5	**38.** 6	**39.** 10	**40.** 6
n	**41.** $n + 1$	**42.** $2n$	**43.** $n + 1$

44. $V - E + F = (n + 1) - 2n + (n + 1)$

$= n - 2n + n + 1 + 1$

$= 2$

45. $V = \frac{1}{3}Bh = \frac{1}{3}(42)(30)(\sqrt{29^2 - 21^2}) = 420\sqrt{400} = 8400 \text{ units}^3$

Since the base is rectangular, the lateral faces have different areas.

The lateral edge of the pyramid has a measure of $\sqrt{15^2 + 29^2} \approx 32.65 \text{ units}$.

The altitude of the front lateral face measures $\sqrt{(32.65)^2 - 21^2} = 25 \text{ units}$.

So, the surface area of the pyramid is given by:

$S = 2\left(\frac{1}{2}\right)(42)(25) + 2\left(\frac{1}{2}\right)(30)(29) + (30)(42) = 3180 \text{ units}^2$

46. $V = \frac{1}{3}Bh = \frac{1}{3}(225)(15) = 1125 \text{ m}^3$

47. $S = \frac{1}{2}\ell p$

$s = \sqrt{225} = 15$

$\ell = \sqrt{15^2 + (7.5)^2} \approx 16.77$

$S = \frac{1}{2}(16.77)(4 \times 15) \approx 503.1 \text{ m}^2$

48. The complement of 51° is 90° − 51° or 39°.
The supplement of 51° is 180° − 51° or 129°.

49.

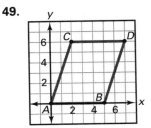

The slope of $\overline{CD}$ is $\frac{6-6}{7-2} = 0$.

The slope of $\overline{AB}$ is $\frac{0-0}{5-0} = 0$.

The slope of $\overline{AC}$ is $\frac{6-0}{2-0} = 3$.

The slope of $\overline{BD}$ is $\frac{6-0}{7-5} = 3$.

By the Parallel Lines Theorem, $\overline{CD}$ is parallel to $\overline{AB}$ and $\overline{AC}$ is parallel to $\overline{BD}$.
Therefore $ABCD$ is a parallelogram.

50. $P = 2(5 + \sqrt{6^2 + 2^2})$
$= 2(5 + \sqrt{40})$
$= 10 + 4\sqrt{10}$ units

51. $A = bh = (5)(6)$
$= 30$ units^2

52. Sample answer: $A(0, 0)$, $B(5, 0)$, $C(0, 6)$, $D(5, 6)$; Using these vertices, the area is still 30 but the perimeter is $2(5) + 2(6) = 22$ units.

53. $A = (10)^2 - \pi\left(\frac{10}{2}\right)^2 = 21.46$ units^2

54. $A = (4)^2 + \frac{1}{2}\pi\left(\frac{4}{2}\right)^2$
$= 16 + 2\pi \approx 22.28$ units^2

55. $A = (6)(4) - [\pi(2)^2 + \pi(1)^2]$
$A = 24 - 5\pi \approx 8.29$ units^2

56. The intersecting plane divides the lateral edges in half.

57. The lengths of the sides of the red triangles are half the lengths of the sides of the bases.

58. By Exercise 57, the areas of the red triangles are $\frac{1}{4}$ the areas of the bases. The red triangles have the same area.

59. Yes and no. The areas of the red triangles will be the same no matter where the plane intersects its the pyramids as long as it intersects the pyramids in a way that is parallel to the plane of the bases of the pyramid. However the comparison between the bases and the red triangles depends on exactly where the plane intersects the pyramids.

60. This is another application of Cavalieri's Principle.

7.4 | **PAGE 456, GUIDED SKILLS PRACTICE**

6. $r = 10.605$ mm
$S = 2\pi rh + 2\pi r^2$
$S = 2\pi(10.605)(1.95) + 2\pi(10.605)^2 \approx 836.58$ mm^2

7. $r = 8.955$ mm
$S = 2\pi rh + 2\pi r^2$
$S = 2\pi(8.955)(1.35) + 2\pi(8.955)^2 \approx 579.82$ mm^2

8. $r = 12.13$ mm
$S = 2\pi rh + 2\pi r^2$
$S = 2\pi(12.13)(1.75) + 2\pi(12.13)^2 \approx 1057.86$ mm^2

9. $V = \pi r^2 h = \pi(10.605)^2(1.95) \approx 688.98$ mm^3

10. $V = \pi r^2 h = \pi(8.955)^2(1.35) \approx 340.11$ mm^3

11. $V = \pi r^2 h = \pi(12.13)^2(1.75) \approx 808.93 \text{ mm}^3$

12.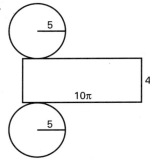

13.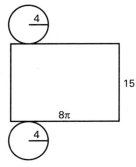

14. $S = 2\pi(5)(4) + 2\pi(5)^2 = 90\pi \approx 282.7 \text{ units}^2$

15. $S = 2\pi(4)(15 + 4) = 152\pi \approx 477.5 \text{ units}^2$

16. $S = 2\pi\left(\frac{1}{2}\right)(1) + 2\pi\left(\frac{1}{2}\right)^2 = \frac{3}{2}\pi \approx 4.7 \text{ units}^2$

17. $S = 2\pi rh + 2\pi r^2 = 72\pi$
$2r(h + r) = 72$
$h = \frac{72}{2r} - r = \frac{72}{6} - 3 = 9 \text{ units}$

18. $S = 2\pi rh + 2\pi r^2 = 550$
$h = \frac{550}{2\pi r} - r = \frac{550}{2\pi(7)} - 7 \approx 5.5 \text{ units}$

19. $S = 2\pi rh + 2\pi r^2 = 70\pi$
$2r(h + r) = 70$
$2r^2 + 2hr - 70 = 0$
$2r^2 + 4r - 70 = 0$
$r^2 + 2r - 35 = 0$
$(r - 5)(r + 7) = 0$
$r - 5 = 0 \quad \text{or} \quad r + 7 = 0$
$\quad r = 5 \quad \text{or} \quad\quad r = -7$
r must be positive, so $r = 5 \text{ units}$

20. $V = \pi r^2 h = \pi(5)^2(4) = 100\pi \text{ units}^3$

21. $V = \pi r^2 h = \pi(4)^2(15) = 240\pi \text{ units}^3$

22. $V = \pi r^2 h = 12\pi$
$h = \frac{12}{r^2} = \frac{12}{2^2} = 3 \text{ units}$

23. $V = \pi r^2 h = 1536$
$h = \frac{1536}{\pi r^2} = \frac{1536}{64\pi} = \frac{24}{\pi} \text{ units}$

24. $V = \pi r^2 h = 54\pi$
$r = \sqrt{\frac{54}{h}} = \sqrt{\frac{54}{6}} = 3 \text{ units}$

25. $V = \pi r^2 h = 80\pi$
$r = \sqrt{\frac{80}{h}} = \sqrt{\frac{(16)(5)}{9}} = \frac{4\sqrt{5}}{3} \text{ units}$

26. The diameter is equal to the height means $h = 2r$.
$200 = 2\pi r(2r + r) = 2\pi r(3r) = 6\pi r^2$
$200 = 6\pi r^2$
$r = \sqrt{\frac{200}{6\pi}} \approx 3.26 \text{ cm}$

27. $360\pi = \pi r^2(10)$
$r^2 = 36$
$r = 6$
$C = 2\pi r = 12\pi \approx 37.7 \text{ mm}$

28. $V = \frac{1}{2}\pi(8)^2(10) \approx 1005.31 \text{ cubic feet}$
$S = 2\left(\frac{1}{2}\right)\pi r^2 + \frac{1}{2}(2\pi rh)$
$\quad = \pi(8)^2 + \pi(8)(10)$
$\quad = 144\pi \approx 452.39 \text{ ft}^2$

29. $V = \pi r^2 h$
Doubling the height of a cylinder doubles the volume of the cylinder.

30. Doubling the radius of a cylinder multiplies the volume by $2^2 = 4$.

31. Doubling the height and radius of a cylinder multiples the volume by $(2)(4) = 8$.

32. $S = 2\pi r^2 + 2\pi rh$
$S = 2\pi(2r)^2 + 2\pi(2r)(2h)$
$\quad = 2\pi(4)r^2 + 2\pi(4)rh$
$\quad = 4(2\pi r^2 + 2\pi rh)$

Doubling the height and radius of a cylinder multiples the surface area by 4.

33. $S = V$
$2\pi r^2 + 2\pi rh = \pi r^2 h$
$\quad 2r + 2h = rh$
$\quad \dfrac{2r + 2h}{rh} = 1$
$\quad \dfrac{2}{h} + \dfrac{2}{r} = 1$

If either h or r is less than or equal to 2,
then $\dfrac{2}{h} \geq 1$ or $\dfrac{2}{r} \geq 1$, so $\dfrac{2}{h} + \dfrac{2}{r} > 1$.

Therefore, both h and r must be greater than 2.

34. $\pi r^2 h = 16\pi$
$\quad h = \dfrac{16}{r^2}$
$S = 2\pi rh + 2\pi r^2$
$\quad = 2\pi r \dfrac{16}{r^2} + 2\pi r^2$
$\quad = \dfrac{32\pi}{r} + 2\pi r^2$

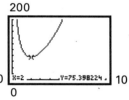

$r = 2$ units
$S \approx 75.40$ units^2
$h = 4$ units

35. $2\pi r = 4.7$ cm
$\quad r = \dfrac{4.7}{2\pi} \approx 0.75$ cm

36. $V = \pi r^2 h$
$\quad r = \sqrt{\dfrac{V}{\pi h}}$
$\quad d = 2r = 2\sqrt{\dfrac{(200,000)(0.134)}{23\pi}}$
$\qquad \approx 38.52$ ft

37. The volume of the cigarette is
$V = \pi r^2 h = \pi(0.08)^2(2.56) \approx 0.051$ in^3
The volume of the cigar is
$V = \pi r^2 h = \pi(0.375)^2(7) \approx 3.093$ in^3
$\dfrac{3.093}{0.051} \approx 60$

60 times more tobacco is contained in the cigar than in the cigarette.

38. 500,000 gallons $= \dfrac{500,000}{7.48} \approx 66,844.92$ cubic feet
Each tank holds $\pi(25)^2(25) \approx 49,087$ cubic feet.
$\dfrac{66,844.92}{49,087} \approx 1.36$
The city will need 2 tanks.

39. $V = \pi r^2 h = 115.5$ in^3
$\quad h = \dfrac{115.5}{\pi r^2}$
$\quad S = 2\pi rh + 2\pi r^2$
$\qquad = 2\pi r \dfrac{115.5}{\pi r^2} + 2\pi r^2$
$\qquad = \dfrac{231}{r} + 2\pi r^2$

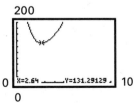

$r_{min} = 2.64$ in.
$S_{min} = 131.29$ in^2
$h_{min} \approx 5.28$ in.

40.

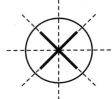

The symbol has 90°, 180°, and 270° rotational symmetry.

41.

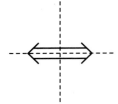

The symbol has 180° rotational symmetry

42.

The symbol has only trivial rotational symmetry.

43.

The symbol has 180° rotational symmetry.

44. $S = (2)(4)(7) + (2)(10)(4) + (2)(10)(7) = 276 \text{ units}^2$
$V = (10)(7)(4) = 280 \text{ units}^3$

45. $S = 6(7.5)^2 = 337.5 \text{ units}^2$
$V = (7.5)^3 \approx 421.88 \text{ units}^3$

46. $S = (12)(3 + 4 + 5) + 2\left(\frac{1}{2}\right)(3)(4) = 156 \text{ units}^2$
$V = \left(\frac{1}{2}\right)(3)(4)(12) = 72 \text{ units}^3$

47. $S = \frac{1}{2}(13)(40) + (10)^2 = 360 \text{ units}^2$
$V = \frac{1}{3}(10)^2(\sqrt{13^2 - 5^2}) = 400 \text{ units}^3$

48. $d^2 = w^2 + h^2$

49. $w^2 = \left(\frac{d}{3}\right)^2 + x^2$
$w^2 = \frac{d^2}{9} + x^2$

50. $h^2 = x^2 + \left(\frac{d}{3} + \frac{d}{3}\right)^2$
$h^2 = x^2 + \frac{4d^2}{9}$

51. $d = 18 \text{ in.}$
$d^2 = w^2 + h^2$

$$h^2 = x^2 + \frac{4d^2}{9}$$
$$- \quad \left(w^2 = \frac{d^2}{9} + x^2\right)$$
$$\overline{h^2 - w^2 = \frac{3d^2}{9} = \frac{d^2}{3}}$$

$h^2 + w^2 = d^2 = (18)^2 = 324$
$h^2 - w^2 = \frac{d^2}{3} = \frac{(18)^2}{3} = 108$

$2h^2 = 432$
$h = \sqrt{216} \approx 14.7 \text{ in.}$
$2w^2 = 216$
$w = \sqrt{108} \approx 10.4 \text{ in.}$

6. The area of the circle is $\pi r^2 = \pi(5)^2 = 25\pi$ units^2.

   The circumference of the circle is $2\pi r = 10\pi$ units.

   The circumference of the larger circle that the sector is part of is

   $2\pi(13) = 26\pi$ units. The sector is $\frac{10\pi}{26\pi} = \frac{5}{13}$ of the larger circle. The area of the

   larger circle is $\pi(13)^2 = 169\pi$, so the area of the sector is $\frac{5}{13}(169\pi) = 65\pi$ units^2.

   Thus the surface area of the cone is $25\pi + 65\pi = 90\pi \approx 282.74$ units^2.

7. $S = \pi r \ell + \pi r^2 = \pi(8)(17) + \pi(8)^2 = 200\pi \approx 628.32$ units^2

8. $S = L + B$ can be rewritten as $S = \pi r \ell + \pi r^2$.

9. $V = \frac{1}{3}Bh = \frac{1}{3}\pi r^2 h = \frac{1}{3}\pi(8)^2(27) = 576\pi \approx 1809.56$ units^3

PAGES 465–467, PRACTICE AND APPLY

10. $S = \pi(3)(5) + \pi(3)^2 = 24\pi \approx 75.40$ cm^2

11. $S = \pi(6)(10) + \pi(6)^2 = 96\pi \approx 301.59$ cm^2

12. $S = \pi(1.2)(2) + \pi(1.2)^2 = 3.84\pi \approx 12.06$ in^2

13. $S = \pi(0.5)(1.3) + \pi(0.5)^2 = 0.9\pi \approx 2.83$ units^2

14. $S = \pi(0.8)(1.7) + \pi(0.8)^2 = 2\pi \approx 6.28$ units^2

15. $S = \pi(4.5)(5.3) + \pi(4.5)^2 = 44.1\pi \approx 138.54$ units^2

16. $\ell = \sqrt{3^2 + 4^2} = 5$ in.

17. $\ell = \sqrt{5^2 + 6^2} = \sqrt{61} \approx 7.81$ in.

18. $\ell = \sqrt{(24)^2 + 7^2} = 25$

    $S = \pi r \ell + \pi r^2 = 175\pi + 49\pi = 224\pi \approx 703.72$ units^2

19. $\ell = \sqrt{40^2 + 9^2} = 41$

    $S = \pi r \ell + \pi r^2 = 369\pi + 81\pi = 450\pi \approx 1413.72$ units^2

20. $\ell = \sqrt{(12)^2 + (15)^2} \approx 19.21$

    $S = \pi r \ell + \pi r^2 \approx 374.51\pi \approx 1176.57$ units^2

21. $\ell = \sqrt{(20)^2 + (23)^2} \approx 30.48$

    $S = \pi r \ell + \pi r^2 = 1009.59\pi \approx 3171.72$ units^2

22. $\ell = \sqrt{(6)^2 + (1.1)^2} = 6.1$

    $S = \pi r \ell + \pi r^2 = 7.92\pi \approx 24.88$ units^2

23. $\ell = \sqrt{(6.3)^2 + (1.6)^2} = 6.5$

    $S = \pi r \ell + \pi r^2 = 12.96\pi \approx 40.72$ units^2

24. $V = \frac{1}{3}\pi r^2 h$

    $= \frac{1}{3}\pi(15)^2(13.3) \approx 3133.74$ units^3

25. $V = \frac{1}{3}\pi r^2 h$

    $= \frac{1}{3}\pi(7)^2(45) \approx 2309.07$ units^3

26. $V = \frac{1}{3}\pi r^2 h$

    $= \frac{1}{3}\pi(6.0)^2(33.2) \approx 1251.61$ units^3

27. $V = \frac{1}{3}\pi r^2 h$

    $= \frac{1}{3}\pi(21)^2(42) \approx 19{,}396.19$ units^3

28. $V = \frac{1}{3}\pi r^2 h$

    $= \frac{1}{3}\pi(13)^2(33.2)$

    ≈ 5875.62 units^3

29. $V = \frac{1}{3}\pi r^2 h$

    $= \frac{1}{3}\pi(9)^2(66) \approx 5598.32$ units^3

30. A cone with $r = 20$ units and slant height $\ell = \sqrt{30^2 + 20^2} = \sqrt{1300}$ units is formed.

    $S = \pi(20)(\sqrt{1300}) + \pi(20)^2 \approx 3522.07$ units^2

31. When rotated about the leg of length 10, $h = 10$ and $r = 24$.

    $V = \frac{1}{3}\pi(24)^2(10) \approx 6031.86$ units^3.

    When rotated about the leg of length 24, $h = 24$ and $r = 10$:

    $V = \frac{1}{3}\pi(10)^2(24) \approx 2513.27$ units^3.

32. The volume of the cone $= \frac{1}{3}\pi(10)^2(10) = \frac{1000\pi}{3}$.

The volume of the pyramid $= \frac{1}{3}(10\sqrt{2})^2(10) = \frac{2000}{3}$.

The ratio of the pyramid's volume to the cone's volume is then $\frac{\frac{2000}{3}}{\frac{1000\pi}{3}} = \frac{2}{\pi}$.

33. $180 = \pi(5)l + \pi(5)^2$

$\ell = \frac{180 - 25\pi}{5\pi} \approx 6.46$ in.

34. $\pi r^2 = 25\pi$ means that $r = 5$ cm

$\ell = \sqrt{5^2 + 13^2} \approx 13.93$ cm

$S = \pi(5)(13.93) + \pi(5^2) \approx 297.33$ cm^2

35. $1000 = \frac{1}{3}\pi r^2(10)$

$300 = \pi r^2$

$r = \sqrt{\dfrac{300}{\pi}} \approx 9.77$ cm

Radius	Height	Volume
1	$\sqrt{99}$	$\frac{1}{3}\pi \times 1^2 \times \sqrt{99} \approx 10.4$
2	$\sqrt{96}$	**36.** $\frac{1}{3}\pi \times 2^2 \times \sqrt{96} \approx 41.04$
3	**37.** $\sqrt{91}$	**38.** $\frac{1}{3}\pi \times 3^2 \times \sqrt{91} \approx 89.91$
4	**39.** $\sqrt{84}$	**40.** $\frac{1}{3}\pi \times 4^2 \times \sqrt{84} \approx 153.56$
5	**41.** $\sqrt{75}$	**42.** $\frac{1}{3}\pi \times 5^2 \times \sqrt{75} \approx 226.72$

43.

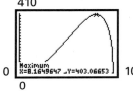

410

0 ··· 10

0

Maximum
X=8.1649647 Y=403.06653

$V = \frac{1}{3}\pi(x^2)(\sqrt{100 - x^2})$

$x_{max} \approx 8.2$ units

44. $V_{max} \approx 403.1$ units^3

45. $V = \frac{1}{3}\pi r^2 h = \frac{1}{3}\pi(30)^2(150) \approx 141{,}372$ ft^3

46. Cylindrical glass:

$r = \frac{7.2}{2} = 3.6$ cm

$V = \pi r^2 h = \pi(3.6)^2(14.6) \approx 594$ cm^3

Conical glass:

$r = \frac{7.5}{2} = 3.75$ cm

$V = \frac{1}{3}\pi r^2 h = \frac{1}{3}\pi(3.75)^2(18.4) \approx 271$ cm^3

47. straight angle:

$S = \frac{1}{2}\pi 6^2 \approx 56.55$ cm^2

angle of 120°:

$S = \frac{1}{3}\pi 6^2 \approx 37.70$ cm^2

48. circumference of the base of the cone is

$\frac{1}{2}(2)\pi(6) = 2\pi r$

$r = 3$ cm

$h^2 + r^2 = 6^2$

$\qquad h = \sqrt{6^2 - 3^3} = \sqrt{27} = 3\sqrt{3}$ cm

$\qquad V = \frac{1}{3}\pi r^2 h = \frac{1}{3}\pi(3)^2(3\sqrt{3}) \approx 48.97$ cm^3

49. circumference of the base of the cone is

$\frac{1}{3}(2\pi(6)) = 2\pi r$

$r = 2$ cm

$h^2 + r^2 = 6^2$

$h = \sqrt{6^2 - 2^2} = 4\sqrt{2}$ cm

$V = \frac{1}{3}\pi r^2 h = \frac{1}{3}\pi(2)^2(4\sqrt{2}) \approx 23.70$ cm^3

PAGE 468, LOOK BACK

50. The slope of $\overleftrightarrow{CD}$ is $\frac{-2-4}{8-(-2)} = \frac{-6}{10} = -\frac{3}{5}$.
So the slope of $\overleftrightarrow{AB}$ is $\frac{5}{3}$, by the
Perpendicular Lines Theorem.

51. The slope of $\overleftrightarrow{CD}$ is $-\frac{3}{5}$. So the slope of $\overleftrightarrow{AB}$ is $-\frac{3}{5}$,
by the Parallel Lines Theorem.

52. Sample answer:
Given: Parallelogram $ABCD$ with diagonal $\overline{AC}$
Prove: $\triangle ADC \cong \triangle CBA$
Proof:

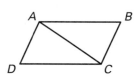

Statements	Reasons
$\overline{AB} \parallel \overline{CD}$ and $\overline{AD} \parallel \overline{BC}$	Definition of parallelogram
$\angle DAC \cong \angle BCA$, $\angle ACD \cong \angle CAB$	Alternate Interior Angles Theorem
$\overline{AC} \cong \overline{AC}$	Reflexive Property of Congruence
$\triangle ADC \cong \triangle CBA$	ASA

53. distance $= \sqrt{7^2 + 4^2} = \sqrt{65} \approx 8.06$

54. The dimensions of the box are x, $18 - 2x$, and $12 - 2x$; students should use their
calculators to graph the equation $y = x(18 - 2x)(12 - 2x)$ from $x = 0$ to $x = 6$.

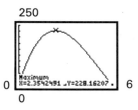

The maximum volume occurs when $x \approx 2.35$. The squares that should be cut from
the corners are approximately 2.35 × 2.35 in. The dimensions of the box are
approximately 2.35 × 13.30 × 7.30 in. The volume is approximately 228 in^3.

55. Since the model is $\frac{1}{87}$ the size of a real engine, the length of the real engine is

$87(5.5) = 478.5$ inches, or 39.875 feet.

7.6 **PAGE 473, GUIDED SKILLS PRACTICE**

6. $V = \frac{4}{3}\pi(40)^3 \approx 268,083 \text{ ft}^3$

7. $r = 24 \text{ ft}$
$S = 4\pi r^2 \approx 7238.23 \text{ ft}^2$
cost of construction is $1.50 \times 7238.23 \approx \$10,857$

PAGES 474–475, PRACTICE AND APPLY

8. $S = 4\pi(4)^2 = 64\pi \approx 201.06 \text{ units}^2$
$V = \frac{4}{3}\pi(4)^3 = \frac{256}{3}\pi \approx 268.08 \text{ units}^3$

9. $S = 4\pi(8)^2 = 256\pi \approx 804.25 \text{ units}^2$
$V = \frac{4}{3}\pi(8)^3 = \frac{2048}{3}\pi \approx 2144.66 \text{ units}^3$

10. $S = 4\pi(41)^2 = 6724\pi = 21,124.07 \text{ units}^2$
$V = \frac{4}{3}\pi(41)^3 = \frac{275,684}{3}\pi \approx 288,695.61 \text{ units}^3$

11. $S = 4\pi(33)^2 = 4356\pi \approx 13,684.78 \text{ units}^2$
$V = \frac{4}{3}\pi(33)^3 = 47,916\pi \approx 150,532.55 \text{ units}^3$

12. $S = 4\pi(13.41)^2 \approx 719.3124\pi \approx 2259.79 \text{ units}^2$
$V = \frac{4}{3}\pi(13.41)^3 \approx 3215.33\pi \approx 10,101.25 \text{ units}^3$

13. $S = 4\pi(12.22)^2 = 597.3136\pi \approx 1876.52 \text{ units}^2$
$V = \frac{4}{3}\pi(12.22)^3 \approx 2433.06\pi \approx 7643.68 \text{ units}^3$

14. $S = 4\pi(9)^2 = 324\pi \approx 1017.88 \text{ units}^2$
$V = \frac{4}{3}\pi(9)^3 = 972\pi \approx 3053.63 \text{ units}^3$

15. $S = 4\pi(8)^2 = 256\pi \approx 804.25 \text{ units}^2$
$V = \frac{4}{3}\pi(8)^3 = \frac{2048}{3}\pi \approx 2144.66 \text{ units}^3$

16. $S = 4\pi(11.17)^2 \approx 499.0756\pi \approx 1567.89 \text{ units}^2$
$V = \frac{4}{3}\pi(11.17)^3 \approx 1858.22\pi \approx 5837.79 \text{ units}^3$

17. $S = 4\pi(5.74)^2 \approx 131.7904\pi \approx 414.03 \text{ units}^2$
$V = \frac{4}{3}\pi(5.74)^3 \approx 252.16\pi \approx 792.18 \text{ units}^3$

18. $S = 4\pi(12.33)^2 \approx 608.1156\pi \approx 1910.45 \text{ units}^2$
$V = \frac{4}{3}\pi(12.33)^3 \approx 2499.36\pi \approx 7851.96 \text{ units}^3$

19. $S = 4\pi(99.98)^2 \approx 39,984.0016\pi \approx 125,613.45 \text{ units}^2$
$V = \frac{4}{3}\pi(99.98)^3 \approx 1,332,533.49\pi \approx 4,186,277.43 \text{ units}^3$

20. $S = 4\pi x^2 = 4x^2\pi$
$V = \frac{4}{3}\pi x^3 = \frac{4}{3}x^3\pi$

21. $S = 4\pi(2y)^2 = 16\pi y^2 = 16y^2\pi$
$V = \frac{4}{3}\pi(2y)^3 = \frac{4}{3}\pi(8y^3) = \frac{32}{3}y^3\pi$

22. $S = 4\pi(6x)^2 = 144x^2\pi$
$V = \frac{4}{3}\pi(6x)^3 = 288x^3\pi$

23. $S = 4\pi(2y)^2 = 16y^2\pi$
$V = \frac{4}{3}\pi(2y)^3 = \frac{32}{3}y^3\pi$

24. $S = 4\pi\left(\frac{x}{2}\right)^2 = x^2\pi$
$V = \frac{4}{3}\pi\left(\frac{x}{2}\right)^3 = \frac{x^3}{6}\pi$

25. $S = 4\pi\left(\frac{y}{4}\right)^2 = \frac{y^2}{4}\pi$
$V = \frac{4}{3}\pi\left(\frac{y}{4}\right)^3 = \frac{y^3}{48}\pi$

26. $A = \pi r^2$
$S = 4\pi r^2 = 4A = 4 \times 225 = 900 \text{ units}^2$

27. $A = \pi r^2$
$S = 4\pi r^2 = 4A = 4 \times 125 = 500 \text{ units}^2$

28. $A = \pi r^2$
$S = 4\pi r^2 = 4A = 4 \times 32.30 = 129.2 \text{ units}^2$

29. $A = \pi r^2$
$S = 4\pi r^2 = 4A = 4 \times 11.22 = 44.88 \text{ units}^2$

30. $A = \pi r^2$
$S = 4\pi r^2 = 4A = 4 \times 16\pi = 64\pi \approx 201.06 \text{ units}^2$

31. $A = \pi r^2$
$S = 4\pi r^2 = 4A = 4 \times 225\pi = 900\pi \approx 2827.43 \text{ units}^2$

32. The largest ball would have a diameter equal in length to the edge of the cube.

$V = \frac{4}{3}\pi(6)^3 = 288\pi \approx 904.78 \text{ in}^3$

33. For the cube: $s^3 = 1000$, so $s = 10$, and $S = 6(100) = 600 \text{ units}^2$

For the sphere: $\frac{4}{3}\pi r^3 = 1000$, so $r = \sqrt[3]{\frac{750}{\pi}} \approx 6.204$

$S = 4\pi(6.204)^2 \approx 483.60 \text{ units}^2$.

34. For the cube: $6s^2 = 864$, so $s = 12$ and $V = (12)^3 = 1728 \text{ units}^3$

For the sphere, $4\pi r^2 = 864$, so $r \approx 8.292$, and $V = \frac{4}{3}\pi r^3 = \frac{4}{3}\pi(8.292)^3 \approx 2388.06 \text{ units}^3$

35. The altitude of the hemisphere is $r = 10$ in.

Volume of the hemisphere is

$\frac{1}{2}\left(\frac{4}{3}\pi r^3\right) = \frac{2}{3}\pi(10)^3 = \frac{2000}{3}\pi$

For the cone, $V = \frac{1}{3}Bh = \frac{1}{3}\pi r^2 h = \frac{2000}{3}\pi$

$h = \dfrac{(3)\left(\frac{2000}{3}\pi\right)}{\pi(10)^2} = 20$ in.

For the cylinder, $V = \pi r^2 h = \frac{2000}{3}\pi$

$h = \dfrac{\frac{2000}{3}\pi}{\pi(10)^2} \approx 6.67$ in.

36. $S = 4\pi r^2 = 4\pi\left(\frac{3.5}{2}\right)^2 \approx 38.48 \text{ in}^2$

$V = \frac{4}{3}\pi r^3 = \frac{4}{3}\pi\left(\frac{3.5}{2}\right)^3 \approx 22.45 \text{ in}^3$

37. $S = 4\pi r^2 = 4\pi\left(\frac{2.9}{2}\right)^2 \approx 26.42 \text{ in}^2$

$V = \frac{4}{3}\pi r^3 = \frac{4}{3}\pi\left(\frac{2.9}{2}\right)^3 \approx 12.77 \text{ in}^3$

38. The volume of each tennis ball $= \frac{4}{3}\pi(1.5)^3 \approx 14.14 \text{ in}^3$

The total volume taken up by the tennis balls $= 3(14.1) \approx 42.4 \text{ in}^3$

Volume of the cylinder $= \pi(1.5)^2 9 \approx 63.6 \text{ in}^3$

The percent of the space occupied by the balls is $\frac{42.4}{63.6} \approx 0.67$, or 67%

39. $S = 4\pi r^2$

Area of the land $= \frac{1}{3}(4\pi 4000^2) \approx 67{,}020{,}643 \text{ mi}^2$

40. Volume of the ice cream $= \frac{4}{3}\pi(1.25)^3 \approx 8.18 \text{ in}^3$

41. Volume of the cone $= \frac{1}{3}\pi(1.25)^2(8) \approx 13.09 \text{ in}^3$

42. The slant height of the cone $= \sqrt{8^2 + (1.25)^2} \approx 8.097$ in.

Total surface area $= \frac{1}{2}$(surface area of the sphere) + lateral area of the cone

$= \frac{1}{2}[4\pi(1.25)^2] + \pi(1.25)(8.097) \approx 41.61 \text{ in}^2$

43. Volume of the solid sphere $= \frac{4}{3}\pi(10)^3$

The thickness is the difference between the outer radius (R) and the inner radius (r). The volume occupied by the metal in the hollow sphere is the same as the volume of the original solid sphere.

$\frac{4}{3}\pi R^3 - \frac{4}{3}\pi r^3 = \frac{4}{3}\pi(10)^3$

$R^3 - r^3 = (10)^3$

$R^3 = r^3 + (10)^3 = (10)^3 + (10)^3 = 2000$

$R \approx 12.60$ in.

The thickness $= 12.60 - 10 = 2.60$ in.

44. $B = \ell w = 30 \text{ in}^2; p = 2(3 + 10) = 26 \text{ in.}$
$L = hp = 5 \cdot 26 = 130 \text{ in}^2$
$S = L + 2B = 130 + 2(30) = 190 \text{ in}^2$

45. $S = \pi(15)(45) + \pi(15)^2 = 900\pi \approx 2827.43 \text{ cm}^2$

46. $S = 2\pi rh + 2\pi r^2 = 180\pi + 162\pi \approx 1074.42 \text{ in}^2$

47. $V = \frac{1}{3}(15)(7)(12) = 420 \text{ ft}^3$

48. $V = \frac{1}{3}\pi r^2 h = \frac{1}{3}\pi(25)(10) \approx 261.80 \text{ in}^3$

49. $V = \pi r^2 h = \pi(7.5)^2(20) = 3534.29 \text{ m}^3$

50. $\dfrac{\frac{4}{3}\pi r^3}{4\pi r^2} = \dfrac{r}{3}$

51. $V = \frac{4}{3}\pi r^3 \quad S = 4\pi r^2; r = \sqrt{\dfrac{S}{4\pi}}$
$V = \frac{4}{3}\pi\left(\sqrt{\dfrac{S}{4\pi}}\right)^3 = \frac{4}{3}\pi\left(\dfrac{S}{4\pi}\right)^{3\backslash 2}$
$= \dfrac{S\sqrt{S}}{6\sqrt{\pi}} = \dfrac{S\sqrt{\pi S}}{6\pi}$

52. The number is $\dfrac{\sqrt{\pi S}}{6\pi} = \dfrac{\sqrt{\pi 4\pi r^2}}{6\pi} = \dfrac{r}{3} = 1.5$ units

53. Volume of the cube $= s^3 = (2r)^3 = 8r^3$
Volume of the sphere $= \frac{4}{3}\pi r^3$
ratio of the volume of the cube to the volume of
the sphere $= \dfrac{8r^3}{\frac{4}{3}\pi r^3} = \dfrac{6}{\pi}$

5.–7.

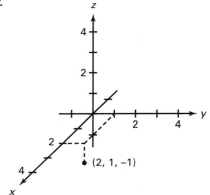

(2, 1, −1)

5. $(2, 1, 1)$ **6.** $(2, -1, -1)$ **7.** $(-2, 1, -1)$

8.

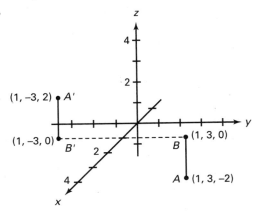

about the *x*-axis

9.

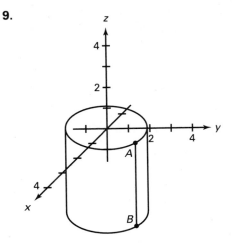

10.

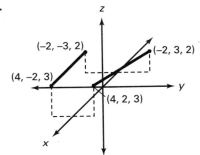

11.

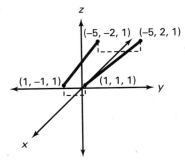

12.

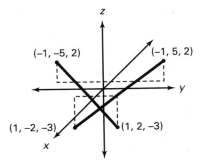

13.

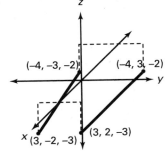

14.

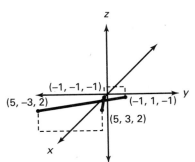

15.

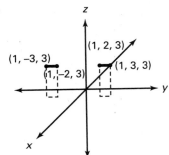

For Exercises 16–21, find the coordinates of the reflected point by multplying the *z*-coordinate of the point by −1.

16. $(6, 5, -8)$

17. $(-2, 3, 1)$

18. $(1, 1, -1)$

19. $(4, -2, -3)$

20. $(-5, 2, -1)$

21. $(1, -2, 3)$

For exercises 22.–27., find the coordinates of the reflected point by multiplying the y-coordinate of the point by −1.

22. front-right-top, $(6, 2, 8)$

23. back-right-bottom, $(-4, 4, -1)$

24. in the xz-plane, $(1, 0, 1)$

25. in the xy-plane $(1, -1, 0)$

26. front-right-top $(4, 4, 1)$

27. front-left-bottom $(2, -2, -8)$

28. The resulting figure is a circle and its interior. Its center is at $(5, 0, 0)$ with radius $= 10$.

29. The area of this circle $= \pi(10)^2 = 100\pi \approx 314.16$ units^2

30. A cylinder with $h = 10$ and $r = 5$, centered on the z-axis.

31. $V = \pi(5)^2(10) = 250\pi \approx 785.40$ units^3

32. A cone with $r = 4$, $h = 4$ and $\ell = 4\sqrt{2}$

33. $V = \frac{1}{3}\pi(4)^2(4) = \frac{64}{3}\pi \approx 67.02$ units^3

34. No. Based on Exercise 25–33 in Lesson 7.5, if the hypotenuse of the right triangle (the slant height of the cone) is kept constant, the volume of the cone is given by the equation $y = \frac{1}{3}\pi(x^2)(\sqrt{\ell^2 - x^2})$, where x is the length of the leg of the triangle that generates the base of the cone. The maximum of this function is not at the point where $x = \sqrt{\ell^2 - x^2}$ (that is, when the triangle is isosceles). For example, for $\ell = 4$ the maximum volume is for $x \approx 3.24$, not $x \approx 2.83$.

35.

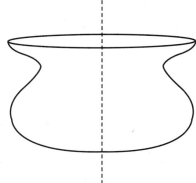

36.

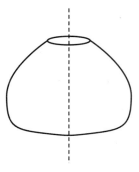

37.

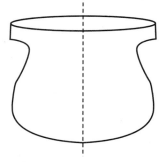

38.

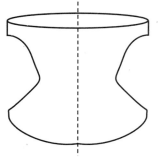

39.

40.

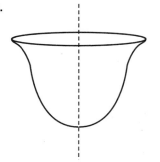

41. 3rd side of base $= \sqrt{11^2 - 7^2} = \sqrt{72}$

$V = \frac{1}{2}(7)(\sqrt{72})(15) \approx 445.48$ units^3

$S = 15(7 + 11 + \sqrt{72}) + 2\left(\frac{1}{2}\right)(7)(\sqrt{72})$

≈ 456.68 units^2

42. $V = \frac{1}{2}(3\sqrt{3})(36)(12) \approx 1122.37$ units^3,

since the apothem is $3\sqrt{3}$

$S = hp + 2B = hp + 2\left(\frac{1}{2}ap\right) = p(h + a)$

$= 36(12 + 3\sqrt{3}) \approx 619.06$ units^2

43. $V = Bh$

7 ft. 10 in = 94 in.

8 ft. 2 in. = 98 in.

24 ft = 288 in.

$V = 94 \times 98 \times 288 = 2,653,056$ in^3 ≈ 1535 ft^3

44. $\ell = \sqrt{3^2 + 5^2} = \sqrt{34}$

$S = \frac{1}{2}\ell p + B = \frac{1}{2}\sqrt{34}(4)(6) + 36 \approx 105.97$ cm^2

$V = \frac{1}{3}Bh = \frac{1}{3}(36)(5) = 60$ cm^3

45. From the photo, the diameter of the smoker is about 2.5 ft and the width is about 3.5 ft.

$V = \frac{1}{2}\pi(1.25)^2(3.5) \approx 8.6$ ft^3

46. The strobe will "freeze" the motion of the fan blades if it flashes at an integer multiple of 90° of rotation of the blades.

$\frac{12 \text{ rev}}{1 \text{ sec}} \cdot \frac{360°}{\text{rev}} \cdot \frac{1 \text{ flash}}{n \cdot 90°} = \frac{48}{n} \frac{\text{flash}}{\text{sec}}$

where $n = 1, 2, 3, \ldots$

So the strobe must flash at

$\frac{48}{1}, \frac{48}{2}, \frac{48}{3}, \frac{48}{4}, \ldots, \frac{48}{n}$ or 48, 24, 12, 6, … times per second.

47. $\frac{x \text{ rev}}{\text{sec}} \cdot \frac{360°}{\text{rev}} \cdot \frac{1 \text{ flash}}{n \cdot 90°} = 36 \frac{\text{flashes}}{\text{sec}}$

$\frac{360x}{90n} = 36$

$x = 9n$, for $n = 1, 2, 3, \ldots$

The fan speed must be an integer multiple of 9 revolutions/second.

CHAPTER REVIEW AND ASSESSMENT

1. The length of each edge of the cube is $\sqrt[3]{64} = 4$ in., so the area of each face is $4 \times 4 = 16$ in^2. There are 6 faces, so the surface area is $6(16) = 96$ in^2. Thus, the ratio of surface area to volume is $\frac{96}{64} = \frac{3}{2}$.

2. The length of each edge of the cube is $\sqrt[3]{100} \approx 4.64$ in., so the area of each face is $4.64 \times 4.64 \approx 21.53$ in^2. There are 6 faces, so the surface area is $6(21.53) \approx 129.18$ in^2. Thus, the ratio of surface area to volume is $\frac{129.18}{100} \approx 1.29$.

3. Cube: $V = 4^3 = 64$ units^3; $S = (6)(16) = 96$ units^2; $\frac{S}{V} = \frac{96}{64} = \frac{3}{2} = 1.5$

Prism: $V = (7)(4)(3) = 84$ units^3; $S = (2)(4)(3) + (2)(7)(4) + (2)(7)(3)$

$= 122$ units^2; $\frac{S}{V} = \frac{122}{84} = \frac{61}{42} \approx 1.45$

The ratio is larger for the cube.

4. s is the base edge; h is the height, $h = 3s$

$$\frac{S}{V} = \frac{hp + 2B}{Bh} = \frac{h(4s) + 2(s^2)}{s^2 h}$$
$$= \frac{(3s)(4s) + 2s^2}{s^2(3s)} = \frac{14s^2}{3s^3} = \frac{14}{3s}$$

5. The lateral area is hp, or
$8(6 \times 6) = 288 \text{ units}^2$

6. The apothem is $3\sqrt{3}$.

The surface area is $hp + 2B = hp + 2\left(\frac{1}{2}ap\right)$
$$= hp + ap = (h + a)p$$
$$= (8 + 3\sqrt{3})(6 \times 6) \approx 475.06 \text{ units}^2$$

7. $V = Bh = \frac{1}{2}aph = \frac{1}{2}(3\sqrt{3})(6 \times 6)(8) \approx 748.25 \text{ units}^3$

8. The volume will be multiplied by a factor of 8.
$V \approx 748.25 \times 8 = 5986 \text{ units}^3$

9. $L = 3\left(\frac{1}{2}\right)(6)(2) = 18 \text{ units}^2$

10. $S = L + B = 18 + \frac{1}{2}(6)(3\sqrt{3}) = 18 + 9\sqrt{3} \approx 33.59 \text{ units}^2$

11. $V = \frac{1}{3}Bh = \frac{1}{3}(9\sqrt{3})(1) = 3\sqrt{3} \approx 5.20 \text{ units}^3$

12. $V = \frac{1}{3}(9\sqrt{3})(2) = 6\sqrt{3} \approx 10.39 \text{ units}^3$

13. The lateral area is $2\pi rh$,
or $2\pi(4)(9) = 72\pi \approx 226.19 \text{ units}^2$.

14. The area of each base is
πr^2, or $\pi(4)^2 = 16\pi \text{ units}^2$.
The surface area is
$S = L + 2B$, or
$72\pi + 2(16\pi) = 104\pi \approx 326.73 \text{ units}^2$.

15. The area of the base is 16π, so the volume is
$V = Bh = 16\pi(9) \approx 144\pi \approx 452.39 \text{ units}^3$.

16. $V = Bh = \pi r^2 h$.
Doubling the height will double the volume of
the cylinder. But, doubling the radius will multiply
the volume by 4. Thus, doubling the radius increases
the volume more than doubling the height.

17. The lateral area is $\pi r \ell$, or
$\pi(20)(29) = 580\pi \approx 1822.12 \text{ units}^2$.

18. The area of the base is πr^2, or
$\pi(20)^2 = 400\pi \text{ units}^2$.
The surface area is $S = L + B$ or
$580\pi + 400\pi = 980\pi \approx 3078.76 \text{ units}^2$.

19. The area of the base is 400π, so the volume is
$V = \frac{1}{3}Bh = \frac{1}{3}400\pi(21) = 2800\pi \approx 8796.12 \text{ units}^3$.

20. If $h = 42$, then
$V = \frac{1}{3}\pi(20^2)(42) = 5600\pi \approx 17{,}592.92 \text{ units}^3$

21. The surface area is
$S = 4\pi r^2 = 4\pi 5^2 = 100\pi \approx 314.16 \text{ units}^2$.

22. The volume is $V = \frac{4}{3}\pi r^3 = \frac{4}{3}\pi(5^3) \approx 166.67\pi \text{ units}^3$.

23. $V = \frac{4}{3}\pi r^3$
So $r = \sqrt[3]{\frac{3}{4\pi}V} = \sqrt[3]{\frac{3(36\pi)}{4\pi}} = 3 \text{ units}$

24. The volume of the sphere is
$V = \frac{4}{3}\pi r^3 = \frac{4}{3}\pi(1^3) = \frac{4}{3}\pi \text{ units}^3$.
The volume of the cube is $V = s^3$,
so $s = \sqrt[3]{V} = \sqrt[3]{\frac{4}{3}\pi} \approx 1.61 \text{ units}$.

25.

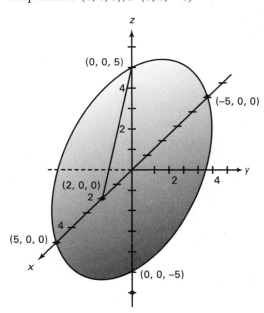

(left figure for 25, showing 3D axes with A'(0, 0, 0) and B'(0, 4, −6))

Multiply each z coordinate by -1 to find the endpoints: $A'(0, 0, 0)$, $B'(0, 4, -6)$

26.

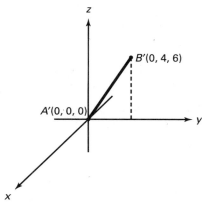

Multiply each x coordinate by -1 to find the endpoints: $A'(0, 0, 0)$, $B'(0, 4, 6)$

27.

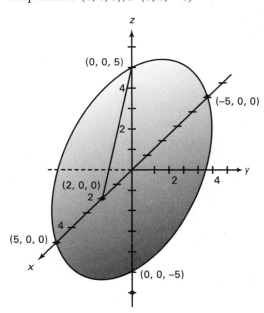

The figure is a flat disk in the xz-plane.

28.

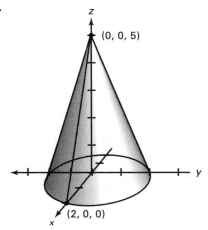

29. Find the volume of paper on the roll.

Volume of outer cylinder $= \pi(1.5)^2(5) = \frac{5}{36}\pi$ ft^3

Volume of hollow core $= \pi\left(\frac{1}{6}\right)^2(5) = \frac{5}{36}\pi$ ft^3

Volume of paper $= 11.25\pi - \frac{5}{36}\pi = \frac{100}{9}\pi$ ft^3

$\frac{100}{9}\frac{\pi \text{ ft}^3}{5 \text{ ft}} = \frac{20}{9}\pi$ ft^2, and 1 ft^2 = 144 in^2

so $\frac{20}{9}\pi$ ft^2 = 320π in^2 $\approx$ 1005.31 in^2

Then the length of paper on the roll is

$\frac{320\pi}{0.0015} \approx 670{,}206$ in., or about 55,850 ft.

30. The surface area is $S = 4\pi r^2 = 4\pi(4.75)^2 = 90.25\pi$

$= 283.53$ in.^{2}

The volume is $V = \frac{4}{3}\pi r^3 = \frac{4}{3}\pi(4.75)^3 \approx 142.90\pi$

≈ 448.92 in.^{3}

The volume of the cube is $V = (9.5)^3 = 857.375$ in.^{3}

The percent of the box not filled by the basketball is

$\frac{857.375 - 448.92}{857.375} \approx 0.48$ or 48%.

31. The volume of the bubble is

$V = \frac{4}{3}\pi r^3 = \frac{4}{3}\pi 5^3 \approx 166.67\pi$

≈ 523.50 cm^3.

Let the radius of the hemisphere be r_h.

$V = \frac{1}{2}\left(\frac{4}{3}\pi r_h^3\right) = 523.50$

$r_h^3 = \frac{\left(\frac{3}{4}\right)(2)(523.50)}{\pi} \approx 249.95$

$r_h \approx 6.30$ cm

CHAPTER TEST

1. $\frac{150}{125} = 1.2$

2. $\frac{4n + 2}{n} = 4 + \frac{2}{n}$

3. $\frac{2s^2 + 4(4s^2)}{4s^3} = \frac{18s^2}{4s^3} = \frac{9}{2s}$

4. Sample answer: The large rectangular shape maximizes surface area to absorb solar power, while the thinness of the panel minimizes volume.

5. $2(56) + 2(40) + 2(35) = 262$ units^2;
   $(5)(8)(7) = 280$ units^3

6. $2(24) + 2(1.5) + 2(1) = 53$ units^2;
   $(6)(4)(0.25) = 6$ units^3

7. $(12)(6)(15) = 1080$ cm^3

8. $2[0.5(4.5)(5)] + 2\left(6\sqrt{2.25^2 + 5^2}\right) + (4.5)(6) \approx$ 115.3 ft^2

9. $4[(0.5)(24)(26)] = 1248$ units^2

10. $1248 + 24^2 = 1824$ units^2

11. $\frac{1}{3}(24^2)(10) = 1920$ units^3

12. $1920(2) = 3840$ units^3

13. $2(6)\pi \cdot 8 = 96\pi \approx 301.59$ in^2

14. $96\pi + 2(6^2)\pi = 168\pi \approx$ 527.79 in^2

15. $6^2\pi \cdot 8 = 288\pi \approx 904.79$ in^3

16. $100\pi \div 4 = 25\pi$; $r = \sqrt{25} = 5$;
    $C = 2 \cdot 5\pi = 10\pi$ cm

17. $S = 216\pi \approx 678.58$ units^2;
    $V = 324\pi \approx 1017.88$ units^3

18. $S = 90\pi \approx 282.74$ units^2;
    $V = 100\pi \approx 314.16$ units^3

19. $S = 4\pi(12^2) = 576\pi \approx 1809.56$ units^2

20. $V = \frac{4}{3}\pi(12^3) = 2304\pi \approx 7238.23$ units^3

21. $r = 3x$; $S = 36\pi x^2$ units^2; $V = 36\pi x^3$ units^3

22. $r = 4.5$ in.; $S = 81\pi \approx 254.47$ in^2;
    $V = 121.5\pi \approx 381.70$ in^3

23. $(1, 5, -9)$

24. $(-3, 2, 5)$

25. $(6, -1, 4)$

26. circle and its interior with center $(3, 0, 0)$ and $r = 6$

27. cylinder centered on the z-axis, with $h = 6$ and $r = 3$

CHAPTERS 1–7 CUMULATIVE ASSESSMENT

1. $V_{\text{cone}} = \frac{1}{3}\pi r^2 h = \frac{1}{3}\pi(3)^2(6) = 18\pi$
   $V_{\text{sphere}} = \frac{4}{3}\pi r^3 = \frac{4}{3}\pi(3)^3 = 36\pi$
   The answer is B.

2. $m\angle C = 90°$
   $m\angle A + m\angle B = 180 - m\angle C = 180 - 90 = 90°$
   The answer is C.

3. The area of $RSTU$ is $(1.5)(3.5) = 5.25$ units^2.
   The area of $RUYV$ is $(2)(3.5) = 7$ units^2.
   The answer is B.

4. It is not clear whether the lines shown are parallel, so no conclusion can be drawn about $\angle 1$ and $\angle 2$. Thus the answer is D.

5. c

6. b

7. b

8. c

9. a

10. $\frac{4.8 + 7.2}{2} = \frac{12}{2} = 6$; The answer is c.

11. No conclusion

12. Squares

13. $A = \pi r^2 = 49\pi$ units^2

14. $\pi r^2 = 7$; $r = \sqrt{\frac{7}{\pi}}$ units

15.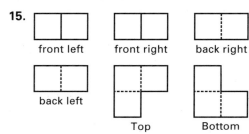
    front left   front right   back right
    back left   Top   Bottom

16. The length is
    $$\sqrt{3^2 + \left(\sqrt{6^2 + 8^2}\right)^2} = \sqrt{9 + 36 + 64}$$
    $$= \sqrt{109} \approx 10.4 \text{ units}$$

17. $S = 2\ell w + 2wh + 2\ell h$
    $= 2(8)(6) + 2(6)(3) + 2(8)(3) = 180$ units^2

18. $V = \frac{1}{3}Bh = \frac{1}{3}\left(\frac{1}{2}ap\right)h$
    $= \frac{1}{3}\left[\frac{1}{2}\left(5\sqrt{3}\right)(6 \times 10)\right](20) \approx 1732$ units^3

CHAPTER 8

Similar Shapes

7. $(3, 15)$

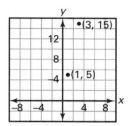

8. $(-2, 8)$

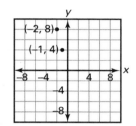

9. $(1.5, -0.5)$

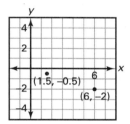

10. $(-4, -6)$

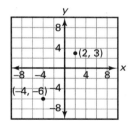

11.

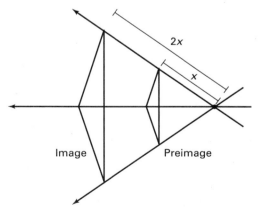

12.

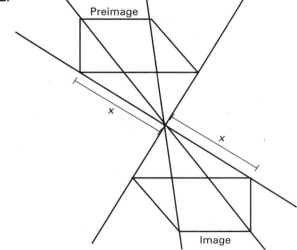

13. The ratio of the size of the image to the size of the preimage is $\frac{1.8 \text{ cm}}{3.6 \text{ cm}} = 0.5$ and the image is opposite the preimage from the center of dilation, so the scale factor is -0.5.

14. $(2, 6), (4, 10), (8, 6)$

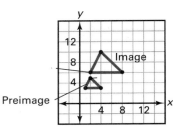

15. $\left(-1, \frac{5}{3}\right), \left(\frac{8}{3}, 3\right), \left(\frac{2}{3}, -2\right)$

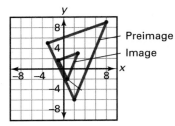

16. $(0, 0), (-3, 0), (-2, -2), \left(-1, -\frac{3}{2}\right)$

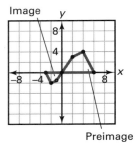

17. $(1.6, 1.6), (4.8, -1.6), (-3.2, -4.8)$

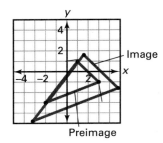

18. 3

19. $\frac{1}{2}$

20. -1

21. $-\frac{1}{3}$

22.

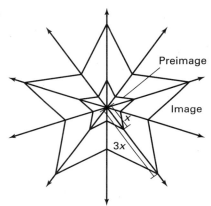

23.

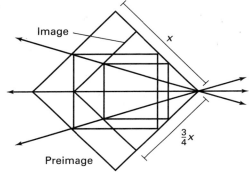

24.

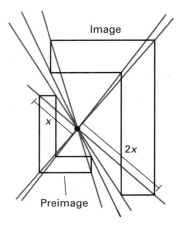

25.

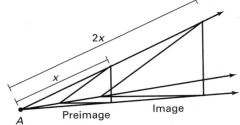

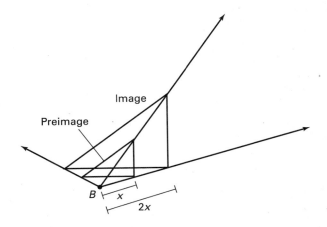

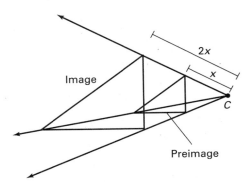

The dimensions of each of the three images are twice as large as the dimensions of the corresponding images. The difference is that each image is placed in a different position relative to the preimage.

26.

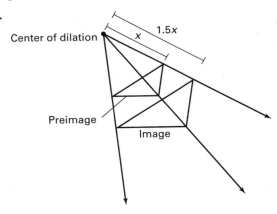

Scale factor = 1.5

27. Slope of preimage: $m = \dfrac{3 - 0}{5 - 1} = \dfrac{3}{4}$

Endpoints of image are $(2, 0)$ and $(10, 6)$.

So, $m = \dfrac{6 - 0}{10 - 2} = \dfrac{6}{8} = \dfrac{3}{4}$.

28. Slope of preimage: $m = \dfrac{1 - 3}{3 - (-2)} = -\dfrac{2}{5}$.

Endpoints of image are $(-10, 15)$ and $(15, 5)$.

So, $m = \dfrac{5 - 15}{15 - (-10)} = \dfrac{-10}{25} = -\dfrac{2}{5}$.

29. Slope of preimage: $m = \dfrac{8 - 4}{4 - (-2)} = \dfrac{4}{6} = \dfrac{2}{3}$.

Endpoints of image are $(1, -2)$ and $(-2, -4)$.

So, $m = \dfrac{-4 - (-2)}{-2 - 1} = \dfrac{-2}{-3} = \dfrac{2}{3}$.

30. Slope of preimage: $m = \frac{4-1}{2-1} = \frac{3}{1} = 3$.

Endpoints of image are $(1.7, 1.7)$ and $(3.4, 6.8)$.

So, $m = \frac{6.8 - 1.7}{3.4 - 1.7} = \frac{5.1}{1.7} = 3$.

31. Image is $(20, 4)$.

So, $m = \frac{4-1}{20-5} = \frac{3}{15} = \frac{1}{5}$.

Then, $y - 1 = \frac{1}{5}(x - 5)$

$y - 1 = \frac{1}{5}x - 1$

$y = \frac{1}{5}x$

Equation of line is $y = \frac{1}{5}x$.

$0 = \frac{1}{5}(0) \Rightarrow 0 = 0$. So, the origin is on this line.

32. Image is $\left(-\frac{5}{3}, \frac{5}{2}\right)$.

So, $m = \frac{\frac{5}{2} - 3}{-\frac{5}{3} - (-2)} = \frac{-\frac{1}{2}}{\frac{1}{3}} = -\frac{3}{2}$.

Then, $y - 3 = -\frac{3}{2}(x - (-2))$

$y - 3 = -\frac{3}{2}x - 3$

$y = -\frac{3}{2}x$

Equation of line is $y = -\frac{3}{2}x$.

$0 = -\frac{3}{2}(0) \Rightarrow 0 = 0$. So the origin is on this line.

33. Image is $(-9, 15)$.

So, $m = \frac{15 - (-5)}{-9 - 3} = -\frac{20}{12} = -\frac{5}{3}$.

Then, $y - (-5) = -\frac{5}{3}(x - 3)$

$y + 5 = -\frac{5}{3}x + 5$

$y = -\frac{5}{3}x$

Equation of line is $y = -\frac{5}{3}x$.

$0 = -\frac{5}{3}(0) \Rightarrow 0 = 0$. So, the origin is on this line.

34. Image $(10, 17.5)$.

So, $m = \frac{17.5 - 7}{10 - 4} = \frac{10.5}{6} = 1.75$.

Then $y - 7 = 1.75(x - 4)$

$y - 7 = 1.75x - 7$

$y = 1.75x$

Equation of line is $y = 1.75x$.

$0 = 1.75(0) \Rightarrow 0 = 0$. So the origin is on this line.

35.

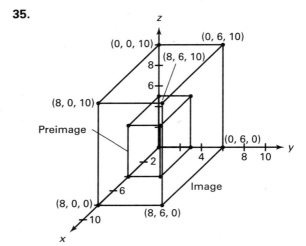

36. 2

37. 4

38. 8

39.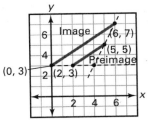

The endpoints of the image are $(0, 3)$ and $(6, 7)$, respectively. Distance of the preimage segment is $\sqrt{13}$. Distance of the image segment is $2\sqrt{13}$. Thus, the scale factor is $\frac{2\sqrt{13}}{\sqrt{13}} = 2$. Equation of the line passing through $(0, 3)$ and $(2, 3)$ is $y = 3$. Equation of the line passing through $(5, 5)$ and $(6, 7)$ is $y = 2x - 5$. Intersection of the two lines is the point $(4, 3) \Rightarrow$ Center of the dilation is $(4, 3)$. A dilation with a center of $(2, 1)$ and a scale factor of 4 is $D(x, y) = (4x - 2, 4y - 1)$.

40. The scale factor is $\frac{12}{4} = 3$. Thus, the coordinates of the image are $(0, 3)$, $(3, 3)$, $(6, 6)$, and $(3, 6)$, respectively.

41. The small opening in the center plate.

42. The scale factor is negative because the image is inverted.

43. The image is inverted because the light from point A travels in a straight line to A'; the light from point B travels in a straight line to point B'; and so on for each point on the preimage.

44. $\frac{2}{5}$

PAGE 505, LOOK BACK

45. Perimeter $= 6 + 2(8) = 22$ meters.
$$h = \sqrt{8^2 - \left(\frac{6}{2}\right)^2} = \sqrt{55}$$
So, Area $= \frac{1}{2}(6)(\sqrt{55}) \approx 22.25$ square meters.

46. Sides opposite congruent angles are congruent. So, $h^2 = 7^2 + 7^2 = 98 \Rightarrow h = 7\sqrt{2} \approx 9.9$ Hypotenuse is about 9.9 cm.

47. 40 ft = 480 in.
Outside surface area $= 4\pi(240)^2 \approx 723,822.95$ in^2.
Inside surface area $= 4\pi(239)^2 \approx 717,803.66$ in^2.
Difference is about 6,019.29 in^2.

48. Outside surface: $\frac{723,822.95 \text{ in}^2}{144 \text{ in}^2/\text{ft}^2} \approx 5026.55$ ft^2
Inside surface: $\frac{717,803.66 \text{ in}^2}{144 \text{ in}^2/\text{ft}^2} \approx 4984.75$ ft^2
$\frac{5026.55 \text{ ft}^2 + 4984.75 \text{ ft}^2}{400 \text{ gal/ft}^2} = \frac{10,011.3 \text{ ft}^2}{400 \text{ gal/ft}^2} \approx 25$ gal

49. $C = 2\pi r \Rightarrow r = \frac{C}{2\pi}$
$r = \frac{40,000}{2\pi}$
$r \approx 6366.2$
Radius is about 6366.2 km.

50. Total radius = Earth's radius + Height of atmosphere
$= 6366.2 + 550 = 6916.2$ km.
$V = \frac{4}{3}\pi r^3 = \frac{4}{3}\pi(6916.2)^3 \approx 1.39 \times 10^{12}$
$\Rightarrow$ Volume is about 1.39×10^{12} km^3.

51.

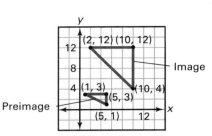

52. The preimage and the image are both right triangles with one vertical leg and one horizontal leg. The image is twice as wide and 4 times as high as the preimage, and the slopes of the hypotenuses are different. The form of the rule, $T(x, y) = (mx, ny)$ is similar to the rule for a dilation, but the scale factors are different for the x- and y-coordinates. In a dilation, the scale factors for the coordinates must be the same.

53. Slope of the segment with endpoints $(1, 3)$ and $(5,1)$ is $\frac{3 - 1}{1 - 5} = -\frac{2}{4} = -\frac{1}{2}$.
Image segment has endpoints $(2, 12)$ and $(10, 4)$.
Slope is $\frac{12 - 4}{2 - 10} = -\frac{8}{8} = -1$. Thus, ratio is $\frac{1}{2}$.

54.

Preimage	Image
$(0, 0)$	$(0, 0)$
$(2, 1)$	$(6, -1)$
$(2, 2)$	$(6, -2)$
$(1, 2)$	$(3, -2)$
$(1, 1)$	$(3, -1)$
$(0, 1)$	$(0, -1)$
$(-1, 1)$	$(-3, -1)$
$(-1, 0)$	$(-3, 0)$
$(-2, -1)$	$(-6, 1)$
$(-2, -2)$	$(-6, 2)$
$(-1, -1)$	$(-3, 1)$
$(2, -2)$	$(6, 2)$
$(2, -1)$	$(6, 1)$
$(1, -1)$	$(3, 1)$

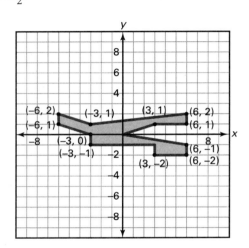

8.2 | **PAGE 511, GUIDED SKILLS PRACTICE**

6. Yes, corresponding angles are congruent, and corresponding sides are proportional: $\frac{20}{25} = \frac{24}{30} = \frac{28}{35}$.

7. No. $\frac{27}{36} \neq \frac{32}{48}$
Corresponding sides are not proportional.

8. $\frac{AD}{EH} = \frac{AB}{EF} \Rightarrow \frac{33}{EH} = \frac{20}{12} \Rightarrow 20EH = 396 \Rightarrow EH = 19.8$

9. $\frac{MN}{JK} = \frac{MP}{JL} \Rightarrow \frac{24}{JK} = \frac{40}{25} \Rightarrow 40JK = 600 \Rightarrow JK = 15$

10. $\frac{x}{7.5} = \frac{42}{3.5} \Rightarrow 3.5x = 315 \Rightarrow x = 90$
Width is 90 ft.

PAGES 512–515, PRACTICE AND APPLY

11. $\triangle SGT \sim \triangle MWR$

12. Rectangle $JDCP \sim$ Rectangle $LEBH$.

13. $\frac{AB}{XY} = \frac{BC}{YZ} = \frac{CA}{ZX}$

14. $\frac{EF}{VW} = \frac{FG}{WX} = \frac{GH}{XY} = \frac{HE}{YV}$

15. $\dfrac{JK}{PQ} = \dfrac{KL}{QR} = \dfrac{LM}{RS} = \dfrac{MN}{ST} = \dfrac{NJ}{TP}$

16. Since every angle is right, then corresponding angles are congruent. $\dfrac{3.6}{2.4} = \dfrac{5.4}{3.6}$, so corresponding sides are proportional.
Thus, quadrilateral $ABCD \sim$ quadrilateral $EFGH$.

17. $\angle J \cong \angle P$, $\angle K \cong \angle M$, and $\angle L \cong \angle N$, so corresponding angles are congruent.
$\dfrac{30}{22.5} = \dfrac{20}{15} = \dfrac{27}{20.25}$,
so corresponding sides are proportional.
Thus, $\triangle JKL \sim \triangle PMN$.

18. $\angle S \cong \angle U$, $\angle Q \cong \angle V$, and $\angle R$, $\angle T$ are right angles, so corresponding angles are congruent.
$\dfrac{30}{19.5} = \dfrac{16}{10.4} = \dfrac{34}{22.1}$, so corresponding sides are proportional. Thus $\triangle RSQ \sim \triangle TUV$.

19. $\dfrac{25}{18} \neq \dfrac{3}{2}$, so corresponding sides are not proportional. Thus, pentagon $KLMNJ \nsim PQRST$.

20. $(3)(20) = (15)(4) \Rightarrow 60 = 60$

21. $\dfrac{2}{5} = \dfrac{4}{10} \Rightarrow \dfrac{2}{5} = \dfrac{2}{5}$

22. $\dfrac{6}{2} = \dfrac{9}{3} \Rightarrow 3 = 3$

23. $\dfrac{x}{y} = \dfrac{4}{8} \Rightarrow \dfrac{x}{y} = \dfrac{1}{2}$

24. Ratio of the sides is $\dfrac{3}{6} = \dfrac{2}{4} = \dfrac{1}{2}$.
Area of first triangle is 6. Area of second triangle is 24. So, ratio of the areas is $\dfrac{6}{24} = \dfrac{1}{4}$.
Observe that $\left(\dfrac{1}{2}\right)^2 = \dfrac{1}{4}$.

25. Ratio of the sides is $\dfrac{24}{8} = \dfrac{51}{17} = \dfrac{45}{15} = 3$.
Area of first triangle is $\dfrac{1}{2}(24)(45) = 540$.
Area of second triangle is $\dfrac{1}{2}(8)(15) = 60$.
So, the ratio of the areas is $\dfrac{540}{60} = 9$.
Observe that $(3)^2 = 9$.

26. $\dfrac{x}{28} = \dfrac{33}{38.5} \Rightarrow 38.5x = 28 \cdot 33 \Rightarrow x = 24$

27. $\dfrac{x}{2} = \dfrac{1.8}{3} \Rightarrow 3x = 3.6 \Rightarrow x = 1.2$

28. $\dfrac{8}{x} = \dfrac{x}{2} \Rightarrow x^2 = 16 \Rightarrow x = 4$

29. $\dfrac{4}{x} = \dfrac{4.8}{x+1} \Rightarrow 4(x+1) = 4.8x \Rightarrow 0.8x = 4 \Rightarrow x = 5$

30. $\angle G \cong \angle P$, $\angle H \cong \angle L$, $\angle I \cong \angle M$, and $\angle K \cong \angle N$, so corresponding angles are congruent. $\overline{GH}$, $\overline{HI}$, $\overline{IK}$, and $\overline{KG}$ are proportional to $\overline{PL}$, $\overline{LM}$, $\overline{MN}$, and $\overline{NP}$, respectively, so corresponding sides are proportional. Thus, quadrilateral $GHIK \sim$ quadrilateral $PLMN$.

31. $6x(9) = (27)(24) \Rightarrow 54x = 648 \Rightarrow x = 12$

32. $6x = (4.8)(8.4) \Rightarrow 6x = 40.32 \Rightarrow x = 6.72$

33. $\dfrac{2}{5}x = \dfrac{56}{10} \Rightarrow \dfrac{2x}{5} = \dfrac{56}{10} \Rightarrow 20x = 280 \Rightarrow x = 14$

34. $x^2 = 900 \Rightarrow x = 30$

35. $3(x + 4) = 7(x - 4) \Rightarrow 4x = 40 \Rightarrow x = 10$

36. $4(5 - 2x) = 8(3x + 1) \Rightarrow 32x = 12 \Rightarrow x = \dfrac{3}{8}$

37. If the ratio of the sides is n, then the ratio of the areas is n^2.

38. False. It is true that $\dfrac{3}{4} = \dfrac{6}{8}$. But, $\dfrac{3+1}{4} = \dfrac{4}{4} = 1$, and $\dfrac{6+1}{8} = \dfrac{7}{8}$. So, $\dfrac{3+1}{4} \neq \dfrac{6+1}{8}$.

39. False. It is true that $\dfrac{3}{4} = \dfrac{6}{8}$. But, $\dfrac{3+1}{4+1} = \dfrac{4}{5}$ and $\dfrac{6+1}{8+1} = \dfrac{7}{9}$. So, $\dfrac{3+1}{4+1} \neq \dfrac{6+1}{8+1}$.

40. True. It is true by the "Add-One" Property.

41. True. If $\frac{a}{b} = \frac{c}{d}$ then $ad = bc \Rightarrow ac + ad = ac + bc$
$\Rightarrow a(c + d) = c(a + b) \Rightarrow \frac{a}{a + b} = \frac{c}{c + d}.$

42. If $\frac{a}{b} = \frac{c}{d} = \frac{e}{f}$, then
$ad = bc$
$af = be$
$ab = ba$
Thus, $ab + ad + af = ba + bc + be$
$a(b + d + f) = b(a + c + e)$
$\frac{a}{b} = \frac{a + c + e}{b + d + f}$

43. Let x be the estimate of the number of fish in the lake. Then, $\frac{8}{100} = \frac{300}{x} \Rightarrow 8x = 30{,}000 \Rightarrow x = 3750$
$\Rightarrow 3750$ fish is the estimate.

44. Sample answer:
Area ≈ 9 cm$^2 \cdot \frac{12.25 \text{ mi}^2}{1 \text{ cm}^2} = 110.25$ mi^2
110.25 mi$^2 \cdot 640 \frac{\text{acres}}{\text{mi}^2} \approx 70{,}500$ acres

Estimates should vary from 110 mi^2 to 120 mi^2 and from $70{,}000$ to $75{,}000$ acres.

45. Center to center, one row of trees is about 6.8 cm long. The distance between them should be
$\frac{6.8}{4} = 1.7$ cm.
$\frac{1.7 \text{ cm}}{x \text{ ft}} = \frac{1 \text{ cm}}{15 \text{ ft}} \Rightarrow x = 1.7 \cdot 15 = 25.5$ ft
The trees should be planted 25.5 ft apart.

46. Let x be the height of the canvas.
Then, $\frac{16}{24} = \frac{15}{x} \Rightarrow 16x = 15 \cdot 24 \Rightarrow x = 22.5$
The canvas should be 22.5 in. tall.

47. The scale factor is $\frac{15}{16}$.
4 in. $\cdot \frac{15}{16} = 3\frac{3}{4}$ in.
6 in. $\cdot \frac{15}{16} = 5\frac{5}{8}$ in.
The face in Brenda's reproduction should be $5\frac{5}{8}$ in. tall and $3\frac{3}{4}$ in. wide.

48. 12 ft = 144 in. and 2 ft 3 in. = 27 in. Let x be the length of a side of the table.
$\frac{27}{144} = \frac{x}{8} \Rightarrow 144x = 216 \Rightarrow x = 1.5$
The sides should be 1.5 in. long.

PAGE 516, LOOK BACK

49. $(x + 5)° + (5x + 12)° + (2x + 3)° = 180°$
$x + 5 + 5x + 12 + 2x + 3 = 180$
$8x = 160 \Rightarrow x = 20.$
So, the measures of the angles are 25°, 112°, and 43°.

50. Valid

51. Invalid

52. Valid

53. Valid

54. $2x + 92 = 180 \Rightarrow 2x = 88 \Rightarrow x = 44$
Base angle is 44°.

55. $h^2 = 5^2 + 7^2 \Rightarrow h^2 = 74 \Rightarrow h \approx 8.6$
Length is about 8.6 cm.

56. $7^2 = 5^2 + \ell^2 \Rightarrow \ell^2 = 24 \Rightarrow \ell = 2\sqrt{6} \approx 4.9$
Length is about 4.9 cm.

PAGE 516, LOOK BEYOND

57. Anthony can note where point A is, and the pace off or measure the distance to it from where he is standing. This distance should be equivalent to the width of the river.

58. $\overline{NA} \cong \overline{NR}$

59. ASA proves that $\triangle SNR \cong \triangle SNA$, because $\angle ASN \cong \angle RSN$, $\angle SNA \cong \angle SNR$ (because both are right angles), and $\overline{SN} \cong \overline{SN}$.

60. Sample answer:
It is difficult to keep your head at the same angle, and it's difficult to know when you are keeping your eyes level to pick out appropriate points.

5. $\triangle ABC \sim \triangle LKJ$; SSS

6. $\triangle EFD \sim \triangle QPR$; AA

7. $\triangle GHI \sim \triangle MOA$; SAS

PAGES 521–523, PRACTICE AND APPLY

8. $\triangle TXZ \sim \triangle IJK$; SSS

9. $\triangle GHF \sim \triangle ECD$; AA

10. Cannot be proven similar.

11. $\triangle LJK \sim \triangle CBA$; SSS

12. Cannot be proven similar.

13. RS and LN can be calculated because the triangles are right triangles. $RS = 5$ and $LN = 3.45$
$\triangle RTS \sim \triangle NLM$; SSS Similarity Theorem

14. Yes, by AA

15. Yes, by SAS (Each has a right angle.)

16. Rectangle $QRST \sim$ rectangle $WXYZ$ because $\frac{36}{20} = \frac{27}{15} = 1.8$. Neither rectangle is similar to $HIJK$ because $\frac{40}{20} \neq \frac{36}{15}$ and $\frac{40}{36} \neq \frac{36}{27}$.

17. $\triangle XYZ \sim \triangle RST$. By the Angle-Sum Theorem, m$\angle S = 32°$. Also, $\angle Z \cong \angle T$ since both are right angles. Then, apply the AA Similarity Postulate.

18. Yes. By the Corresponding Angles Postulate $\angle D \cong \angle B$. Also, $\angle A$ is shared by both $\triangle ABC$ and $\triangle ADE$, so by the AA Similarity Postulate, $\triangle ABC \sim \triangle ADE$.

19. Yes. AB and AC are proportional to AD and AE, respectively. Also, the included angle, $\angle A$, is shared by both $\triangle ABC$ and $\triangle ADE$, so by the SAS Similarity Theorem, $\triangle ABC \sim \triangle ADE$.

20. Yes. Since $AD = DB$ and $AE = EC$, then $AB = 2AD$ and $AC = 2AE$. Thus, AB and AC are proportional to AD and AE, respectively. Also, the included angle, $\angle A$, is shared by both $\triangle ABC$ and $\triangle ADE$, so by the SAS Similarity Theorem, $\triangle ABC \sim \triangle ADE$.

21. Not enough. Need $\angle D \cong \angle B$ to ensure that $\triangle ABC \sim \triangle ADE$.

22. Yes. $\frac{10}{5} = \frac{8}{4} = \frac{6}{3} = 2$. Thus, by the SSS Similarity Theorem, $\triangle NOP \sim \triangle KLM$.

23. No. Because $\triangle QRS$ is not a right triangle; and thus, $\triangle QRS$ is not congruent to a dilation of $\triangle KLM$. So, they are not similar.

24. Since $\frac{TU}{KL} = \frac{3}{2}$ then $2TU = 3KL \Rightarrow TU = \frac{3}{2}KL$. Similarly, $UV = \frac{3}{2}LM$ and $VT = \frac{3}{2}MK$. Thus,
$$\frac{TU + UV + VT}{KL + LM + MK} = \frac{\frac{3}{2}(KL + LM + MK)}{KL + LM + MK} = \frac{3}{2}.$$
So, the ratio is $\frac{3}{2}$.

25. Since $\overline{GH} \parallel \overline{BC}$, $\angle AGH \cong \angle ABC$ by the Corresponding Angles Postulate. Also, $\angle A$ is shared by both triangles, so by the AA Similarity Postulate, $\triangle AGH \sim \triangle ABC$.

26. By Exercise 25, $\triangle AGH \sim \triangle ABC$. This means that the sides of $\triangle AGH$ are proportional to the sides of $\triangle ABC$. But, it is given that the sides of $\triangle ABC$ are proportional to the sides of $\triangle DEF$. Thus, the sides of $\triangle AGH$ are proportional to the sides of $\triangle DEF$, or $\frac{AG}{DE} = \frac{GH}{EF} = \frac{HA}{FD}$. But, $AG = DE$ which means that $1 = \frac{GH}{EF} = \frac{HA}{FD}$ or $EF = GH$ and $HA = FD$. So, it follows that $\triangle AGH \cong \triangle DEF$ by SSS.

27. Since $\triangle AGH \cong \triangle DEF$, then $\triangle AGH \sim \triangle DEF$. But, $\triangle AGH \sim \triangle ABC$, so $\triangle ABC \sim \triangle DEF$ by the Polygon Similarity Postulate.

28. Since $\overline{ST} \parallel \overline{VW}$, $\angle UST \cong \angle UVW$ by the Corresponding Angles Postulate. Also, $\angle U$ is shared by both triangles, so by the AA Similarity Postulate, $\triangle UST \sim \triangle UVW$.

29. By Exercise 28, $\triangle UST \sim \triangle UVW$. This means that $\overline{US}$ and $\overline{UT}$ are proportional to $\overline{UV}$ and $\overline{UW}$, respectively. That is, $\frac{US}{UV} = \frac{UT}{UW}$. But, it is given that $\frac{UV}{XY} = \frac{UW}{XZ}$. So, $\frac{US}{XY} = \frac{UT}{XZ}$. But, $US = XY$ which means that $UT = XZ$. Also $\angle U \cong \angle X$, so $\triangle UST \cong \triangle XYZ$ by SAS.

30. Since $\triangle UST \cong \triangle XYZ$, then $\triangle UST \sim \triangle XYZ$. But, $\triangle UST \sim \triangle UVW$ so $\triangle UVW \sim \triangle XYZ$ by the Polygon Similarity Postulate.

31. $\angle A \cong \angle A$ and $\angle B \cong \angle B$ by the Symmetric Property of Congruence, so $\triangle ABC \sim \triangle ABC$ by the AA Similarity Postulate.

32. If $\triangle ABC \sim \triangle DEF$ then $\angle A \cong \angle D$ and $\angle B \cong \angle E$ by the Polygon Similarity Postulate. But, this means that $\angle D \cong \angle A$ and $\angle E \cong \angle B$ by the Reflexive Property of Congruence. So, $\triangle DEF \sim \triangle ABC$ by the AA Similarity Postulate.

33. If $\triangle ABC \sim \triangle DEF$, then $\angle A \cong \angle D$ and $\angle B \cong \angle E$, and if $\triangle DEF \sim \triangle GHI$, then $\angle D \cong \angle G$ and $\angle E \cong \angle H$. So, it follows that $\angle A \cong \angle G$ and $\angle B \cong \angle H$ by the Transitive Property of Congruence. Thus, by the AA Similarity Postulate, $\triangle ABC \sim \triangle GHI$.

34. Since all four sides of $STUV$ are proportional to all four sides of $WXYZ$, then WX and XY are proportional to ST and TU, respectively. $\angle T$ and $\angle X$ are each included angles, so by the SAS similarity postulate, $\triangle WXY \sim \triangle STU$. Similarly, it follows that $\triangle WYZ \sim \triangle SUV$. Appending $\triangle WXY$ and $\triangle STU$ to $\triangle WYZ$ and $\triangle SUV$, respectively, two similar quadrilaterals are formed.

35. Yes. Since Tony copied two angles from the original triangle, then by the AA Similarity Postulate, the two triangles are similar.

36. Yes. Since Biata multiplied each length by 5 to get the sides of her triangle, then by the SSS Similarity Theorem, the two triangles are similar.

37. Yes. Since Miki copied an angle which is adjacent to two sides of her triangle, of which the lengths are 5 times the lengths of two sides of the original triangle, then by the SAS Similarity Theorem, the two triangles are similar.

38. No. Even though the lengths of two sides of George's triangle are 5 times the lengths of two sides of the original triangle, and he copied an angle from the original triangle, the angle is not adjacent to those two sides. So, the SAS Similarity Theorem cannot be applied.

PAGE 524, LOOK BACK

39. Since $j \parallel k$, corresponding angles are congruent, so the angle supplementary to x is 45°. So, $x + 45 = 180 \Rightarrow x = 135$.

40. y is supplementary to x, so $x + y = 180$ $\Rightarrow 135 + y = 180 \Rightarrow y = 45$.

41. $\angle CAB$ is supplementary to a 120° angle, so $m\angle CAB = 60°$. But $\overline{AB} \cong \overline{BC}$, so $\angle CAB \cong \angle ACB$. Thus $m\angle ACB = 60°$. By the Triangle Sum Theorem, $m\angle CAB + m\angle ACB + z = 180$ $\Rightarrow 60 + 60 + z = 180 \Rightarrow z = 60$.

42. Since $j \parallel k$ then the 120° angle and the angle with mesaure $z + v$ are congruent corresponding angles. Thus $z + v = 120 \Rightarrow 60 + v = 120 \Rightarrow v = 60$.

43.

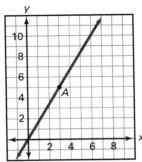

$$m = \frac{5 - 0}{3 - 0} = \frac{5}{3} \Rightarrow \text{Slope is } \frac{5}{3}.$$

44. $d = \sqrt{(3 - 0)^2 + (5 - 0)^2} = \sqrt{3^2 + 5^2} = \sqrt{34} \approx 5.83$

45.

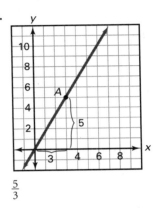

$\frac{5}{3}$

46. Sample answer.

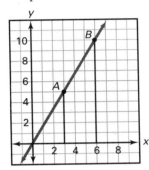

Let B be the point $(6, 10)$. Then the ratio is $\frac{10}{6} = \frac{5}{3}$.

47. Yes. Both have a right angle and share an angle with its vertex at the origin. Thus, by the AA Similarity Postulate, both are similar.

PAGE 524, LOOK BEYOND

48.

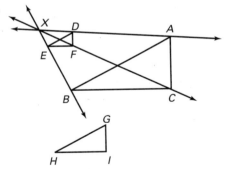

49.

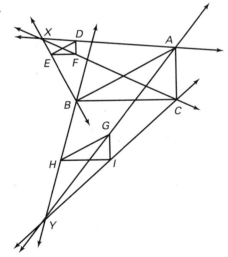

50.

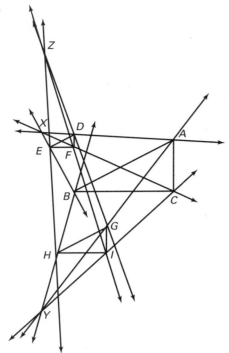

51.

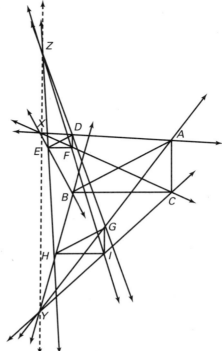

They are collinear.

8.4   **PAGE 528, GUIDED SKILLS PRACTICE**

6. $\frac{2}{10} = \frac{3}{x} \Rightarrow 2x = 30 \Rightarrow x = 15.$

7. $\frac{9}{x} = \frac{6}{4} \Rightarrow 6x = 36 \Rightarrow x = 6.$

8. $\frac{5+4}{4} = \frac{x}{2} \Rightarrow \frac{9}{4} = \frac{x}{2} \Rightarrow 4x = 18 \Rightarrow x = 4.5.$

9. $\frac{DF}{DA} = \frac{EG}{EA} \Rightarrow \frac{DF}{4} = \frac{1}{5} \Rightarrow 5DF = 4 \Rightarrow DF = \frac{4}{5}.$

10. $\frac{DF}{FB} = \frac{EG}{GC} \Rightarrow \frac{\frac{4}{5}}{FB} = \frac{1}{2} \Rightarrow FB = \frac{8}{5}.$

11. $\triangle ADE \sim \triangle AFG$, so $\frac{AD}{DE} = \frac{AF}{FG} \Rightarrow \frac{4}{3} = \frac{\frac{24}{5}}{FG}$

$\Rightarrow 4FG = \frac{72}{5} \Rightarrow FG = \frac{18}{5}.$

12. $\triangle ADE \sim \triangle ABC$, so $\frac{AD}{DE} = \frac{AB}{BC} \Rightarrow \frac{4}{3} = \frac{\frac{32}{5}}{BC} \Rightarrow 4BC = \frac{96}{5} \Rightarrow BC = \frac{24}{5}$

PAGES 528–530, PRACTICE AND APPLY

13. $\frac{x}{24} = \frac{8}{20} \Rightarrow 20x = 192 \Rightarrow x = \frac{48}{5} = 9.6$

14. $\frac{3}{4} = \frac{6}{x} \Rightarrow 3x = 24 \Rightarrow x = 8$

15. $\frac{7}{7+x} = \frac{6}{15} \Rightarrow 6(7+x) = 105 \Rightarrow 42 + 6x = 105$
$\Rightarrow 6x = 63 \Rightarrow x = \frac{21}{2} = 10.5$

16. $\frac{8}{8+3} = \frac{6}{x} \Rightarrow \frac{8}{11} = \frac{6}{x} \Rightarrow 8x = 66 \Rightarrow x = \frac{33}{4} = 8.25$

17. $\frac{4}{x} = \frac{x}{9} \Rightarrow x^2 = 36 \Rightarrow x = 6$

18. $\frac{x+1}{x-2} = \frac{x-1}{x-3} \Rightarrow (x+1)(x-3) = (x-1)(x-2)$
$\Rightarrow x^2 - 2x - 3 = x^2 - 3x + 2$
$\Rightarrow -2x - 3 = -3x + 2 \Rightarrow x = 5$

19. $\dfrac{3x + 6}{2x} = \dfrac{9}{x + 1} \Rightarrow (3x + 6)(x + 1) = 18x$

$\Rightarrow 3x^2 + 9x + 6 = 18x$

$\Rightarrow x^2 - 3x + 2 = 0 \Rightarrow (x - 1)(x - 2) = 0$

$\Rightarrow x = 1$ or $x = 2$.

20. $\dfrac{5}{x + 1} = \dfrac{2x + 4}{3x - 1} \Rightarrow (2x + 4)(x + 1) = 5(3x - 1)$

$\Rightarrow 2x^2 + 6x + 4 = 15x - 5$

$\Rightarrow 2x^2 - 9x + 9 = 0 \Rightarrow (2x - 3)(x - 3) = 0$

$\Rightarrow x = \dfrac{3}{2}$ or $x = 3$

21. By the Side Splitting Theorem and the SAS Similarity Theorem, $\triangle BED \sim \triangle BAC$. Similarly, $\triangle ACB \sim \triangle FCD$. Then by the Transitive Property, $\triangle EBD \sim \triangle FDC$.

22. $\dfrac{6}{4} = \dfrac{3}{2}$ and $\dfrac{6 + 12}{4 + 8} = \dfrac{18}{12} = \dfrac{3}{2}$, so $\dfrac{NK}{KO} = \dfrac{PK}{KQ}$. The included angle $\angle K$, is shared by both $\triangle NKO$ and $\triangle PKQ$. Thus, $\triangle NKO \sim \triangle PKQ$ by the SAS Similarity Theorem.

$\dfrac{6}{4} = \dfrac{3}{2}$ and $\dfrac{6 + 12 + 3}{4 + 8 + 2} = \dfrac{21}{14} = \dfrac{3}{2}$, so $\dfrac{NK}{KO} = \dfrac{LK}{KM}$. Thus, by the SAS Similarity Theorem, $\triangle NKO \sim \triangle LKM$.

$\dfrac{6 + 12}{4 + 8} = \dfrac{18}{12} = \dfrac{3}{2}$ and $\dfrac{6 + 12 + 3}{4 + 8 + 2} = \dfrac{21}{14} = \dfrac{3}{2}$ so $\dfrac{PK}{KQ} = \dfrac{LK}{KM}$. Thus, by the SAS Similarity Theorem, $\triangle PKQ \sim \triangle LKM$.

23. Since $QRST$ is a parallelogram, $QR = ST$ and $SA = 10$.

$\dfrac{RA}{AB} = \dfrac{SA}{AT} \Rightarrow \dfrac{y}{12} = \dfrac{10}{8} \Rightarrow 8y = 120 \Rightarrow y = 15$.

Since $\overline{RQ} \parallel \overline{TS}$, then by the Side Splitting Theorem,

$\dfrac{20}{x} = \dfrac{y}{12} \Rightarrow \dfrac{20}{x} = \dfrac{15}{12} \Rightarrow 15x = 240 \Rightarrow x = 16$.

24. $\dfrac{DE}{EA} = \dfrac{BC}{AB} \Rightarrow \dfrac{3}{x} = \dfrac{4}{12} \Rightarrow 4x = 36 \Rightarrow x = 9$.

$\triangle EAB \sim \triangle DAC$, so $\dfrac{AE}{AD} = \dfrac{EB}{DC} \Rightarrow \dfrac{x}{x + 3} = \dfrac{y}{20}$

$\Rightarrow \dfrac{9}{12} = \dfrac{y}{20} \Rightarrow 12y = 180 \Rightarrow y = 15$.

25. By the Two-Transversal Proportionality Corollary,

$\dfrac{7}{2.5} = \dfrac{6x + 3y}{3}$ and

$\dfrac{2.5}{5} = \dfrac{3}{5x + 2y}$, or $15x + 7.5y = 21$ and

$12.5x + 5y = 15$. Solving this set of equations,

$x = \dfrac{2}{5}$ and $y = 2$.

26. Sample answer:

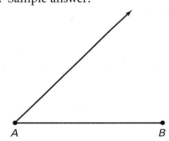

A $\quad$ B

27. Sample answer:

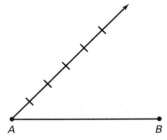

A $\quad$ B

28. Sample answer:

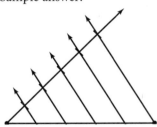

29. By the Two-Transversal Proportionality Corollary, the parallel lines divide $\overline{AB}$ and the ray proportionally. But the ray was divided into five congruent parts, so $\overline{AB}$ is divided into five congruent parts.

30. Sample answer:

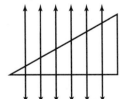

31.

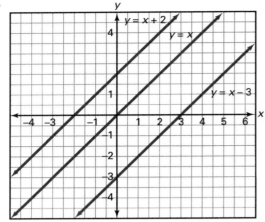

They all have a slope of 1.

32.

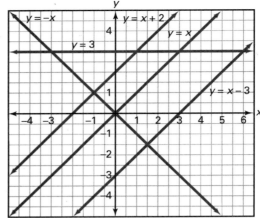

The slope of $y = -x$ is -1, the negative reciprocal of the slope of the other lines, which is 1.

33. $y = -x$ and $y = x$ intersect at $(0, 0)$ $y = -x$ and $y = x + 2$ intersect at $(-1, 1)$.
$d = \sqrt{(0 - (-1))^2 + (0 - 1)^2} \Rightarrow d = \sqrt{1^2 + 1^2}$
$\Rightarrow d = \sqrt{2} \approx 1.41$

34.

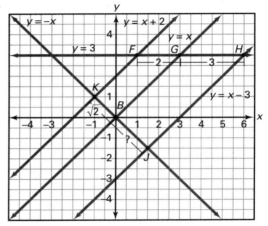

$\frac{KB}{BJ} = \frac{FG}{GH} \Rightarrow \frac{\sqrt{2}}{BJ} = \frac{2}{3} \Rightarrow 2BJ = 3\sqrt{2} \Rightarrow BJ = \frac{3\sqrt{2}}{2}$
$\Rightarrow$ The distance between $y = x$ and $y = x - 3$ is $\frac{3\sqrt{2}}{2}$ or about 2.12 units.

35. Given: $\frac{DB}{AD} = \frac{BC}{AE}$

Prove: $\overline{DE} \parallel \overline{BC}$

Proof:

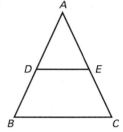

Statements	Reasons
$\frac{DB}{AD} = \frac{EC}{AE}$	Given
$1 + \frac{DB}{AD} = 1 + \frac{EC}{AE}$	Addition Property
$\frac{AD}{AD} + \frac{DB}{AD} = \frac{AE}{AE} + \frac{EC}{AE}$	Substitution Property
$\frac{AD + DB}{AD} = \frac{AE + EC}{AE}$	Addition of fractions
$AD + DB = AB$ $AE + EC = AC$	Segment Addition Postulate
$\frac{AB}{AD} = \frac{AC}{AE}$	Substitution Property
$\angle A \cong \angle A$	Reflexive Property
$\triangle ABC \sim \triangle ADE$	SAS Similarity Theorem
$\angle ADE \cong \angle B$	Polygon Similarity Postulate
$\overline{DE} \parallel \overline{BC}$	Converse of the Corresponding Angles Postulate

36. Since $\overline{FE} \parallel \overline{CB}$, $\frac{AB}{AE} = \frac{AC}{AF}$ by the Side Splitting Theorem. $\overline{GF} \parallel \overline{DC}$, so $\frac{AC}{AF} = \frac{AD}{AG}$ by the Side-Splitting Theorem. By the Transitive Property of Equality, $\frac{AB}{AE} = \frac{AD}{AG}$. By the Exchange Property, $\frac{AB}{AD} = \frac{AE}{AG}$. Since $\frac{AE}{AG} = 1.3$ and $AD = 10$, $\frac{AB}{10} = 1.3$ so $AB = 13$ cm.

37. Extend the sides of the trapezoid to form a triangle, as shown.

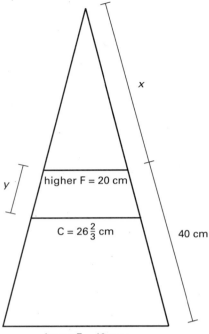

The string for the higher F is the midsegment of the triangle, so $x = 40$ cm.

Since the strings are all parallel, the triangles formed are all similar, so

$\frac{40}{40 + y} = \frac{20}{26\frac{2}{3}} \Rightarrow 1066\frac{2}{3} = 800 + 20y \Rightarrow 266\frac{2}{3} = 20y \Rightarrow y = 13\frac{1}{3} \Rightarrow$ The

string should be placed $13\frac{1}{3}$ cm from the higher F.

38. The length of the string for the note G is $\frac{8}{9} \cdot 40 = 35\frac{5}{9}$ cm. If the distance

from the string for the higher F to the string for the G is y, then $\frac{40}{40 + y}$

$= \frac{20}{35\frac{5}{9}} \Rightarrow 1422\frac{2}{9} = 800 + 20y \Rightarrow 622\frac{2}{9} = 20y \Rightarrow y = 31\frac{1}{9} \Rightarrow$ The string

should be placed $31\frac{1}{9}$ cm from the higher F.

39.

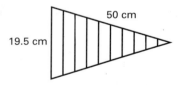

By the Side-Splitting Theorem, the lengths of the inner segments are 1.95, 3.9, 5.85, 7.8, 9.75, 11.7, 13.65, 15.6, and 17.55 cm. So the total length of all the cross-segments of one triangular sector is 107.25 cm. Each triangular sector also has a radial segment with a length of 50 cm, so the amount of silk for each triangular sector is 107.25 + 50 = 157.25 cm. Thus, the amount of silk used to build the web is 16(157.25) = 2516 cm.

40.

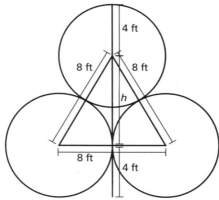

$h^2 + 4^2 = 8^2 \Rightarrow h^2 = 48 \Rightarrow h = \sqrt{48} = 4\sqrt{3}.$

Height of stack $= 4 + h + 4 = 4 + 4\sqrt{3} + 4 = 8 + 4\sqrt{3} \approx 14.93$

$\Rightarrow$ The height of the stack is about 14.93 ft.

41.

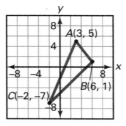

42. $d(A, B) = \sqrt{(3 - 6)^2 + (5 - 1)^2} = \sqrt{(-3)^2 + (4)^2} = \sqrt{9 + 16} = \sqrt{25} = 5$

$d(B, C) = \sqrt{(6 - (-2))^2 + (1 - (-7))^2} = \sqrt{8^2 + 8^2} = \sqrt{64 + 64} = \sqrt{128} = 8\sqrt{2}$

$d(C, A) = \sqrt{(-2 - 3)^2 + (-7 - 5)^2} = \sqrt{(-5)^2 + (-12)^2} = \sqrt{25 + 144} = \sqrt{169} = 13$

So, the perimeter of $\triangle ABC$ is $5 + 8\sqrt{2} + 13 = 18 + 8\sqrt{2} \approx 29.31$ units.

43. Midpoint of $A(3, 5)$ and $B(6, 1)$ is $\left(\frac{3 + 6}{2}, \frac{5 + 1}{2}\right) = \left(\frac{9}{2}, 3\right) \Rightarrow D\left(\frac{9}{2}, 3\right)$

Midpoint of $B(6, 1)$ and $C(-2, -7)$ is $\left(\frac{6 - 2}{2}, \frac{1 - 7}{2}\right) = (2, -3) \Rightarrow E(2, -3)$

Midpoint of $C(-2, -7)$ and $A(3, 5)$ is $\left(\frac{-2 + 3}{2}, \frac{-7 + 5}{2}\right) = \left(\frac{1}{2}, -1\right) \Rightarrow F\left(\frac{1}{2}, -1\right)$

44. $D\left(\frac{9}{2}, 3\right), E(2, -3), F\left(\frac{1}{2}, -1\right)$

$d(D, E) = \sqrt{\left(\frac{9}{2} - 2\right)^2 + (3 - (-3))^2} = \sqrt{\left(\frac{5}{2}\right)^2 + (6)^2} = \sqrt{\frac{25}{4} + 36} = \sqrt{\frac{169}{4}} = \frac{13}{2}$

$d(E, F) = \sqrt{\left(2 - \frac{1}{2}\right)^2 + (-3 - (-1))^2} = \sqrt{\left(\frac{3}{2}\right)^2 + (-2)^2} = \sqrt{\frac{9}{4} + 4} = \sqrt{\frac{25}{4}} = \frac{5}{2}$

$d(F, D) = \sqrt{\left(\frac{1}{2} - \frac{9}{2}\right)^2 + (-1 - 3)^2} = \sqrt{(-4)^2 + (-4)^2} = \sqrt{16 + 16} = \sqrt{32} = 4\sqrt{2}$

So, the perimeter of $\triangle DEF$ is $\frac{13}{2} + \frac{5}{2} + 4\sqrt{2} = 9 + 4\sqrt{2} \approx 14.66$ units.

45.

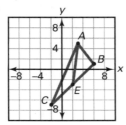

Midpoint of $B(6, 1)$ and $C(-2, -7)$ is $E(2, -3)$, so

$d(A, E) = \sqrt{(3 - 2)^2 + (5 - (-3))^2} = \sqrt{(1)^2 + (8)^2} = \sqrt{1 + 64} = \sqrt{65} \approx 8.06$

46.

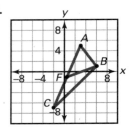

Midpoint of $C(-2, -7)$ and $A(3, 5)$ is $F\left(\frac{1}{2}, -1\right)$, so

$$d(B, F) = \sqrt{\left(6 - \frac{1}{2}\right)^2 + (1 - (-1))^2} = \sqrt{\left(\frac{11}{2}\right)^2 + (2)^2} = \sqrt{\frac{121}{4} + 4} = \sqrt{\frac{137}{2}} = \frac{\sqrt{137}}{2} \approx 5.85$$

47.

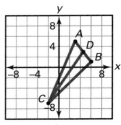

Midpoint of $A(3, 5)$ and $B(6, 1)$ is $D\left(\frac{9}{2}, 3\right)$, so

$$d(C, D) = \sqrt{\left(-2 - \frac{9}{2}\right)^2 + (-7 - 3)^2} = \sqrt{\left(-\frac{13}{2}\right)^2 + (-10)^2} = \sqrt{\frac{169}{4} + 100} = \sqrt{\frac{569}{4}} = \frac{\sqrt{569}}{2} \approx 11.93$$

PAGE 534, LOOK BEYOND

48. Yes. Each triangle has a right angle and a 30° angle. Thus, by the AA Similarity Postulate, they are similar.

49. Since the sum of the angles in any triangle is 180°, the remaining angle in both triangles measures $180° - 90° - 30° = 60°$. Then by the 30–60–90 Triangle Theorem, the length of the hypotenuse is 2 times the length of the shorter leg. The ratio is $\frac{1}{2}$ for both triangles.

50. As in exercise 49, the triangle is a 30–60–90 triangle, so the ratio is also $\frac{1}{2}$.

8.5 PAGE 537, GUIDED SKILLS PRACTICE

5. Since each triangle has a right angle and vertical angles are congruent, the triangles are similar by the AA Similarity Postulate. Let x be the width of the lake. Then, $\frac{x}{10} = \frac{5.4}{3} \Rightarrow 3x = 54 \Rightarrow x = 18$
The width of the lake is 18 km.

6. Since each triangle has a right angle and shares a non-right angle, then the triangles are similar by the AA Similarity Postulate.

Let x be the width of the lake. Then, $\frac{x}{x + 10} = \frac{9}{14}$
$\Rightarrow 9(10 + x) = 14x \Rightarrow 90 + 9x = 14x$
$\Rightarrow 90 = 5x \Rightarrow x = 18$
The width of the lake is 18 km.

7. Let x be the height of the street light. Then,
$\frac{x}{170} = \frac{840}{310} \Rightarrow 310x = 142,800 \Rightarrow x \approx 460.65$
The street light is about 460.65 cm tall.

8. $\frac{x}{2.4} = \frac{7.2}{3.6} \Rightarrow 3.6x = 17.28 \Rightarrow x = 4.8$

9. $\frac{1}{x} = \frac{1.5}{3.3} \Rightarrow 1.5x = 3.3 \Rightarrow x = 2.2$

10. $\frac{h}{22.5} = \frac{81 + 27}{27} \Rightarrow \frac{h}{22.5} = \frac{108}{27} \Rightarrow 27h = 2430$
$\Rightarrow h = 90 \Rightarrow$ Height is 90 ft

11. $\frac{h}{20} = \frac{10}{12} \Rightarrow 12h = 200 \Rightarrow h = 16\frac{2}{3}$
$\Rightarrow$ Height is $16\frac{2}{3}$ ft

12. $\frac{h}{18} = \frac{20}{12} \Rightarrow 12h = 360 \Rightarrow h = 30 \Rightarrow$ Height is 30 ft

13. $\frac{6}{14} = \frac{9}{h} \Rightarrow 6h = 126 \Rightarrow h = 21$
$\Rightarrow$ Height is 21 ft

14. $\frac{x}{15} = \frac{115}{23} \Rightarrow 23x = 1725 \Rightarrow x = 75$

15. $\frac{12}{x} = \frac{15.6}{13} \Rightarrow 15.6x = 156 \Rightarrow x = 10$

16. $\frac{7.2}{x} = \frac{8}{10} \Rightarrow 8x = 72 \Rightarrow x \Rightarrow 9$

17. $\frac{x}{40} = \frac{35}{56} \Rightarrow 56x = 1400 \Rightarrow x = 25$

18. $\frac{RW}{AE} = \frac{TW}{CE} \Rightarrow \frac{RW}{20} = \frac{10}{16}$
$\Rightarrow 16RW = 200 \Rightarrow RW = 12.5$

19. $\frac{RW}{AE} = \frac{SW}{BE} \Rightarrow \frac{12.5}{20} = \frac{9}{BE}$
$\Rightarrow 12.5BE = 180 \Rightarrow BE = 14.4$

20. $\frac{WV}{ED} = \frac{WT}{EC} \Rightarrow \frac{WV}{28} = \frac{10}{16}$
$\Rightarrow 16WV = 280 \Rightarrow WV = 17.5$

21. $\frac{MK}{JK} = \frac{16}{12} = \frac{4}{3}, \frac{KJ}{KL} = \frac{12}{9} = \frac{4}{3}$, and $\angle MKJ \cong \angle JKL$
since both are right angles. Thus, $\triangle MKJ \sim \triangle JKL$
by the SAS Similarity Theorem.

22. $LJ^2 = LK^2 + KJ^2 \Rightarrow LJ^2 = 9^2 + 12^2$
$\Rightarrow LJ^2 = 81 + 144 \Rightarrow LJ^2 = 225 \Rightarrow LJ = 15$
$\Rightarrow$ Length of hypotenuse of $\triangle JKL$ is 15.
$JM^2 = JK^2 + KM^2 \Rightarrow JM^2 = 12^2 + 16^2$
$\Rightarrow JM^2 = 144 + 256 \Rightarrow JM^2 = 400 \Rightarrow JM = 20$
$\Rightarrow$ Length of hypotenuse of $\triangle MKJ$ is 20.

23. $\frac{1}{2}(LK)(JK) = \frac{1}{2}(9)(12) = 54 \Rightarrow$ Area of $\triangle JKL$ is 54.
$\frac{1}{2}(JK)(KM) = \frac{1}{2}(12)(16) = 96$
$\Rightarrow$ Area of $\triangle MKJ$ is 96.

24. In $\triangle JKL$, use $\overline{LJ}$ as the base in the area formula
$A = \frac{1}{2}bh$ and solve for h.
$A = \frac{1}{2}bh \Rightarrow 54 = \frac{1}{2}(LJ)h \Rightarrow 54 = \frac{1}{2}(15)h \Rightarrow h = 7.2$
In $\triangle MKJ$, use $\overline{MJ}$ as the base in the area formula
$A = \frac{1}{2}bh$ and solve for h.
$A = \frac{1}{2}bh \Rightarrow 96 = \frac{1}{2}(MJ)h \Rightarrow 96 = \frac{1}{2}(20)h \Rightarrow h = 9.6$

25. Yes, $\frac{9.6}{7.2} = \frac{4}{3} = \frac{MK}{JK}$.

26. Since $\triangle ABC \sim \triangle XYZ$, $\angle ABC \cong \angle XYZ$ by the
Polygon Similarity Postulate. $\angle ABD$ and $\angle XYW$ are
supplementary to $\angle ABC$ and $\angle XYZ$, respectively,
so $\angle ABD \cong \angle XYW$. $\angle ADB \cong \angle XWY$, because
both are right angles. Thus, $\triangle ABD \sim \triangle XYW$
by the AA Similarity Postulate. So $\frac{AD}{XW} = \frac{AB}{XY}$ by the
Polygon Similarity Postulate.

27. Given

28. Polygon Similarity Postulate

29. Polygon Similarity Postulate

30. $m\angle BAC$
$m\angle YXZ$

31. $m\angle BAD = m\angle YXW$

32. $\triangle ABD \sim \triangle XYW$

33. $\frac{AD}{XW} = \frac{AB}{XY}$

34.

Statements	Reasons
$\overline{KN} \parallel \overline{MJ}$	By construction
$\dfrac{KM}{LM} = \dfrac{NJ}{JL}$	Side Splitting Theorem
$\angle 1 \cong \angle 4$	Alternate Interior Angles Theorem
$\angle 2 \cong \angle 3$	Corresponding Angles Postulate
$\angle 1 \cong \angle 2$	Given (Definition of angle bisector)
$\angle 3 \cong \angle 4$	Transitive Property of Congruence
$NJ = JK$	Converse of the Isosceles Triangle Theorem
$\dfrac{KM}{LM} = \dfrac{JK}{JL}$	Substitution Property

35. Definition of median

36. Definition of midpoint

37. Substitution Property and Division Property

38. Polygon Similarity Postulate

39. Substitution Property

40. $\triangle EFH \sim \triangle STV$

41. $\dfrac{EH}{SV} = \dfrac{EF}{ST}$

42. $\dfrac{\text{Height of person}}{\text{Distance from person to mirror}} = \dfrac{\text{Height of dinosaur}}{\text{Distance from dinosaur to mirror}}$

$\dfrac{5}{3} = \dfrac{x}{12} \Rightarrow 3x = 60 \Rightarrow x = 20$

The dinosaur is 20 ft tall.

43. Since vertical angles are congruent, $\angle BAC \cong \angle B'AC'$. Since $\overline{BC} \parallel \overline{B'C'}$, $\angle B \cong \angle B'$ by the Alternate Interior Angles Theorem. Thus, by the AA Similarity Postulate, $\triangle ABC \sim \triangle AB'C'$.

44. $\dfrac{6}{15} = \dfrac{2}{5}$.

45. Let x be the height of the image. Then, $\dfrac{10}{x} = \dfrac{25}{5} \Rightarrow 25x = 50 \Rightarrow x = 2$
The image was 2 cm tall.

46. Arrange them so that the distance from the object to the lens is the same as the distance from the image to the lens.

47. Arrange them so that the distance from the image to the lens is 20 times greater than the distance from the object to the lens.

PAGE 541, LOOK BACK

48. $A = 4^2 + 4\left(\dfrac{\pi(2)^2}{2}\right) = 16 + 8\pi \approx 41.13$

49. $A = \dfrac{1}{2}(2)(2) + 1 + \dfrac{1}{2}(1)(2) = 2 + 1 + 1 = 4$

50. The diagonal of the rectangle has length 5 units. Both the shaded triangles and the unshaded triangles are similar to the 3–4–5 triangle by the AA Similarity Postulate. Let x = length of the shorter legs of the shaded triangles. Then, $\dfrac{x}{3} = \dfrac{3}{5}$ $\Rightarrow 5x = 9 \Rightarrow x = \dfrac{9}{5}$. Let y = length of the longer legs of the shaded triangles. Then, $\dfrac{y}{4} = \dfrac{3}{5}$ $\Rightarrow 5y = 12 \Rightarrow y = \dfrac{12}{5}$.

$A = 2\left[\dfrac{1}{2}\left(\dfrac{12}{5}\right)\left(\dfrac{9}{5}\right)\right] = \dfrac{108}{25} = 4.32$

51. Let h be the height of the triangle with length of base equal to 3. Then $\dfrac{h}{3} = \dfrac{h+2}{5} \Rightarrow 5h = 3(h+2)$ $\Rightarrow h = 3$. So, $A = \dfrac{1}{2}(3)(3) + 2\left[\dfrac{1}{2}(1)(2)\right] = \dfrac{9}{2} + 2$ $= \dfrac{13}{2}$

52. $V = \dfrac{4}{3}\pi(7)^3 \Rightarrow V = \dfrac{1372\pi}{3} \approx 1436.76$

53. Area of base $= \dfrac{1}{2}(3)(4) = 6$
$V = \dfrac{1}{3}Bh \Rightarrow V = \dfrac{1}{3}(6)(5) = 10$

54. The angles are alternate interior angles where the parallel lines are the sun's rays and the transversal is the radius from the center of Earth to Alexandria.

56. relative error $= \dfrac{\text{estimate} - \text{actual value}}{\text{actual value}}$

$\dfrac{27,600 - 24,900}{24,900} = \dfrac{2700}{24,900} \approx 0.11$

The relative error is 0.11 or 11%.

55. Let $x =$ circumference of Earth.
$\dfrac{7.5°}{360°} = \dfrac{575 \text{ mi}}{x} \Rightarrow 7.5x = 207,000 \Rightarrow x = 27,600$
Eratosthenes' estimate was 27,600 miles.

8.6 | PAGE 547, GUIDED SKILLS PRACTICE

6. The ratio of the sides of the larger parallelogram to the sides of the smaller parallelogram is $\frac{2}{1}$, so the ratio of the areas is $\left(\frac{2}{1}\right)^2 = \frac{4}{1}$.

7. The ratio of the linear dimensions of the larger pyramid to the smaller pyramid is $\frac{2}{1}$, so the ratio of the volumes is $\left(\frac{2}{1}\right)^3 = \frac{8}{1}$.

8. The volume of the animal would increase by a factor of $(3)^3$ or 27, and the cross-sectional area would increase by a factor of 9.

PAGES 548–550, PRACTICE AND APPLY

9. Since the ratio of the corresponding sides is $\frac{7+3}{3} = \frac{10}{3}$, the ratio of the perimeters is $\frac{10}{3}$.

10. $\left(\frac{10}{3}\right)^3 = \frac{100}{9}$

11. The ratio of the perimeters is $\frac{5}{4}$, so the perimeter of $VWXYZ$ is $24 \cdot \frac{5}{4} = 30$ m.

12. The ratio of the areas is $\left(\frac{5}{4}\right)^2 = \frac{25}{16}$, so the area of $VWXYZ$ is $50 \cdot \frac{25}{16} = 78.125$ m².

13. The ratio of the perimeters $\frac{4}{6} = \frac{2}{3}$, so the perimeter of $PQRSTU$ is $42 \cdot \frac{2}{3} = 28$ m. The ratio of the areas is $\left(\frac{2}{3}\right)^2 = \frac{4}{9}$, so the area of $PQRSTU$ is $96 \cdot \frac{4}{9} = \frac{128}{3} \approx 42.67$ m².

14. $\frac{7}{5}$

15. $\left(\frac{7}{5}\right)^2 = \frac{49}{25}$

16. $\left(\frac{7}{5}\right)^3 = \frac{343}{125}$

17. $\frac{5}{7}$

18. $\left(\frac{5}{7}\right)^2 = \frac{25}{49}$

19. $\left(\frac{5}{7}\right)^3 = \frac{125}{343}$

20. $\frac{7}{9}$

21. $\left(\frac{7}{9}\right)^3 = \frac{343}{729}$

22. $\left(\frac{7}{9}\right)^2 = \frac{49}{81}$

23. $\sqrt{\frac{16}{25}} = \frac{4}{5}$

24. $\sqrt{\frac{16}{25}} = \frac{4}{5}$

25. $\left(\frac{4}{5}\right)^3 = \frac{64}{125}$

26. $\sqrt{\frac{144}{169}} = \frac{12}{13}$

27. $\left(\frac{12}{13}\right)^3 = \frac{1728}{2197}$

28. $\sqrt{\frac{144}{169}} = \frac{12}{13}$

29. Ratio of the heights is $\sqrt{\frac{16}{49}} = \frac{4}{7}$. Let h be the height of the smaller cylinder, so $\frac{4}{7} = \frac{h}{21} \Rightarrow 7h = 84$ $\Rightarrow h = 12$ cm

30. Ratio of the heights is $\sqrt{\frac{225}{441}} = \frac{15}{21} = \frac{5}{7}$. Let h be the height of the smaller cone, so $\frac{5}{7} = \frac{h}{12} \Rightarrow$ $7h = 60 \Rightarrow h = 8\frac{4}{7}$ cm

31. Ratio of the volumes is $\left(\frac{5}{7}\right)^3 = \frac{125}{343}$
Let V be the volume of the larger cone, so $\frac{125}{343} = \frac{250}{V} \Rightarrow 125V = 85{,}750 \Rightarrow V = 686$ cm^3

32. $\sqrt[3]{\frac{64}{125}} = \frac{4}{5} \Rightarrow \left(\frac{4}{5}\right)^2 = \frac{16}{25}$

33. $\sqrt[3]{\frac{64}{125}} = \frac{4}{5}$

34. $\sqrt[3]{\frac{64}{125}} = \frac{4}{5}$

35. $\left(\frac{4}{5}\right)^2 = \frac{16}{25}$

36. Ratio of the lengths is $\frac{2}{8} = \frac{1}{4}$, so the ratio of the volumes is $\left(\frac{1}{4}\right)^3 = \frac{1}{64}$. Let V be the volume of the smaller cone, so $\frac{1}{64} = \frac{V}{288} \Rightarrow 64V = 288$ $\Rightarrow V = 4.5$ cm^3

37. Ratio of the areas is $\left(\frac{1}{4}\right)^2 = \frac{1}{16}$. Let A be the area of the base of the larger cone, so $\frac{1}{16} = \frac{3.6}{A}$ $\Rightarrow A = 57.6$ cm^2.

38. $A = \pi r^2 \Rightarrow 19.6 = \pi r^2 \Rightarrow r = \sqrt{\frac{19.6}{\pi}} \Rightarrow r \approx 2.5$ $\Rightarrow$ Diameter is about 5 cm.

39. Since the height scale factor is 2 then the cross-sectional radius scale factor is $\sqrt{8}$. So, radius of the leg bone should be $(2.5)\sqrt{8} \approx 7.07$ cm $\Rightarrow$ Diameter should be 14.14 cm.

40. Let r be the radius of the 350-foot tall tower. Then, $\frac{100}{26} = \frac{350}{r} \Rightarrow 100r = 9100 \Rightarrow r = 91$ $\Rightarrow$ Radius is 91 ft.

41. Ratio of diameters is $\frac{40{,}200}{10{,}000} = \frac{201}{50}$
Ratio between volumes is $\left(\frac{201}{50}\right)^3 = \frac{8{,}120{,}601}{125{,}000}$.

42. The area of the 8-inch pizza is 16π, so the ratio of the pizza area to its price is $\frac{16\pi}{4} = 4\pi$. The area of the 16-inch pizza is 64π, so the ratio of this pizza area to its price is $\frac{64\pi}{8} = 8\pi$. Thus, since the 16-inch pizza has the bigger ratio, then this one is a better deal.

43. No. Doubling all the dimensions would increase the volume of the box by a factor of $2^3 = 8$. To double the volume, all the dimensions would be increased by a factor of $\sqrt[3]{2} \approx 1.26$ or any one dimension could be doubled.

44. The ratio of the radii of the openings is $\frac{3}{2}$. If people use the same length of toothpaste, then one of the linear measures in the volume of toothpaste used remains the same. Thus, the amount of toothpaste used will increase by a factor of $\left(\frac{3}{2}\right)^2 = \frac{9}{4} = 2.25$.

45. Let x be the original length and y be the original width. Then $xy = 400$. So,
$(1.75)(x)(1.5)(y) = (1.75)(1.5)xy$
$\qquad\qquad\quad = (1.75)(1.5)(400)$
$\qquad\qquad\quad = 1050$
$\Rightarrow$ Area of expanded lot is 1050 m^2.

46. The ratio of the diameter of the standard basketball to the diameter of the promotional basketball is $\frac{9.5}{5}$, so the ratio of the surface areas is $\left(\frac{9.5}{5}\right)^2 = \frac{90.25}{25}$. Let x be the cost of the materials for the promotional basketball. Then, $\frac{90.25}{1.40} = \frac{25}{x}$ $\Rightarrow 90.25x = 35 \Rightarrow x \approx 0.39 \Rightarrow$ cost is about 39¢

47. The ratio of the volume of the smaller dome to the larger dome is $\left(\frac{3}{4}\right)^3 = \frac{27}{64}$. Let x be the storage capacity of the smaller (new) dome. Thus, $\frac{27}{64} = \frac{x}{3366} \Rightarrow 64x = 90{,}882 \Rightarrow x \approx 1420$ $\Rightarrow$ Estimated storage capacity is 1420 tons.

48. The measure of an interior angle of a regular polygon gives $165° = 180° - \frac{360°}{n}$
$\Rightarrow -15° = -\frac{360°}{n} \Rightarrow -15°n = -360° \Rightarrow n = 24$

49. Since the sum of the measures of the exterior angles of a polygon is 360°, then if n = the number of sides of the polygon, $40n = 360 \Rightarrow n = 9$. The polygon has 9 sides.

50. a. $5x + 4y = 18$
$4y = -5x + 18$
$y = -\frac{5}{4}x + 18$

b. $-10x + 8y = 21$
$8y = 10x + 21$
$y = \frac{5}{4}x + \frac{21}{8}$

c. $30x - 24y = 45$
$-24y = -30x + 45$
$y = \frac{5}{4}x + \frac{15}{8}$

Lines **b** and **c** are parallel.

51. Slope of the line $5x + 4y = 18$ is $-\frac{5}{4}$. So, equation of the line parallel to $5x + 4y = 18$ through the point $(3, 8)$ is $y - 8 = -\frac{5}{4}(x - 3)$.
$4y - 32 = -5(x - 3)$
$4y - 32 = -5x + 15$
$y = -\frac{5}{4}x + \frac{47}{4}$.

52. Slope of the line $5x + 4y = 18$ is $-\frac{5}{4}$. So, slope of the perpendicular line is $\frac{4}{5}$. So, equation of the line perpendicular to $5x + 4y = 18$ through the point $(3, 8)$ is $y - 8 = \frac{4}{5}(x - 3)$.
$5y - 40 = 4(x - 3)$
$5y - 40 = 4x - 12$
$y = \frac{4}{5}x + \frac{28}{5}$.

53. $\frac{x + 1}{4} = \frac{3x}{8}$
$(x + 1)(8) = (3x)(4)$
$8x + 8 = 12x$
$4x = 8$
$x = 2$

54. $\frac{2}{x + 2} = \frac{x - 1}{2}$
$(2)(2) = (x - 1)(x + 2)$
$x^2 + x - 6 = 0$
$(x + 3)(x - 2) = 0$
$x = -3$ or $x = 2$

55. $\frac{x - 1}{x + 1} = \frac{x + 2}{2x + 2}$
$(x - 1)(2x + 2) = (x + 2)(x + 1)$
$2x^2 - 2 = x^2 + 3x + 2$
$x^2 - 3x - 4 = 0$
$(x - 4)(x + 1) = 0$
$x = 4$ or $x = -1$
But, $x = -1$ is an impossibility. So, $x = 4$.

56. Let x be the length of the long side. Then
$\frac{5}{3} = \frac{x}{9}$
$3x = 45$
$x = 15$

57. $\frac{x - 1}{x - 9} = \frac{x + 13}{x - 2} \Rightarrow (x - 1)(x - 2) = (x + 13)(x - 9)$
$\Rightarrow x^2 - 3x + 2 = x^2 + 4x - 117$
$\Rightarrow 119 = 7x \Rightarrow x = 17$

58. The perimeter of each square is $\frac{1}{\sqrt{2}}$ times the perimeter of its preceding larger square. The sum of the perimeters of the first eight squares is about 12.8.
No, the sum of the perimeters is:

$$4 + \frac{4}{\sqrt{2}} + 2 + \frac{2}{\sqrt{2}} + 1 + \frac{1}{\sqrt{2}} + \frac{1}{2} + \frac{1}{2\sqrt{2}} + \ldots$$
$$= \left(4 + 2 + 1 + \frac{1}{2} + \ldots\right) + \left(\frac{4}{\sqrt{2}} + \frac{2}{\sqrt{2}} + \frac{1}{\sqrt{2}} + \frac{1}{2\sqrt{2}} + \ldots\right)$$
$$= 4\left(1 + \frac{1}{2} + \frac{1}{4} + \ldots\right) + \frac{4}{\sqrt{2}}\left(1 + \frac{1}{2} + \frac{1}{4} + \ldots\right)$$

Since the sum $1 + \frac{1}{2} + \frac{1}{4} + \ldots$ is 2(see pages 84–85), the sum is

$$4(2) + \frac{4}{\sqrt{2}}(2) \approx 13.66$$

CHAPTER REVIEW AND ASSESSMENT

1. The endpoints of the image are $(-6, 3)$ and $(9, 12)$.

2. The endpoints of the image are $(2, -1)$ and $(-3, -4)$.

3.

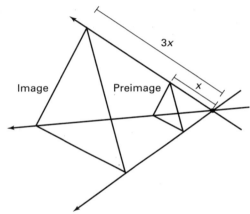

4.

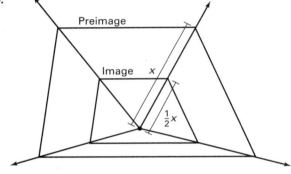

5. If the triangles were similar, the proportion $\frac{30}{18} = \frac{36}{21}$ would be true. By cross-multiplying, we obtain $30 \cdot 21 = 36 \cdot 18$ or $630 = 648$, so the triangles are not similar.

6. If the rectangles were similar, the proportion $\frac{4.2}{5.4} = \frac{5.6}{7.2}$ would be true. By cross-multiplying, we obtain $4.2 \cdot 7.2 = 5.6 \cdot 5.4$ or $30.24 = 30.24$, so the rectangles are similar.

7. $\frac{3}{x} = \frac{2}{3.6} \Rightarrow 2x = 10.8 \Rightarrow x = 5.4$

8. $\frac{18}{9} = \frac{x}{18} \Rightarrow 9x = 324 \Rightarrow x = 36$

9. $\frac{28}{21} = \frac{32}{24} \Rightarrow 28 \cdot 24 = 32 \cdot 21 \Rightarrow 672 = 672$
Also, the included angles are congruent, so the triangles are similar by the SAS Similarity Theorem.

10. $\frac{22}{11} = \frac{20}{10} \Rightarrow 22 \cdot 10 = 20 \cdot 11 \Rightarrow 220 = 220$
So, the triangles are similar by the SSS Similarity Theorem.

11. Two pairs of corresponding angles are congruent. So, the triangles are similar by the AA Similarity Postulate.

12. $\frac{18}{42} = \frac{12}{28} \Rightarrow 18 \cdot 28 = 12 \cdot 42 \Rightarrow 504 = 504$. Also the included right angles are congruent, so the triangles are similar by the SAS Similarity Theorem.

13. $\frac{15}{x} = \frac{20}{16} \Rightarrow 20x = 240 \Rightarrow x = 12$

14. $\frac{13}{7} = \frac{x}{21} \Rightarrow 7x = 273 \Rightarrow x = 39$

15. $\frac{15}{12} = \frac{12}{x} \Rightarrow 15x = 144 \Rightarrow x = 9.6$

16. $\frac{8.4}{x} = \frac{40}{14} \Rightarrow 40x = 117.6 \Rightarrow x = 2.94$

17. Let h be the height of the house. Then, $\frac{1.6}{3.5} = \frac{h}{17.5}$
$\Rightarrow 3.5h = 28 \Rightarrow h = 8 \Rightarrow$ House is 8 m tall.

18. Vertical angles are congruent, so $\angle WVX \cong \angle ZVY$ and alternate interior angles are congruent, so $\angle XWV \cong \angle VYZ$. So, $\triangle WXV \cong \triangle YZV$ by the AA Similarity Postulate. So,

$$\frac{XV}{ZV} = \frac{WX}{YZ}$$
$$\frac{1750}{750} = \frac{WX}{1250}$$
$$750WX = 2{,}187{,}500$$
$$WX = 2916\frac{2}{3}$$

Distance is $2916\frac{2}{3}$ m.

19. Since the triangles are similar, their altitudes are proportional, so $\frac{21}{12} = \frac{42}{x} \Rightarrow 21x = 504 \Rightarrow x = 24$.

20. Since the triangles are similar, their medians are proportional, so $\frac{30}{x} = \frac{32}{16} \Rightarrow 32x = 480 \Rightarrow x = 15$.

21. The ratio of their areas is $\left(\frac{5}{1}\right)^2 = \frac{25}{1}$.

22. The ratio of their areas is $\left(\frac{1}{2}\right)^2 = \frac{1}{4}$.

23. The ratio of their volumes is $\left(\frac{1}{10}\right)^3 = \frac{1}{1000}$.

24. The ratio of their volumes is $\left(\frac{1}{2}\right)^3 = \frac{1}{8}$.

25. Since the design is being enlarged by 120%, each of the linear dimensions is being scaled by a factor of 1.2. Thus the area is scaled by a factor of $(1.2)^2 = 1.44$. The area of the new design is $(10)(1.44) = 14.4$ in^2.

26. Let h be the height of the tree. Then, $\frac{1.8}{h} = \frac{1}{15}$
$\Rightarrow h = 27 \Rightarrow$ Height is 27 m. Since the tree is 30 m away from his house, then the tree will miss the house by 3 m.

27. 100 ft 9 in. = 1209 in. So, the ratio of the length of the sculpture to the length of the bat is $\frac{1209}{35}$. Thus, the ratio of the volumes is $\left(\frac{1209}{35}\right)^3 = \frac{1{,}767{,}172{,}329}{42{,}875} \approx 41{,}216.85$

1. $D(6, -1) = (-2 \cdot 6, -2 \cdot -1) = (-12, 2)$
$D(-4, 2) = (-2 \cdot -4, -2 \cdot 2) = (8, -4)$

2. $D(6, -1) = (0.5 \cdot 6, 0.5 \cdot -1) = (3, -0.5)$
$D(-4, 2) = (0.5 \cdot -4, 0.5 \cdot 2) = (-2, 1)$

3.

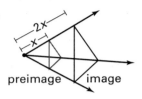

preimage   image

4. Yes; every angle is a right angle and $\frac{2.6}{7.8} = \frac{1.8}{5.4}$, so the corresponding side lengths are proportional.

5. No; $\frac{30}{15} \neq \frac{38}{18}$, so the corresponding side lengths are not proportional.

6. $\frac{4.6}{23} = \frac{x}{29}$
$x = \frac{(4.6)(29)}{23}$
$x = 5.8$

7. $\frac{x}{3} = \frac{18}{5}$
$x = \frac{(3)(18)}{5}$
$x = 10.8$ inches

8. AA Similarity Postulate

9. SAS Similarity Theorem

10. The triangles cannot be proven similar.

11. $\frac{12}{8} = \frac{x}{9}$
$x = \frac{(9)(12)}{18}$
$x = 6$

12. $\frac{x - 8}{x + 2} = \frac{x - 2}{x + 13}$
$x^2 + 5x - 104 = x^2 - 4$
$5x = 100$
$x = 20$

13. $\frac{1.2}{1.5} = \frac{x}{25}$
$x = \frac{(1.2)(25)}{1.5}$
$x = 20$ meters

14. $\frac{42}{28} = \frac{x}{18}$
$x = \frac{(42)(18)}{28}$
$x = 27$

15. $\frac{16}{20} = \frac{x}{45}$
$x = \frac{(16)(45)}{20}$
$x = 36$

16. $\frac{8^2}{1^2} = \frac{64}{1}$

17. $\frac{2^3}{5^3} = \frac{8}{125}$

18. $\frac{4^3}{3^3} = \frac{64}{27}$

1. $d(A, B) = |-1 - 2| = |-3| = 3$
   $d(B, A) = |2 - (-1)| = |2 + 1| = |3| = 3$
   C

2. The sum of first 15 odd numbers $= 15^2 = 225$.
   A

3. The number of lateral edges is 6.
   The number of faces is 8.
   B

4. The lateral area is $\pi\sqrt{2} \approx 4.44$.
   The volume is $\frac{1}{3}\pi \approx 1.05$.
   A

5. Parallel lines have the same slope, perpendicular lines have slopes that are negative reciprocals, and vertical lines have undefined slope. Since $0.2 = \frac{1}{5} \neq -\frac{1}{5}$, the lines are neither parallel, perpendicular, nor vertical.
   Choose **d.**

6. In congruent polygons, corresponding sides and angles are congruent. $\angle C$ corresponds to $\angle J$, $\angle A$ corresponds to $\angle M$, and $\overline{AE}$ corresponds to $\overline{ML}$.
   Choose **b.**

7. All rectangles have congruent diagonals. A trapezoid has congruent diagonals only if it is isosceles, a parallelogram has congruent diagonals only if it is a rectangle, and a rhombus has congruent diagonals only if it is a square.
Choose **b.**

8.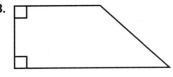

The trapezoid above has exactly two right angles. A rectangle has four right angles. If a parallelogram or a rhombus has a right angle, then it must have four right angles.
Choose **a.**

9. $\left(\frac{5+1}{2}, \frac{1+(-3)}{2}\right) = (3, -1)$
Choose **c.**

10. $x = 3, y = -1, z = 2$
Positive x is in the front, negative y is on the left, and positive z is on the top.
Choose **c.**

11. Let $x = 0$ and $z = 0$. Then, $5(0) + 2y - 0 = 6$
$\Rightarrow 2y = 6 \Rightarrow y = 3$. So, y-intercept is $(0, 3, 0)$.

12. $\frac{2+x}{3} = \frac{x}{2} \Rightarrow (2+x)(2) = 3x \Rightarrow 4 + 2x = 3x$
$\Rightarrow x = 4$

13. The AA Similarity Postulate as well as the SSS and SAS Similarity Theorems can be used to show that triangles are similar. An SSA combination does not guarantee that two tringles are similar.
Choose **c.**

14. $\left(\frac{3}{6}\right)^3 = \left(\frac{1}{2}\right)^3 = \frac{1}{8}$

15. The number of sides is 5, so the sum of the angles is $180(5-2) = 180(3) = 540°$.

16. The sum of the exterior angles of any polygon is $360°$.

17. $x^2 = 54^2 + 36^2 \Rightarrow x^2 = 4212 \Rightarrow x \approx 64.9$

18. By the Side Splitting Theorem,
$\frac{AD}{BD} = \frac{AE}{CE} \Rightarrow \frac{24}{BD} = \frac{32}{20} \Rightarrow 32BD = 480 \Rightarrow BD = 15$

CHAPTER 9

Circles

PAGE 569, GUIDED SKILLS PRACTICE

8. $m\widehat{AB} = 70°$
$m\widehat{BC} = 360° - (70° + 130°) = 160°$
$m\widehat{CA} = 130°$

9. The measure of $\angle APB = 70°$, and the radius of the circle is 3.
Thus the length of $\widehat{AB}$ is
$L = \dfrac{70°}{360°}(2\pi \cdot 3) \approx 3.67$

10.
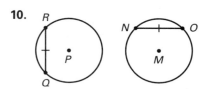

It is given that $\odot P$ and $\odot M$ are congruent, and that $\overline{QR}$ and $\overline{NO}$ are congruent. Form two triangles by adding radii $\overline{PR}$, $\overline{PQ}$, $\overline{MN}$, and $\overline{MO}$. Then $\triangle PQR$ and $\triangle MNO$ are congruent by SSS. Thus, $\angle QPR$ and $\angle NMO$ are congruent because CPCTC, and so the arcs $\widehat{QR}$ and $\widehat{NO}$ are congruent.

PAGES 570–571, PRACTICE AND APPLY

11. point P

12. Sample answer: $\overline{PB}$

13. $\overline{ED}$ or $\overline{AC}$

14. $\overline{AC}$

15. Sample answer: $\angle BPC$

16. $\widehat{ABC}$, $\widehat{ADC}$, or $\widehat{AEC}$

17. Sample answer: $\widehat{AB}$ and $\widehat{BC}$

18. Sample answer: $\widehat{DAB}$ and $\widehat{CEB}$

19. $\overline{AP}$ is a radius.

20. $\overline{AC}$ is a diameter.

21. $\overline{ED}$ is a chord.

22. $\angle APB$ is a central angle.

23. $m\widehat{TU} = 130°$

24. $m\widehat{TSU} = 360° - 130° = 230°$

25. $m\widehat{RT} = m\widehat{RS} + m\widehat{ST} = 90° + 36° = 126°$

26. $m\widehat{UR} = m\widehat{UV} + m\widehat{VR} = 78° + 26° = 104°$

27. $m\widehat{VS} = m\widehat{VR} + m\widehat{RS} = 26° + 90° = 116°$

28. $m\widehat{US} = m\widehat{UT} + m\widehat{TS} = 130° + 36° = 166°$

29. $m\widehat{SUV} = m\widehat{ST} + m\widehat{TU} + m\widehat{UV}$
$= 36° + 130° + 78°$
$= 244°$

30. $m\widehat{VTR} = 360° - \widehat{VR}$
$= 360° - 26°$
$= 334°$

31. $L = \dfrac{90°}{360°}(2\pi \cdot 10) = 5\pi \approx 15.71$

32. $L = \dfrac{60°}{360°}(2\pi \cdot 3) = \pi \approx 3.14$

33. $L = \dfrac{30°}{360°}(2\pi \cdot 120) = 20\pi \approx 62.83$

34. $M = \dfrac{360 \cdot L}{2\pi r}$

$M = \dfrac{360° \cdot 14°}{2\pi \cdot 70} = \dfrac{36°}{\pi} \approx 11.46°$

35. $M = \dfrac{360 \cdot L}{2\pi r}$

$M = \dfrac{360° \cdot 20°}{2\pi \cdot 100} = \dfrac{36°}{\pi} \approx 11.46°$

36. $M = \dfrac{360 \cdot L}{2\pi r}$

$M = \dfrac{360° \cdot 3°}{2\pi \cdot 15} = \dfrac{36°}{\pi} \approx 11.46°$

37. $M = \dfrac{360 \cdot L}{2\pi r}$

$M = \dfrac{360° \cdot 5°}{2\pi \cdot 25} = \dfrac{36°}{\pi} \approx 11.46°$

38. Since $\overline{AR}$, $\overline{BR}$, $\overline{XR}$, and $\overline{YR}$ are all radii of $\odot R$, $\overline{AR} \cong \overline{YR}$ and $\overline{BR} \cong \overline{XR}$. $\angle ARB \cong \angle YRX$ by the Vertical Angles Theorem, so $\triangle ARB \cong \triangle YRX$ by SAS. Thus, $XY = 5$ because CPCTC.

39. $m\angle AQB = \dfrac{360°}{5} = 72°$

40. $m\overarc{AE} = m\angle AQE = 72°$

41. $m\overarc{ACE} = 360° - 72° = 288°$

42. $L_{\overarc{AE}} = \dfrac{72°}{360°}(2\pi \cdot 2) = \dfrac{4\pi}{5} \approx 2.51$

43. $L_{\overarc{ACE}} = \dfrac{288°}{360°}(2\pi \cdot 2) = \dfrac{16\pi}{5} \approx 10.05$

44. In a circle, or in congruent circles, the chords of congruent arcs are congruent.
Proof:

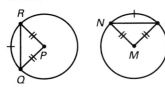

Since $\overarc{QR}$ is congruent to $\overarc{NO}$,
$\angle QPR$ is congruent to $\angle NMO$.
Since $\overline{RP}$ is congruent to $\overline{OM}$
and $\overline{QP}$ is congruent to $\overline{NM}$,
$\triangle QPR$ is congruent to $\triangle NMO$ by SAS.
Therefore $\overline{QR}$ is congruent to $\overline{NO}$ because CPCTC.

45. The radius of the fountain is 10 ft, so the tulips are planted in a circle of radius 11 with an angle measure of 6° between each tulip. The arc length between each tulip is

$L = \dfrac{6°}{360°}(2\pi \cdot 11)$

$= \dfrac{11\pi}{30} \approx 1.15$ ft.

46. Total number of students $= 450 + 375 + 400 + 325 = 1550$

Freshmen: $\frac{450}{1550} \approx 29\% \Rightarrow (0.29)(360°) = 104.4°$

Sophomores: $\frac{375}{1550} \approx 24\% \Rightarrow (0.24)(360°) = 86.4°$

Juniors: $\frac{400}{1550} \approx 26\% \Rightarrow (0.26)(360°) = 93.6°$

Seniors: $\frac{325}{1550} \approx 21\% \Rightarrow (0.21)(360°) = 75.6°$

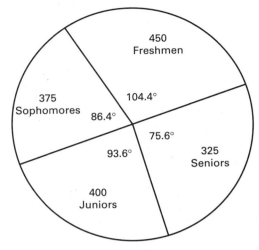

47. Sample answer:
Using the 45° angle shown, the dashed arc measures about 135°, or 3 times the length of the 45° arc. There are about 5 cars that would fit along the 45° arc, thus the length of the section of road is about $5 \cdot 16 \cdot 3 = 240$ ft.

PAGE 571, LOOK BACK

48. A sphere with radius r has volume

$$V_1 = \frac{4}{3}\pi r^3$$

A sphere with radius $2r$ has volume

$$V_2 = \frac{4}{3}\pi(2r)^3$$
$$= 8\left(\frac{4}{3}\pi r^3\right)$$
$$= 8V_1$$

Doubling the radius of a sphere has the effect of increasing its volume by a factor of 8.

49. $V = \frac{4}{3}\pi(1.45)^3 \approx 12.77$ in^3

50. $V = \ell \cdot w \cdot h = 59 \cdot 25 \cdot 2$
$\qquad = 2950$ m^3

51. $C = \pi d$
$\qquad = \pi \cdot 26$
$\qquad \approx 81.68$ in.

52. Distance $=$ (no. of revolutions) $\cdot$ (circumference)
$\qquad = 100 \cdot 81.68$
$\qquad = 8168$ in.
$\qquad = 680$ ft, 8 in.

PAGE 572, LOOK BEYOND

53. $A(1, 0)$, $B(0, 1)$, $C(-1, 0)$, $D(0, -1)$

54. $C = 2\pi r = 2\pi(1) = 2\pi$

55. $m\widehat{AB} = 90°$
$\qquad L_{\widehat{AB}} = \frac{90°}{360°}(2\pi \cdot 1) = \frac{\pi}{2}$

56. $m\widehat{ABC} = 180°$
$\qquad L_{\widehat{ABC}} = \frac{180°}{360°}(2\pi \cdot 1) = \pi$

57. $\stackrel{\frown}{mABD} = 360° - 90° = 270°$

$L_{\stackrel{\frown}{ABD}} = \dfrac{270°}{360°}(2\pi \cdot 1) = \dfrac{3\pi}{2}$

9.2 | **PAGES 576–577, GUIDED SKILLS PRACTICE**

6. By the Pythagorean Theorem
$$(KL)^2 + (KM)^2 = (LM)^2$$
$$(KL)^2 + 1 = 4$$
$$(KL)^2 = 3$$
$$KL = \sqrt{3}$$

7. Since $\overleftrightarrow{PR} \perp \overline{QS}$, $\overline{RS} = 3$ by the Radius and Chord Theorem.

8. By the Pythagorean Theorem
$$(XW)^2 + (YW)^2 = (XY)^2$$
$$5^2 + (YW)^2 = 13^2$$
$$(YW)^2 = 144$$
$$YW = 12$$
$WZ = 12$ by the Radius and Chord Theorem
$YZ = YW + WZ = 12 + 12 = 24$
$YZ = 24$

9. $10^2 = 8^2 + 6^2$
$(AC)^2 = (BC)^2 + (AB)^2$
Thus, $\triangle ABC$ is a right triangle and $\overline{BC} \perp \overline{AB}$. Thus $\overleftrightarrow{AB}$ is tangent to $\odot C$ by the Converse of the Tangent Theorem.

PAGES 577–578, PRACTICE AND APPLY

10. $\overline{XW} \cong \overline{WZ}$ by the Radius and Chord Theorem.

11. $RY = 7$. Thus $XR = RZ = RY = 7$.
By the Pythagorean Theorem $2^2 + (XW)^2 = 7^2$.
Thus, $RW = \sqrt{45} = 3\sqrt{5}$ and $WZ = \sqrt{45} = 3\sqrt{5}$.

12. $RY = 3$. Thus $XR = RZ = RY = 3$.
By the Pythagorean Theorem,
$$2^2 + (RW)^2 = 3^2$$
Thus, RW and $\sqrt{5}$ and $WZ = \sqrt{5}$.

13. Draw a radius from O to one of the points identified, forming a right triangle. Then
$$6^2 + 8^2 = c^2$$
$$100 = c^2$$
$$10 = c.$$
The radius is 10.

14. The triangle formed has sides of length 3, 4, and 5 units. Since $3^2 + 4^2 = 9 + 16 = 25 = 5^2$, the triangle is a right triangle and is perpendicular to the radius at its endpoint on the circle, so line ℓ is tangent to the circle.

15. Let $\overline{PQ}$ be a chord in $\odot M$ and consider the line ℓ which is the perpendicular bisector of $\overline{PQ}$, intersecting $\overline{PQ}$ at point X. Let $\overline{MN}$ be the radius of $\odot M$ which is perpendicular to $\overline{PQ}$. By the Radius and Chord Theorem, $\overline{MN}$ bisects $\overline{PQ}$, so $\overline{MN}$ also passes through point X. Thus, both $\overline{MN} \perp \overline{PQ}$ and $\ell \perp \overline{PQ}$, so $\ell \parallel \overline{MN}$. Since both ℓ and $\overline{MN}$ also pass through point X, they must coincide. Hence, the perpendicular bisector of $\overline{PQ}$, ℓ, passes through the center of the circle, M.

16. Sample answer:

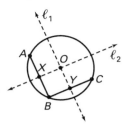

The center of the circle is the intersection of the two perpendicular bisectors.

17. Sample answer:

The tangent line at the point A must be perpendicular to the radius $\overline{AP}$. To complete the construction, extend $\overline{AP}$ and construct a line perpendicular to $\overleftrightarrow{AP}$ at A.

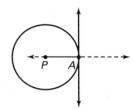

18. Suppse that a line is perpendicular to the radius of $\odot O$ at point P. If Q is any point on the line other than P, then $\triangle OPQ$ is a right triangle with hypotenuse $\overline{OQ}$. The hypotenuse is the longest side of a right triangle, so $\overline{OQ}$ is longer than the radius, $\overline{OP}$, so Q must be outside of the circle. This means that every point on the line is outside the circle except for point P, so the line touches the circle at exactly one point, which is the definition of tangent.

19. $BA = BT = 2$

20. $(BS)^2 + (SA)^2 = (BA)^2$
$1^2 + (SA)^2 = 2^2$
$(SA)^2 = 3$
$SA = \sqrt{3} \approx 1.73$

21. $SN = SA = \sqrt{3} \approx 1.73$

22. $(BT)^2 + (WT)^2 = (BW)^2$
$2^2 + 5^2 = (BW)^2$
$BW = \sqrt{29} \approx 5.39$

23. $XT = XB + BT = 2 + 2 = 4$

24. $1500 \text{ ft} = \frac{1500}{5280} \approx 0.284 \text{ miles}$
$d^2 + 4000^2 = 4000.284^2$
$\Rightarrow d = \sqrt{(4000.284)^2 - (4000)^2} \approx 47.67 \text{ mi}$

25. Let d be the distance to the horizon.
$d^2 + 4000^2 = 4155^2$
$d = \sqrt{4155^2 - 4000^2} \approx 1124.29 \text{ mi}$

26. The center of the circle is the intersection of the segments joining the midpoints. The segments joining the midpoints are diameters of the circle. The circle is tangent to the sides of the square at the midpoints. As in the proof of the Converse of the Tangent Theorem, every point on a side of the square except the midpoint is outside of the circle, so no part of the circle lies outside of the square. Any larger circle would have diameters longer than the sides of the square, so the ends of the diameters perpendicular to sides of the square would have to lie outside the square.

PAGES 578–579, LOOK BACK

27. $P = 24(2) = 48 \text{ cm}$
$A = 24(2)^2 = 96 \text{ cm}^2$

28. $P = 22\left(\frac{8}{3}\right) = 58\frac{2}{3} \text{ ft}$
$A = 22\left(\frac{8}{3}\right)^2 = 156\frac{4}{9} \text{ ft}^2$

29. $V = (12)(8)(15) = 1440 \text{ in}^3$

new volume $= 1440\left(\frac{5}{3}\right)^3$

$= 6666\frac{2}{3} \text{ in}^3$

30. $V = \pi r^2 h$

$= \pi(30)^2\, 100$

$= 90{,}000\pi \text{ ft}^3$

new volume $= 90{,}000\pi \left(\frac{7}{5}\right)^3$

$= 246{,}960\pi \approx 775{,}848 \text{ ft}^3$

PAGE 579, LOOK BEYOND

31. $\overline{OA}$ is the hypotenuse of $\triangle OBA$.
Since the hypotenuse is the longest side of a right triangle, $OA > OB$.

32. Since B is exterior to the circle, $OB > OA$

33. The answers to 31 and 32 are contradictory.

34. Since a contradiction is obtained, we must reject the assumption that m is not perpendicular to $\overline{OA}$. We conclude that $m \perp \overline{OA}$.

9.3 | PAGE 585, GUIDED SKILLS PRACTICE

7. $m\angle OMP = \frac{1}{2}m\overset{\frown}{OP} = 35°$

8. $m\angle ONP = \frac{1}{2}m\overset{\frown}{OP} = 35°$

9. $m\angle GFH = m\overset{\frown}{GH} = 50°$

$m\angle GKH = \frac{1}{2}m\overset{\frown}{GH} = 25°$

$m\angle JFK = m\overset{\frown}{JK} = 90°$

$m\angle KHJ = \frac{1}{2}m\overset{\frown}{JK} = 45°$

10. $m\angle KGH = 90°$, since it intercepts a diameter.

PAGES 585–586, PRACTICE AND APPLY

11. $m\angle XYW = \frac{1}{2}m\overset{\frown}{XZ} = 30°$

12. $m\angle WXY = 30°$ since $\triangle WXY$ is isosceles.

13. $m\angle XWY = 180° - 30° - 30° = 120°$

14. $m\angle XWZ = m\overset{\frown}{XZ} = 60°$

15. $m\overset{\frown}{YXZ} = 180°$ since $\overline{YZ}$ is a diameter.

16. $m\angle YVZ = \frac{1}{2}m\overset{\frown}{YXZ} = 90°$

17. $m\overset{\frown}{XY} = m\angle XWY = 120°$

18. $m\overset{\frown}{VZ} = 2 \cdot m\angle ZYV = 80°$

19. $m\overset{\frown}{VY} = 360° - 120° - 60° - 80° = 100°$

20. $m\angle VZY = 180° - m\angle VYZ - m\angle ZVY$ by the Triangle Sum Theorem.

$= 180° - 40° - 90°$

$= 50°$

21. $m\angle A = \frac{1}{2}m\overset{\frown}{BC} = \frac{1}{2}(58°) = 29°$

22. $m\angle B = 90°$ since it intercepts a diameter

23. $m\angle BCA = 180° - 90° - 29° = 61°$

24. $m\overset{\frown}{AB} = 2m\angle BCA = 2(61°) = 122°$

25. $m\angle PCD = m\angle D = 50°$

26. $m\angle CPD = 180° - 50° - 50° = 80°$

27. $m\overset{\frown}{DC} = m\angle CPD = 80°$

28. $m\angle AD = 180° - 80° = 100°$

29. $m\angle C = \frac{1}{2}(68°) = 34°$

$m\angle D = \frac{1}{2}(68°) = 34°$

30. $m\overset{\frown}{AB} = 2 \cdot m\angle D = 60°$

$m\angle C = m\angle D = 30°$

31. $m\angle B = \frac{1}{2}(87°) = 43.5°$

$m\angle A = m\angle B = 43.5°$

32. $m\overset{\frown}{CD} = 2m\angle B = (2a)°$

$m\angle A = m\angle B = a°$

33. $m\angle\overset{\frown}{QDA} = 2m\angle U = 2 \cdot 100° = 200°$

34. $m\overset{\frown}{QUA} = 360° - m\overset{\frown}{QDA} = 360° - 200° = 160°$

35. $m\angle D = \frac{1}{2}m\overset{\frown}{QUA} = \frac{1}{2} \cdot 160° = 80°$

36. $m\angle U + m\angle D = 100° + 80° = 180°$

37. $m\angle Q + m\angle A = 360° - (m\angle U + m\angle D)$
$= 360° - 180° = 180°$

38. $m\angle U = \frac{1}{2}m\overset{\frown}{QDA} = \frac{1}{2} \cdot 160° = 80°$

39. $m\overset{\frown}{QUA} = 360° - m\overset{\frown}{QDA} = 360° - 160° = 200°$

40. $m\angle D = \frac{1}{2}m\overset{\frown}{QUA} = \frac{1}{2}(200°) = 100°$

41. $m\angle U + m\angle D = 100° + 80° = 180°$

42. $m\angle Q + m\angle A = 360° - (m\angle U + m\angle D) = 360° - 180° = 180°$

43. $m\angle\overset{\frown}{QDA} = 2 \cdot m\angle U = 2 \cdot x° = (2x)°$
by the Inscribed Angle Theorem.

44. $m\overset{\frown}{QUA} = 360° - m\overset{\frown}{QDA} = (360 - 2x)°$
because the sum of the measures of the arcs in a circle is 360°.

45. $m\angle D = \frac{1}{2}m\overset{\frown}{QUA} = \frac{1}{2}(360° - 2x°) = (180 - x)°$
by the Inscribed Angle Theorem.

46. $m\angle U + m\angle D = x° + 180° - x° = 180°$
This is the sum of the values from the first and fourth columns of the table.

47. $m\angle Q + m\angle A = 360° - (m\angle U + m\angle D)$
$= 360° - 180° = 180°$
The sum of the measures of the interior angles of a quadrilateral is 360°.

48. If a quadrilateral is inscribed in a circle, then the opposite angles are supplementary.

49. The longest sides of the triangles intersect at the center of the circle.
Both right angles intercept an arc that measures 180°, so the intercepted arcs are semicircles. Thus, the longest sides of the triangles are diameters of the circles, which intersect at the center of the circle.

50. $m\angle A = 40°$, since it should be half of the measure of the intercepted arc.
$m\angle B = 80°$, since $\angle B$ is the central angle corresponding to the intercepted arc.

PAGE 587, LOOK BACK

51. line m is the angle bisector of $\angle CXD$

52. Vertical Angles Theorem

53. $m\angle 2 = m\angle 5$

54. Addition Property

55. Angle Addition Postulate

56. Substitution Property of Equality

57. Linear Pair Property

58. $2 \cdot m\angle MXL = 180° \Rightarrow m\angle MXL = 90°$
$\Rightarrow m\angle MXN = 90°$

59. Substitution Property and Division Property

60. $\ell \perp m$

61. Definition of perpendicular

PAGE 587, LOOK BEYOND

62. Each of the four areas is adjacent to the other three areas.

63. Sample answer:

5. $m\angle WYX = \frac{1}{2}m\widehat{YW} = \frac{1}{2}(138°) = 69°$

6. $m\angle 1 = \frac{1}{2}(m\widehat{MN} + m\widehat{OP})$
$= \frac{1}{2}(60° + 150°) = 105°$

7. $m\angle BFC = \frac{1}{2}(m\widehat{BC} - m\widehat{ED})$
$= \frac{1}{2}(84° - 40°) = 22°$

8. a. $m\angle ABC = \frac{1}{2}m\widehat{BC} = \frac{1}{2}(70°) = 35°$

b. $m\angle 1 = \frac{1}{2}(m\widehat{BC} + m\widehat{EF})$
$= \frac{1}{2}(70° + 130°) = 100°$

c. $m\widehat{CE} = 360° - (70° + 100° + 130°) = 60°$
$m\angle 2 = \frac{1}{2}(m\widehat{CE} + \widehat{BF})$
$= \frac{1}{2}(60° + 100°) = 80°$

d. $m\angle BDF = \frac{1}{2}(m\widehat{BF} - m\widehat{CE})$
$= \frac{1}{2}(100° - 60°) = 20°$

9. The ship is outside the circle of danger since $m\angle BXA < m\angle BCA$.

PAGES 593–597, PRACTICE AND APPLY

10. $m\widehat{WY} = 360° - (m\widehat{WT} + m\widehat{TU} + m\widehat{UY})$
$= 360° - (254°) = 106°$
$m\angle SYV = 180° - \frac{1}{2}m\widehat{WY}$
$= 180 - \frac{1}{2}(106°) = 180° - 53° = 127°$

11. $m\angle VSY = \frac{1}{2}(m\widehat{YU} - m\widehat{WT})$
$= \frac{1}{2}(70° - 36°)$
$= \frac{1}{2}(34°)$
$= 17°$

12. $m\angle SVY = 180° - (m\angle SYV + m\angle VSY)$
$= 180° - (127° + 17°)$
$= 36°$

13. $m\angle 1 = \frac{1}{2}(m\widehat{CB} + m\widehat{AD}) = \frac{1}{2}(60° + 110°) = 85°$

14. $m\angle 2 = 180° - m\angle 1 = 180° - 85° = 95°$

15. $m\angle 3 = \frac{1}{2}(m\widehat{AD} + m\widehat{BC}) = \frac{1}{2}(110° + 60°) = 85°$

16. $m\angle 4 = 180° - m\angle 3 = 180° - 85° = 95°$

17. $m\angle AVC = \frac{1}{2}m\widehat{AC} = \frac{1}{2} \cdot 150° = 75°$

18. $m\widehat{VC} = 2 \cdot m\angle AVC = 2 \cdot 80° = 160°$

19. $m\angle AVC = \frac{1}{2}m\widehat{AC} = \frac{1}{2}(2x + 4) = (x + 2)°$

20. $m\widehat{VC} = 2 \cdot m\angle AVC = 2(3x - 1) = (6x - 2)°$

21. $m\widehat{AB} = m\widehat{CD} = 105°$

22. $m\widehat{AD} = 360° - m\widehat{AB} - m\widehat{BC} - m\widehat{CD}$
$= 360° - 105° - 47° - 105°$
$= 103°$

23. $m\angle AED = \frac{1}{2}(m\widehat{AD} - m\widehat{BC}) = \frac{1}{2}(103° - 47°) = \frac{1}{2}(56°) = 28°$

24. $m\angle CAF = \frac{1}{2}(m\widehat{AC}) = \frac{1}{2}(m\widehat{AD} + m\widehat{DC})$
$= \frac{1}{2}(103° + 105°) = 104°$

25. $m\angle CQD = \frac{1}{2}(m\widehat{CD} + m\widehat{AB})$
$= \frac{1}{2}(105° + 105°) = 105°$

26. $m\angle BQC = \frac{1}{2}(m\widehat{BC} + m\widehat{AD})$
$= \frac{1}{2}(47° + 103°) = 75°$

27. $m\angle 1 = m\angle 2 + m\angle AVC$

28.

m$\widehat{BC}$	m$\widehat{AC}$	m∠1	m∠2	m∠AVC
250°	60°	**a.** 125°	**b.** 30°	**c.** 95°
200°	40°	**d.** 100°	**e.** 20°	**f.** 80°
130°	40°	**g.** 65°	**h.** 20°	**i.** 45°
70°	30°	**j.** 35°	**k.** 15°	**l.** 20°
$x_1°$	$x_2°$	**m.** $\frac{x_1°}{2}$ m∠1 = $\frac{1}{2}$m$\widehat{BC}$	**n.** $\frac{x_2°}{2}$ m∠2 = $\frac{1}{2}$m$\widehat{AC}$	**o.** $\frac{1}{2}(x_1° - x_2°)$ m∠AVC = m∠1 − m∠2

29. m∠AVC = $\frac{1}{2}$(m$\widehat{BC}$ − m$\widehat{AC}$)

30. one-half the difference of the measures of the intercepted arcs

31. m∠1 = m∠AVC + m∠2

32. $\widehat{AXC}$

33. $\widehat{AC}$

34. m$\widehat{AC}$ = 360° − m$\widehat{AXC}$ = 360° − 260° = 100°

35.

m$\widehat{AXC}$	m$\widehat{AC}$	m∠1	m∠2	m∠AVC
300°	**a.** 60°	**b.** 150°	**c.** 30°	**d.** 120°
250°	**e.** 110°	**f.** 125°	**g.** 55°	**h.** 70°
220°	**i.** 140°	**j.** 110°	**k.** 70°	**l.** 40°
200°	**m.** 160°	**n.** 100°	**o.** 80°	**p.** 20°
$x°$	**q.** 360° − $x°$ m$\widehat{AC}$ = 360° − m$\widehat{AXC}$	**r.** $\frac{x°}{2}$ m∠1 = $\frac{1}{2}$m$\widehat{AXC}$	**s.** 180° − $\frac{x°}{2}$ m∠2 = $\frac{1}{2}$m$\widehat{AC}$	**t.** $x°$ − 180° m∠AVC = m∠1 − m∠2

36. m∠AVC = $\frac{1}{2}$(m$\widehat{AXC}$ − m$\widehat{AC}$)

37. one-half the difference of the measures of the intercepted arcs

38. $\frac{\text{m}\widehat{AXC} - \text{m}\widehat{AC}}{2} = \frac{\text{m}\widehat{AXC} + \text{m}\widehat{AXC} - \text{m}\widehat{AXC} - \text{m}\widehat{AC}}{2}$

$= \frac{2 \cdot \text{m}\widehat{AXC} - (\text{m}\widehat{AXC} + \text{m}\widehat{AC})}{2}$

$= \frac{2 \cdot \text{m}\widehat{AXC} - 360°}{2}$

$= \text{m}\widehat{AXC} - 180°$

39. The measure of a tangent-tangent angle with its vertex outside the circle is the measure of the major arc minus 180°.

40. interior of circle

41. two secants or chords

42. m∠AVC = $\frac{1}{2}$(m$\widehat{AC}$ + m$\widehat{DB}$)

43. on the circle

44. m∠AVC = $\frac{1}{2}$m$\widehat{AC}$

45. on the circle

46. secant and tangent

47. m∠AVC = $\frac{1}{2}$m$\widehat{AV}$

48.

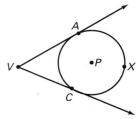

49. $m\angle AVC = \frac{1}{2}(m\widehat{AXC} - m\widehat{AC})$
$= m\widehat{AXC} - 180°$

50. exterior of circle

51. two secants

52. $m\angle AVC = \frac{1}{2}(m\widehat{BD} - m\widehat{AC})$

53. exterior of circle

54. secant and tangent

55. $m\angle AVC = \frac{1}{2}(m\widehat{BC} - m\widehat{AC})$

56. The measure of the tangent angle is $2 \cdot 89.5° = 179°$.
Let $\widehat{AXC}$ be the major intercepted arc on Earth's surface.
$179° = m\widehat{AXC} - 180°$
$m\widehat{AXC} = 359°$
$m\widehat{AC} = 1°$
$\frac{1°}{360°} \cdot 2\pi \cdot 4000 = \frac{8000\pi}{360}$ miles
≈ 69.8 miles

PAGE 597, LOOK BACK

57. $180° - (60° + 50°) = 180° - 110° = 70°$

58. $360° - (120° + 80° + 84°) = 360° - 284° = 76°$

59. $540° - (100° + 130° + 110° + 105°) = 540° - 445° = 95°$

60. $V = \frac{1}{3}s^2h = \frac{1}{3}(775.75)^2(481.4) \approx 96{,}566{,}924 \text{ ft}^3$

61. Yes. Everything was reduced by a common factor.

62. $\frac{A'B'}{AB} = \frac{15}{20} = \frac{3}{4}$
$\frac{A'C'}{12} = \frac{C'B'}{CB} = \frac{3}{4}$ since the triangles are similar.

63. $\frac{A'C'}{AC} = \frac{3}{4} \Rightarrow \frac{A'C'}{12} = \frac{3}{4} \Rightarrow 4A'C' = 36 \Rightarrow A'C' = 9$

PAGE 597, LOOK BEYOND

64. Tilt the plane so that it is no longer perpendicular to the axis of the cone, but so that it still intersects both 'sides' of the top or the bottom of the double cone.

65. Tilt the plane so that it only intersects one side of either the top or the bottom of the double cone, but not the vertex.

66. The plane should intersect the double cone only at the vertex.

67. The plane should contain the axis of the double cone.

68. The plane should be parallel to the axis of the double cone, but not contain the axis.

69. Tilt the plane as in Exercise 65, then move it up or down until it intersects the vertex of the double cone.

9.5 PAGES 605–606, GUIDED SKILLS PRACTICE

6. By Theorem 9.5.1, $BX = AX = 7$.

7. By Theorem 9.5.2,
$AX \cdot CX = BX \cdot DX$
$16 \cdot 9 = 18 \cdot DX$
$\Rightarrow DX = \frac{144}{18} = 8.$

8. By Theorem 9.5.3,
$$AX \cdot CX = (BX)^2$$
$$8 \cdot 2 = (BX)^2 \Rightarrow BX = \sqrt{16} = 4.$$

9. By Theorem 9.5.4,
$$AX \cdot CX = BX \cdot XD$$
$$6 \cdot CX = 3 \cdot 8$$
$$CX = \frac{24}{6} = 4.$$

10. Let H be the other point where $\overleftrightarrow{FG}$ intersects the circle, then $\overline{GH}$ is a diameter of the circle. Since $\overleftrightarrow{FG}$ is the perpendicular bisector of $\overline{DE}$, $DF = FE = 24$.
By Theorem 9.5.4, $DF \cdot FE = GF \cdot FH \Rightarrow 24 \cdot 24 = 12 \cdot FH \Rightarrow FH = 48.$
Thus $GH = GF + FH = 12 + 48 = 60.$

PAGES 606–608, PRACTICE AND APPLY

11. $\overline{AC}$ or $\overline{BC}$

12. $\overline{EF}$

13. $\overline{DE}$

14. Sample answer:
$\angle SRT$ and $\angle SUT$; $\angle RVS$ and $\angle TVU$

15. $\triangle SVR \sim \triangle TVU$ or $\triangle USW \sim \triangle RTW$

16. $16x = 8 \cdot 10 \Rightarrow x = \frac{80}{16} = 5$

17. $6x = 9 \cdot 4 \Rightarrow x = \frac{36}{6} = 6$

18. $VB = VA = 6$ cm

19. $AP = 3$ cm

20. By the Pythagorean Theorem,
$$(AP)^2 + (AV)^2 = (PV)^2 \Rightarrow 3^2 + 6^2 = (PV)^2$$
$$\Rightarrow (PV)^2 = 9 + 36$$
$$\Rightarrow PV = \sqrt{45} = 3\sqrt{5} \approx 6.71 \text{ cm}$$

21. $XV = PV - 3 \approx 3.71$ cm

22. $\angle BVP$

23. $\angle BPV$

24. $\overparen{BX}$

25. $WA \cdot WB = WC \cdot WD$
$$4 \cdot 10 = 5 \cdot WD$$
$$\Rightarrow WD = \frac{40}{5} = 8 \Rightarrow CD = 8 - 5 = 3$$

26. $WA \cdot WB = WC \cdot WD$
$$6 \cdot x = 5(x + 3)$$
$$6x = 5x + 15$$
$$x = 15$$

27. $WA \cdot WB = WC \cdot WD$
$$x(x - 16) = 8 \cdot 5$$
$$x^2 - 16x - 40 = 0$$
$$x = \frac{16 \pm \sqrt{256 + 160}}{2}$$
$$= \frac{16 \pm \sqrt{416}}{2}$$
$$\Rightarrow x = \frac{16 + \sqrt{416}}{2} \approx 18.198$$

28. $AE \cdot EB = CE \cdot CD$
$$x(x - 2) = 3 \cdot 8$$
$$x^2 - 2x - 24 = 0$$
$$x = \frac{2 \pm \sqrt{4 + 96}}{2}$$
$$= \frac{2 \pm 10}{2}$$
$$\Rightarrow x = 6$$

29. Let $AE = x$, then $EB = 10 - x$ since
$AB = AE + EB = 10.$
$ED = CD - CE = 12 - 2 = 10$
By Theorem 9.5.4,
$AE \cdot EB = CE \cdot ED$
$$x(10 - x) = 10 \cdot 2$$
$$10x - x^2 = 20$$
$$0 = x^2 - 10x + 20$$
$$x = \frac{10 \pm \sqrt{100 - 80}}{2}$$
$$= \frac{10 \pm \sqrt{20}}{2}$$
$$\Rightarrow x = AE = \frac{10 \pm \sqrt{20}}{2} \approx 2.76 \text{ or } 7.24$$

30. exterior of circle

32. $AV \cdot BV = CV \cdot DV$

31. 2 secant segments

33.

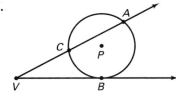

34. $VC \cdot VA = (VB)^2$

36. 2 chords

35. interior of circle

37. $AV \cdot VB = CV \cdot VD$

38. By the Inscribed Angle Theorem,
$m\angle BAC = \frac{1}{2}m\widehat{BC}$. By Theorem 9.4.1,
$m\angle CBX = \frac{1}{2}m\widehat{BC}$, thus by the Transitive Property
of Equality, $m\angle BAC = m\angle CBX$, so
$\angle BAC \cong \angle CBX$. $\angle AXB \cong \angle BXC$ by the Reflexive
Property of Congruence. Thus, by the AA Similarity
Postulate, $\angle AXB \sim \triangle BXC$ and so $\frac{AX}{BX} = \frac{BX}{CX}$. Thus,
$AX \cdot CX = (BX)^2$

39. $CL \cdot BL = AL \cdot XL$
$CL = CB + BL = 32 + 20 = 52$
$20 \cdot 52 = 23 \cdot LX$
$\frac{1040}{23} = LX$
$\Rightarrow LX \approx 45.22$ km

40. Since $\overline{EF}$ is the perpendicular bisector of $\overline{BD}$, we know $\overleftrightarrow{EF}$ passes through the center.
Let x be the distance from F to the opposite side of the circle along the diameter
through E and F.
$x \cdot EF = BF \cdot FD$
$x \cdot 2.5 = 5.05^2$
$x = \frac{25.5025}{2.5} = 10.201 \approx 10.2$ cm
Thus the diameter is $x + EF = 10.2 + 2.5 = 12.7$ cm

PAGE 608, LOOK BACK

41.

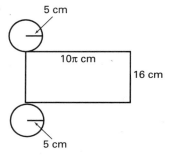

$SA = 2 \cdot \pi r^2 + 2\pi r \cdot h$
$= 2 \cdot \pi 5^2 + 2\pi 5 \cdot 16$
$= 50\pi + 160\pi$
$= 210\pi \approx 659.73$ cm^2

$V = \pi r^2 h$
$= \pi 5^2 \cdot 16$
$= 400\pi \approx 1256.64$ cm^3

42.

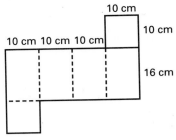

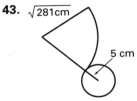

$$SA = 2(\ell w + wh + \ell h)$$
$$= 2(10 \cdot 10 + 10 \cdot 16 + 10 \cdot 16)$$
$$= 2(420)$$
$$= 840 \text{ cm}^2$$

$$V = \ell wh$$
$$= 10 \cdot 10 \cdot 16$$
$$= 1600 \text{ cm}^3$$

43.

$$SA = \pi r \ell + \pi r^2$$
$$= \pi \cdot 5 \cdot \sqrt{281} + \pi \cdot 5^2$$
$$= 5\pi\sqrt{281} + 25\pi \approx 341.85 \text{ cm}^2$$

$$V = \frac{1}{3}\pi r^2 h$$
$$= \frac{\pi}{3} 25 \cdot 16$$
$$= \frac{400\pi}{3} \approx 418.88 \text{ cm}^3$$

44. Opposite sides of a parallelogram are congruent.

45. Definition of a parallelogram.

46. Alternate Interior Angles Theorem

47. Alternate Interior Angles Theorem

48. ASA Postulate

49. CPCTC

PAGE 609, LOOK BEYOND

50. By Theorem 9.5.4,
$$(c - a)(c + a) = b \cdot b$$
$$c^2 + ac - ac - a^2 = b^2$$
$$c^2 - a^2 = b^2$$
$$c^2 = a^2 + b^2$$

9.6 PAGE 614, GUIDED SKILLS PRACTICE

6. For the x-intercepts, set $y = 0$:
$$x^2 + 0^2 = 100$$
$$x^2 = 100$$
$$x = \pm\sqrt{100}$$
$$x = \pm 10$$

x-intercepts: $(10, 0)$ and $(-10, 0)$

For the y-intercepts, set $x = 0$:
$$0^2 + y^2 = 100$$
$$y^2 = 100$$
$$y = \pm\sqrt{100}$$
$$y = \pm 10$$

y-intercepts: $(0, 10)$ and $(0, -10)$

7.

x	y	Points on Graph
0	$y^2 = 100 \Rightarrow y = \pm 10$	$(0, 10), (0, -10)$
$x^2 = 100 \Rightarrow x = \pm 10$	0	$(10, 0), (-10, 0)$
6	$36 + y^2 = 100 \Rightarrow y^2 = 64 \Rightarrow y = \pm 8$	$(6, 8), (6, -8)$
-6	$36 + y^2 = 100 \Rightarrow y^2 = 64 \Rightarrow y = \pm 6$	$(-6, 8), (-6, -8)$
8	$64 + y^2 = 100 \Rightarrow y^2 = 36 \Rightarrow y = \pm 6$	$(8, 6), (8, -6)$
-8	$64 + y^2 = 100 \Rightarrow y^2 = 36 \Rightarrow y = \pm 6$	$(-8, 6), (-8, -6)$

8.

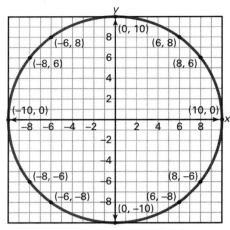

9. For the x-intercepts, set $y = 0$:

$$(x - 4)^2 + (0 - 3)^2 = 25$$
$$(x - 4)^2 + 9 = 25$$
$$(x - 4)^2 = 16$$
$$x - 4 = \pm 4$$

$x - 4 = 4$ or $x - 4 = -4$
$x = 8$ $x = 0$

x-intercepts: $(8, 0)$; $(0, 0)$

For the y-intercepts, set $x = 0$:

$$(0 - 4)^2 + (y - 3)^2 = 25$$
$$16 + (y - 3)^2 = 25$$
$$(y - 3)^2 = 9$$
$$y - 3 = \pm 3$$

$y - 3 = 3$ or $y - 3 = -3$
$y = 6$ $y = 0$

y intercepts: $(0, 6)$; $(0, 0)$

10.

x	y	Points on graph
0	6, 0	$(0, 6), (0, 0)$
8, 0	0	$(8, 0), (0, 0)$
1	7, -1	$(1, 7), (1, -1)$
-1	3	$(-1, 3)$
4	8, -2	$(4, 8), (4, -2)$
7	7, -1	$(7, 7), (7, -1)$
8	6, 0	$(8, 6), (8, 0)$
9	3	$(9, 3)$

11.

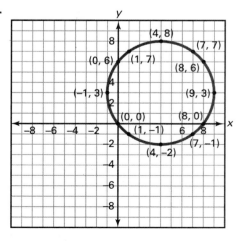

12. x-intercepts: $x^2 + 0 = 64$
$\qquad x = \pm 8$
$\quad (8, 0); (-8, 0)$

y-intercepts: $0^2 + y^2 = 64$
$\qquad y = \pm 8$
$\quad (0, 8); (0, -8)$

13. x-intercepts: $x^2 + 0^2 = 50$
$\qquad x = \pm\sqrt{50} = \pm 5\sqrt{2}$
$\quad (5\sqrt{2}, 0); (-5\sqrt{2}, 0)$

y-intercepts: $0^2 + y^2 = 50$
$\qquad y = \pm\sqrt{50} = \pm 5\sqrt{2}$
$\quad (0, 5\sqrt{2}); (0, -5\sqrt{2})$

14. x-intercepts: $x^2 + (0 - 4)^2 = 25$
$\qquad x^2 + 16 = 25$
$\qquad x^2 = 9$
$\qquad x = \pm 3$
$\quad (3, 0); (-3, 0)$

y-intercepts: $0^2 + (y - 4)^2 = 25$
$\qquad (y - 4)^2 = 25$
$\qquad y - 4 = \pm 5$
$y - 4 = 5$ or $y - 4 = -5$
$\quad y = 9 \qquad y = -1$
$\quad (0, 9); (0, -1)$

15. x-intercepts: $(x - 2)^2 + 0^2 = 9$
$\qquad (x - 2)^2 = 9$
$\qquad x - 2 = \pm 3$
$x - 2 = 3$ or $x - 2 = -3$
$\quad x = 5$ or $\quad x = -1$
$\quad (5, 0); (-1, 0)$

y-intercepts: $(0 - 2)^2 + y^2 = 9$
$\qquad 4 + y^2 = 9$
$\qquad y^2 = \pm\sqrt{5}$
$\quad (0, \sqrt{5}); (0, -\sqrt{5})$

16. x-intercepts: $(x - 6)^2 + (0 - 8)^2 = 100$
$\qquad (x - 6)^2 + 64 = 100$
$\qquad (x - 6)^2 = 36$
$\qquad x - 6 = \pm 6$
$x - 6 = 6$ or $x - 6 = -6$
$\quad x = 12 \qquad x = 0$
$\quad (12, 0); (0, 0)$

y-intercepts: $(0 - 6)^2 + (y - 8)^2 = 100$
$\qquad 36 + (y - 8)^2 = 100$
$\qquad (y - 8)^2 = 64$
$\qquad y - 8 = \pm 8$
$y - 8 = 8$ or $y - 8 = -8$
$\quad y = 16 \qquad y = 0$
$\quad (0, 16); (0, 0)$

17. $(x - 0)^2 + (y - 0)^2 = 6^2 \Rightarrow x^2 + y^2 = 36$

18. $(x - 0)^2 + (y - 0)^2 = (2.5)^2 \Rightarrow x^2 + y^2 = 6.25$

19. $(x - 0)^2 + (y - 0)^2 = (\sqrt{13})^2 \Rightarrow x^2 + y^2 = 13$

20. $(x - 2)^2 + (y - 3)^2 = 4^2 \Rightarrow (x - 2)^2 + (y - 3)^2 = 16$

21. $(x - 0)^2 + (y - 6)^2 = 5^2 \Rightarrow x^2 + (y - 6)^2 = 25$

22. $(x - 4)^2 + (y - (-5))^2 = 7^2 \Rightarrow (x - 4)^2 + (y + 5)^2 = 49$

23. $(x - 1)^2 + (y - (-7))^2 = 10^2 \Rightarrow (x - 1)^2 + (y + 7)^2 = 100$

24. $(x - 4)^2 + (y - (-3))^2 = (\sqrt{7})^2 \Rightarrow (x - 4)^2 + (y + 3)^2 = 7$

25. $(x - 0)^2 + (y - 0)^2 = 10^2$
center: $(0, 0)$
radius: 10

26. $(x - 0)^2 + (y - 0)^2 = 6^2$
center: $(0, 0)$
radius: 6

27. $(x - 0)^2 + (y - 0)^2 = (\sqrt{101})^2$
center: $(0, 0)$
radius $= \sqrt{101}$ units

28. $(x - 6)^2 + (y - 0)^2 = 3^2$
center: $(6, 0)$
radius: 3

29. $(x - 0)^2 + (y - 3)^2 = 2^2$
center: $(0, 3)$
radius: 2

30. $(x - (-5))^2 + (y - 2)^2 = 4^2$
center: $(-5, 2)$
radius: 4

31. $(y - 0)^2 + (x - (-3))^2 = 7^2$
center: $(-3, 0)$
radius: 7

32. $(x - (-1))^2 + (y - (-3))^2 = (\sqrt{19})^2$
center: $(-1, -3)$
radius: $\sqrt{19}$

33. $(x - 0)^2 + (y - 0)^2 = 4^2 \Rightarrow x^2 + y^2 = 16$

34. $(x - 2)^2 + (y - 1)^2 = (2 - (-1))^2$
$\Rightarrow (x - 2)^2 + (y - 1)^2 = 9$

35. $(x - (-2))^2 + (y - 3)^2 = (0 - (-2))^2$
$\Rightarrow (x + 2)^2 + (y - 3)^2 = 4$

36. $x^2 + y^2 = 9$

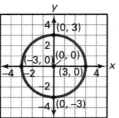

37. Sample answer:
$(x - 4)^2 + y^2 = 4$

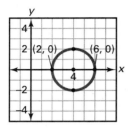

38. $x^2 + (y - 4)^2 = 16$

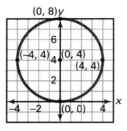

39. Sample answer:
$(x - 1)^2 + (y - 5)^2 = 1$

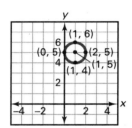

40. Sample answer:
$(x - 2)^2 + (y - 2)^2 = 1$

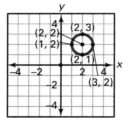

41.

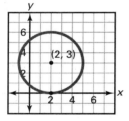

42.

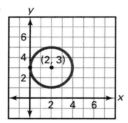

Since the circle is tangent to the x-axis, the x-axis is perpendicular to the radius which ends on the x-axis. Thus the radius of the circle is 3 and the equation is $(x - 2)^2 + (y - 3)^2 = 3^2$ or $(x - 2)^2 + (y - 3)^2 = 9$

Since the circle is tangent to the y-axis, the y-axis is perpendicular to the radius which ends on the y-axis. Thus the radius of the circle is 2 and the equation is $(x - 2)^2 + (y - 3)^2 = 2^2$ or $(x - 2)^2 + (y - 3)^2 = 4$

43.

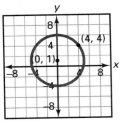

With center $(0, 1)$, the circle has equation
$(x - 0)^2 + (y - 1)^2 = r^2$ or $x^2 + (y - 1)^2 = r^2$.
Substitute the values for x and y to get r.
$$4^2 + (4 - 1)^2 = r^2$$
$$16 + 9 = r^2$$
$$25 = r^2$$
$$5 = r$$
The equation is $x^2 + (y - 1)^2 = 5^2$ or
$x^2 + (y - 1)^2 = 25$

44.

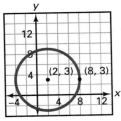

With center $(2, 3)$, the circle has equation
$(x - 2)^2 + (y - 3)^2 = r^2$. Substitute the values for
x and y to get r.
$$(8 - 2)^2 + (3 - 3)^2 = r^2$$
$$36 + 0 = r^2$$
$$6 = r$$
The equation is $(x - 2)^2 + (y - 3)^2 = 6^2$ or
$(x - 2)^2 + (y - 3)^2 = 36$

45.

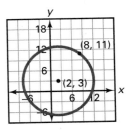

With center $(2, 3)$, the circle has equation
$(x - 2)^2 + (y - 3)^2 = r^2$. Substitute the values for
x and y to get r.
$$(8 - 2)^2 + (11 - 3)^2 = r^2$$
$$36 + 64 = r^2$$
$$100 = r^2$$
$$10 = r$$
The equation is $(x - 2)^2 + (y - 3)^2 = 10^2$ or
$(x - 2)^2 + (y - 3)^2 = 100$

47. $(x - 3)^2 + (y + 5)^2 = 4$

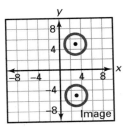

46.

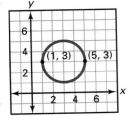

The center of the circle is $\left(\dfrac{1 + 5}{2}, \dfrac{3 + 3}{2}\right) = (3, 3)$,
the midpoint of the diameter. The radius is
$r = \sqrt{(1 - 3)^2 + (3 - 3)^2} = \sqrt{(-2)^2 + 0^2}$
$= \sqrt{4} = 2$, so the equation is
$(x - 3)^2 + (y - 3)^2 = 2^2$ or
$(x - 3)^2 + (y - 3)^2 = 4$

48. $(x + 4)^2 + (y - 2)^2 = 1$

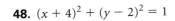

49. $(x - 8)^2 + y^2 = 9$

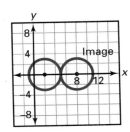

50. $(x - 8)^2 + (y - 3)^2 = 9$

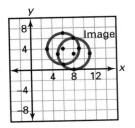

51. $(x + 5)^2 + (y + 4)^2 = 9$

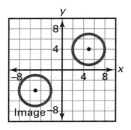

52. The radius from $(0, 0)$ to $(-6, 8)$ has slope $-\frac{4}{3}$. The tangent line is perpendicular to the radius, therefore the slope of the tangent line is $\frac{3}{4}$ and passes through $(-6, 8)$.

$$y - 8 = -\frac{3}{4}[x - (-6)]$$
$$y - 8 = \frac{3}{4}x + \frac{9}{2}$$
$$y = \frac{3}{4}x + \frac{25}{2}$$

53.

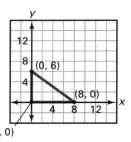

Since the circle circumscribes the triangle, $(0, 0)$, $(0, 6)$, and $(8, 0)$ are on the circle and the line segments from $(0, 0)$ to $(0, 6)$ and from $(0, 8)$ to $(8, 0)$ are chords of the circle. The perpendicular bisector of the chord between $(0, 0)$ and $(0, 6)$ is the line $y = 3$. The perpendicular bisector of the chord between $(0, 0)$ and $(8, 0)$ is the line $x = 4$. The intersection of these bisectors occurs at the point $(4, 3)$, the center of the circle. The radius of the circle is $r = \sqrt{(4 - 0)^2 + (3 - 0)^2} \Rightarrow r = \sqrt{25} = 5$. Therefore the equation of the circle is $(x - 4)^2 + (y - 3)^2 = 25$.

54. Sample answer:
$(x - 320)^2 + (y + 240)^2 = 240^2$ or
$(x - 320)^2 + (y + 240)^2 = 57{,}600$

PAGES 616–617, LOOK BACK

55. Area of larger circle $= \pi \cdot 10^2 = 100\pi$
Area of smaller circle $= \pi \cdot 5^2 = 25\pi$
$\frac{25\pi}{100\pi} = \frac{1}{4}$

56. Let $d =$ the size of the new image.
$\frac{50}{x} = \frac{100}{d} \Rightarrow 50d = 100x \Rightarrow d = 2x$
$\frac{50}{x} = \frac{25}{d} \Rightarrow 50d = 25x \Rightarrow d = \frac{1}{2}x$
$\frac{50}{x} = \frac{45}{d} = \Rightarrow 50d = 45x \Rightarrow d = 0.9x$

57. $\frac{45°}{360°} \cdot \pi(24.5) = \frac{1}{8} \cdot (24.5\pi) \approx 3.06\pi$ in.
≈ 9.6 in.

58. 94 ft = 1128 in.
$24.5\pi \cdot R = 1128$
$R = \frac{1128}{24.5\pi} \approx 14.7$ rotations

59. $AC \cdot AB = AE \cdot AD$
$14 \cdot 8 = (7 + DE) \cdot 7 \Rightarrow 112 = 49 + 7DE$
$\Rightarrow 63 = 7DE$
$\Rightarrow 9 = DE$

60. $FE \cdot DE = BE \cdot CE$
$7 \cdot DE = 6 \cdot 8 \Rightarrow DE = \frac{48}{7} \approx 6.86$

61. The equation of the circle $x^2 + y^2 = r^2$. When $x = p$, then substitute p for x and solve for y to get the y-coordinate.

$$p^2 + y^2 = r^2$$
$$y^2 = r^2 - p^2$$
$$y = \pm\sqrt{r^2 - p^2}$$

The coordinates are $(p, \sqrt{r^2 - p^2})$. Note that this is the top half of the circle, so the positive value is used for y.

62. For the right-hand segment:

$$m_1 = \frac{\sqrt{r^2 - p^2} - 0}{p - r} = \frac{\sqrt{r^2 - p^2}}{p - r}$$

For the left-hand segment:

$$m_2 = \frac{\sqrt{r^2 - p^2} - 0}{p - (-r)} = \frac{\sqrt{r^2 - p^2}}{p + r}$$

63. $m_1 \cdot m_2 = \dfrac{\sqrt{r^2 - p^2}}{p - r} \cdot \dfrac{\sqrt{r^2 - p^2}}{p + r} = \dfrac{(\sqrt{r - p})^2}{(p - r)(p + r)} = \dfrac{r^2 - p^2}{p^2 - r^2} = -1$

Since the product of the slopes is -1, the segments are perpendicular. That is, the inscribed angle measures $90°$.

CHAPTER REVIEW AND ASSESSMENT

1. Sample answer: $\overline{BC}$, $\overline{AP}$, $\angle APC$, $\overparen{CDA}$

2. Given that $\overline{BC}$ is a diameter,
$\mathrm{m}\angle APB = 180° - 72° = 108°$

3. $\mathrm{m}\overparen{AC} = \mathrm{m}\angle APC = 72°$

4. $AC = \dfrac{72°}{360°} \cdot 2\pi \cdot 40 = \dfrac{1}{5} \cdot 80\pi = 16\pi \approx 50$ cm

5. $(NO)^2 + (MO)^2 = (MN)^2$
$12^2 + 5^2 = (MN)^2$
$169 = (MN)^2 \Rightarrow MN = 13$

6. $(ML)^2 = (NL)^2 + (MN)^2$
$(ML)^2 = (31.2)^2 + (13)^2$
$(ML)^2 = 973.44 + 169 = 1142.44 \Rightarrow ML = 33.8$

7. True: $\overline{QX} \cong \overline{RX}$, $\overline{PX} \cong \overline{PX}$, and $\overline{PQ} \cong \overline{PQ}$, so $\triangle PXQ \cong \triangle PXR$ by SSS.
Thus, $\mathrm{m}\angle PXQ = \mathrm{m}\angle PXR = 90°$.

8. False: If $\overline{ST}$ were tangent to $\odot P$ at S, then $\triangle PST$ would be a right triangle, so $(PS)^2 + (ST)^2 = (PT)^2$. However, $(PS)^2 + (ST)^2 = 7^2 + 10^2 = 149$ and $(PT)^2 = 12^2 = 144$

9. $\mathrm{m}\angle F = \dfrac{1}{2}\mathrm{m}\overparen{HE} = \dfrac{1}{2}(40°) = 20°$

10. $\mathrm{m}\overparen{FG} = 2\mathrm{m}\angle FHG = 2(25°) = 50°$

11. $\mathrm{m}\angle E = \dfrac{1}{2}\mathrm{m}\overparen{FG} = \dfrac{1}{2}(50°) = 25°$

12. $\mathrm{m}\angle I = \dfrac{1}{2}\mathrm{m}\overparen{HEFG} = \dfrac{1}{2}(180°) = 90°$

13. $\mathrm{m}\angle KMN = \dfrac{1}{2}(\mathrm{m}\overparen{HP} - \mathrm{m}\overparen{KN}) = \dfrac{1}{2}(136° - 42°) = \dfrac{1}{2}(94°) = 47°$

14. $\mathrm{m}\angle KHJ = \mathrm{m}\angle KMN = 47°$

15. $\mathrm{m}\overparen{KH} = 2 \cdot \mathrm{m}\angle KHJ = 2 \cdot 47° = 94°$

16. $\mathrm{m}\overparen{NP} = 360° - (136° + 42° + 94°)$
$= 360° - 272° = 88°$

17. $SP \cdot SR = (SQ)^2$
$(60 + 75) \cdot 60 = (SQ)^2$
$135 \cdot 60 = (SQ)^2$
$8100 = (SQ)^2 \Rightarrow SQ = 90$

18. $ST = SQ = 90$

19. $SU \cdot SY = SV \cdot SZ$
$SU \cdot 12 = 4 \cdot 18$
$\Rightarrow SU = \dfrac{72}{12} = 6$

20. $WY \cdot WT = VW \cdot WZ$
$WY \cdot 5 = 8 \cdot 6$
$\Rightarrow WY = \dfrac{48}{5} = 9.6$

21. $x^2 + y^2 = 49$; Center $(0, 0)$; radius = 7

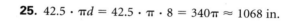

22. $(x - 1)^2 + (y + 2)^2 = 25$; Center $(1, -2)$; radius = 5

23. $(x - 0)^2 + (y - 0)^2 = 1^2$
$x^2 + y^2 = 1$

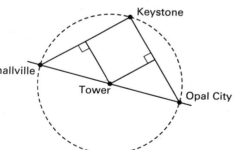

24. $(x - 6)^2 + (y - (-2))^2 = 8^2$
$(x - 6)^2 + (y + 2)^2 = 64$

25. $42.5 \cdot \pi d = 42.5 \cdot \pi \cdot 8 = 340\pi \approx 1068$ in.

26. Take the perpendicular bisectors of two chords; their intersection is the center of the circle.

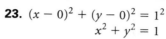

27.

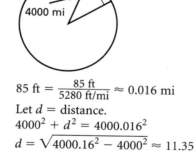

$85 \text{ ft} = \dfrac{85 \text{ ft}}{5280 \text{ ft/mi}} \approx 0.016 \text{ mi}$
Let d = distance.
$4000^2 + d^2 = 4000.016^2$
$d = \sqrt{4000.16^2 - 4000^2} \approx 11.35$
They are 11.35 miles away.

28. Sector area $= \dfrac{330°}{360°} \times \pi r^2$
$= \dfrac{11}{12} \times \pi \cdot \left(\dfrac{1}{2}\right)^2$
$= \dfrac{11}{48}\pi$
$\approx 0.72 \text{ mi}^2$
The irrigated sector is 72% of the entire field.

CHAPTER TEST

1. Sample answer: chord $\overline{QS}$, radius $\overline{CR}$, central angle $\angle QCR$, major arc $\overarc{RSQ}$

2. $m\angle QCR = 180° - 85° = 95°$

3. $m\overarc{RS} = 85°$

4. length $\overarc{RS} = \dfrac{85°}{360°}2(72)\pi = 34\pi \approx 107$ cm

5. $FG^2 + EF^2 = EG^2$
   $FG^2 + 6^2 = 10^2$
   $FG = \sqrt{10^2 - 6^2} = 8$

6. $KJ = HK = 6$
   $HJ = HK + KJ$
   $HJ = 6 + 6 = 12$

7. $\overline{BD}$

8. If $CE = 6$, then $AC = 6$ because they are both radii.
   $AB^2 + BC^2 = AC^2$
   $AB^2 + 4^2 = 6^2$
   $AB = \sqrt{6^2 - 4^2}$
   $AB = \sqrt{20} = 2\sqrt{5}$

9. If $AD = 48$, then $BD = 24$ because a radius that is perpendicular to a chord bisects the chord.
   $CD^2 = BC^2 + BD^2$
   $CD^2 = 7^2 + 24^2$
   $CD = \sqrt{7^2 + 24^2} = 25$
   The radius is 25 units long.

10. $m\angle KJL = \frac{1}{2}(80°) = 40°$

11. $m\angle JMK = 180° - 80° = 100°$, so $m\overarc{JK} = 100°$

12. $m\overarc{LN} = 2(m\angle LJN) = 2(25°) = 50°$

13. $m\overarc{JN} = m\overarc{JNL} - m\overarc{LN}$
    $m\overarc{JN} = 180° - 50°$
    $m\overarc{JN} = 130°$

14. $m\angle C = 120°$
    $m\angle D = \frac{1}{2} \cdot 120° = 60°$

15. $m\overarc{AF} = 2(m\angle ADF) = 2(52°) = 104°$

16. $m\overarc{BD} = 2(m\angle BFD) = 2(36°) = 72°$

17. $m\angle DEF = 180° - 36° - 52° = 92°$ so
    $m\angle AEF = 180° - 92° = 88°$

18. $m\angle BAG = \frac{1}{2}(m\overarc{BA}) = \frac{1}{2}(130°) = 65°$

19. $12x = (8)(9)$
    $x = 6$

20. $8(8 + x) = 15^2$
    $64 + 8x = 225$
    $8x = 161$
    $x = 20.125$ units

21. $HC^2 = CE^2 + EH^2$
    $HC^2 = 12^2 + 9^2$
    $HC = \sqrt{12^2 + 9^2}$
    $HC = 15$

22. Check students' graphs.

23. Check students' graphs.

24. Check students' graphs.

25. $x^2 + y^2 = 4$

26. $(x - 1)^2 + (y + 5)^2 = 16$

1. For A, surface area $= 7 + 6 + 3 + 2 + 5 + 4 + 4 + 1 + 1 + 5 = 38$ units^2
   For B, surface area $= 7 + 8 + 3 + 2 + 5 + 4 + 2 + 1 + 1 + 1 + 4 = 38$ units^2
   The answer is C.

2. $\dfrac{\pi \cdot 5^2}{\pi \cdot 6^2} = \dfrac{25}{36}$ $\qquad$ $\dfrac{\pi \cdot 6^2}{\pi \cdot 5^2} = \dfrac{36}{25}$
   $\dfrac{25}{36} < \dfrac{36}{25}$
   The answer is B.

3. slope of $\ell < 0$
   slope of $m > 0$
   The answer is B.

4. No measurements of $\triangle ABC$ are given.

   The answer is D.

5. The polygons are trapezoids. Not enough information is given to show whether they are congruent or similar.
   Choose b.

6. $\overset{\frown}{BC} = \text{m}\angle BOC = 180° - 120° = 60°$
   Choose d.

7. The back triangular face is an isosceles triangle with sides of length 5, 5, and 2 units. Its height is $h = \sqrt{5^2 - 1^2} = \sqrt{24} = 2\sqrt{6}$ which is the height of the pyramid. The unmarked side of the rectangular base is one leg of a right triangle, hence it has length $\sqrt{(7.81)^2 - 5^2} \approx 6$ units.
   The volume of the pyramid is $V = \frac{1}{3}Bh$ $= \frac{1}{3}(2 \cdot 6)(2\sqrt{6}) = 8\sqrt{6} \approx 19.6$ units^3.

8. Since Volumes have ratio $27 : 1$
   Radii have ratio $\sqrt[3]{27} : \sqrt[3]{1} = 3 : 1$.
   The smaller radius is $15 \Rightarrow$ larger radius is $3 \cdot 15 = 45$ in.

9.

10. $(0, 0)$ and $(-4, -3)$

11. $\overline{PX} \cong \overline{PW}$ since both are radii of $\odot P$.
    $\angle PYX \cong \angle PYW$ since both are right angles.
    $\overline{PY} \cong \overline{PY}$ by the Reflexive Property. Thus, $\triangle PYX \cong \triangle PYW$ by HL.

12. The height of the triangle is $3 + \frac{1}{2}(3) = 4.5$ and a base has length $2 \cdot \frac{1}{2}(3)\sqrt{3} = 3\sqrt{3}$. The area of the circle is $\pi \cdot 3^2 = 9\pi \approx 28.27$. The area of the triangle is $\frac{1}{2}(3\sqrt{3})(4.5) \approx 11.69$
    Probability $= 1 - \dfrac{11.69}{28.27} \approx 0.59$

13. $V = \pi r^2 h$
    $= \pi \cdot 2^2 \cdot 7$
    $= 28\pi$
    ≈ 88.0

14. $m = \dfrac{\triangle y}{\triangle x} = \dfrac{12 - 6}{9 - 2} = \dfrac{6}{7}$

15. The height of the parallelogram is the longer leg of a 30-60-90 right triangle with hypotenuse of length 2, so $h = \frac{1}{2}(\sqrt{3})(2) = \sqrt{3}$. $A = b \cdot h = 4 \cdot \sqrt{3}$ $= 4\sqrt{3} \approx 6.93$

CHAPTER 10

Trigonometry

In Exercises 6 and 7, answers may vary due to inaccuracies in measurement.

6. Sample answer: $\tan A = \dfrac{\text{opposite}}{\text{adjacent}} \approx \dfrac{2.9}{4.65} \approx 0.62.$ **7.** Sample answer: $\tan A \approx 1.45.$

8. $\tan 20° \approx 0.36$ **9.** $\tan 40° \approx 0.84$ **10.** $\tan 70° \approx 2.75$

11. $\tan x = \dfrac{3}{4} \Rightarrow x = \tan^{-1}\dfrac{3}{4} \approx 37°$

PAGE 636, PRACTICE AND APPLY

12. $\dfrac{12}{21} = \dfrac{4}{7} \approx 0.5714$ **13.** $\dfrac{3}{1.2} = \dfrac{5}{2} = 2.5$ **14.** $\dfrac{10}{15} = \dfrac{2}{3} \approx 0.6667$ **15.** $\dfrac{24}{10} = \dfrac{12}{5} = 2.4$

16. $\dfrac{6}{8} = \dfrac{3}{4} = 0.75$ **17.** Since the length of the opposite and adjacent sides are equal, $\tan \angle A = 1.$

18. $\tan 25° \approx 0.47$ **19.** $\tan 67° \approx 2.36$ **20.** $\tan 19° \approx 0.34$ **21.** $\tan 53° \approx 1.33$

22. $\tan 75° \approx 3.73$ **23.** $\tan 89° \approx 57.29$ **24.** $\tan^{-1}\dfrac{3}{8} \approx 21°$ **25.** $\tan^{-1}\dfrac{7}{5} \approx 54°$

26. $\tan^{-1} 3 \approx 72°$ **27.** $\tan^{-1} 9.5 \approx 84°$ **28.** $\tan^{-1} 1 = 45°$ **29.** $\tan^{-1} 0 = 0°$

30. $\tan 37° = \dfrac{x}{18} \Rightarrow x = (18)(\tan 37°) \approx 13.56$ **31.** $\tan 45° = \dfrac{x}{2} \Rightarrow x = (2)(\tan 45°) = 2$

32. $\tan 51° = \dfrac{46}{x} \Rightarrow (x)(\tan 51°) = 46 \Rightarrow x = \dfrac{46}{\tan 51°} \approx 37.25$

33. $\tan 28° = \dfrac{12}{x} \Rightarrow (x)(\tan 28°) = 12 \Rightarrow x = \dfrac{12}{\tan 28°} \approx 22.57$

34. $\tan 40° = \dfrac{y}{x} \Rightarrow y = x \cdot \tan 40°$

Using the Pythagorean theorem, $x^2 + y^2 = 15^2.$

Substituting $y = x \cdot \tan 40°$ into the Pythagorean theorem gives:

$$x^2 + (x \cdot \tan 40°)^2 = 15^2$$
$$x^2 + x^2 \cdot (\tan 40°)^2 = 225$$
$$x^2 + 0.70x^2 = 225$$
$$1.70x^2 = 225$$
$$x^2 = \dfrac{225}{1.70}$$
$$x = \sqrt{\dfrac{225}{1.70}} \approx 11.5$$
$$y = x \cdot \tan 40° \approx (11.5)(0.84) \approx 9.7$$

35. $\tan 43° = \frac{AB}{530} \Rightarrow AB = (530)(\tan 43°) = 494.2$ m

36. Railway: $\tan x = \frac{18}{100} \Rightarrow x = \tan^{-1}\left(\frac{18}{100}\right) \approx 10.2°$

37. $\tan x = \frac{25}{100} \Rightarrow x = \tan^{-1}\left(\frac{25}{100}\right) \approx 14.0°$

38. $\tan 42° = \frac{S}{20} \Rightarrow x = S = (20)(\tan 42°) = 18.0$ m

39. $\tan 75° = \frac{PQ}{300} \Rightarrow PQ = (300)(\tan 75°) \approx 1119.6$ ft

PAGE 637, LOOK BACK

40. $V = (4)(6)(10) = 240$ units^3
$S = (2)(10)(6) + (2)(4)(6) + (2)(4)(10) = 248$ units^2

41. $V = (\pi)(5^2)(14) = 350\pi \approx 1099.56$ units^3
$S = (2)(\pi)(5^2) + (\pi)(10)(14) = 190\pi \approx 596.90$ units^2

42. $V = \frac{1}{3}(\pi)(7^2)(14.2) \approx 728.64$ units^3

$\ell = \sqrt{14.2^2 + 7^2} \approx 15.83$ units

$S = (\pi)(7^2) + (\pi)(7)(15.83) \approx 502.09$ units^2

43. $V = \frac{4}{3}(\pi)(25^3) \approx 65,449.85$ units^3
$S = (4)(\pi)(25^2) = 25\pi \approx 7853.98$ units^2

44. NQ

45. $(MQ)^2 + 3^2 = 8^2 \Rightarrow MQ = \sqrt{8^2 - 3^2}$
$= \sqrt{55} \approx 7.42$
$QN = \sqrt{55} \approx 7.42$

46. $(MQ)^2 + 4^2 = 12^2 \Rightarrow MQ = \sqrt{12^2 - 4^2}$
$= \sqrt{128} \approx 11.31$
$QN = \sqrt{128} \approx 11.31$

PAGE 637, LOOK BEYOND

47. $\frac{\text{run}}{\text{rise}} = \frac{180 \text{ cubits}}{250 \text{ cubits}} = 0.72$

48. The tangent ratio is $\frac{\text{rise}}{\text{run}}$, so this is the reciprocal of the tangent ratio.

49. seked $= 0.72 \times 7 = 5.04$

10.2 PAGE 643, GUIDED SKILLS PRACTICE

5. $\sin 34° = \frac{h}{25} \Rightarrow h = 25 \cdot \sin 34° \approx 14$ ft

6. $\sin 60° = \frac{h}{8} \Rightarrow h = 8 \cdot \sin 60° \approx 7$ ft

7. $\sin 24° = \frac{h}{32} \Rightarrow h = 32 \cdot \sin 24° \approx 13$ ft

8. $\cos \theta = \frac{72}{87} \Rightarrow \theta = \cos^{-1}\left(\frac{72}{87}\right) \approx 34°$

9. $\cos \theta = \frac{42}{82} \Rightarrow \theta \cos^{-1}\left(\frac{42}{82}\right) \approx 59°$

10. $\frac{12}{13} \approx 0.9231$

11. $\frac{5}{13} \approx 0.3846$

12. $\frac{5}{13} \approx 0.3846$

13. $\frac{12}{13} \approx 0.9231$

14. $\frac{12}{5} = 2.4$

15. $\frac{5}{12} \approx 0.4167$

16. $\text{m}\angle C = \sin^{-1}\left(\frac{12}{13}\right) = \cos^{-1}\left(\frac{5}{13}\right) = \tan^{-1}\left(\frac{12}{5}\right) \approx 67.38°$

17. $\text{m}\angle D = 90° - 67.38° \approx 22.62°$

18. X

19. Y

20. Y

21. X

22. X

23. Y

24. 0.57

25. 0.31

26. 0.84

27. 0.62

28. 0.71

29. 0.71

30. 17°

31. 61°

32. 56°

33. 83°

34. 30°

35. 18°

36. Let h be the height of the triangle.

$$\sin 35° = \frac{h}{6} \Rightarrow h = 6 \cdot \sin 35° \approx 3.44$$

$$A = \frac{1}{2}(10)(3.44) = 17.2 \text{ units}^2$$

37. Let h be the height of the parallelogram.

$$\sin 40° = \frac{h}{7} \Rightarrow h = 7 \cdot \sin 40° \approx 4.5$$

$$A = (15)(4.5) = 67.5 \text{ units}^2$$

38. $\sin^{-1}(\sin \theta) = \theta$. If you find the sine of an angle and then the inverse sine of the result, you end up with the original angle.

39. $\cos^{-1}(\cos \theta) = \theta$. If you find the cosine of an angle and then the inverse cosine of the result, you end up with the original angle.

40. $\tan^{-1}(\tan \theta) = \theta$. If you find the tangent of an angle and then the inverse tangent of the result, you end up with the original angle.

41. $\sin 70° = 0.9397$; $\cos 20° = 0.9397$; $\sin 70° = \cos 20°$; They are the same because the side opposite the 70° angle is the side adjacent to the 20° angle.

42. $\sin 30° = \cos 60°$ $\sin 65° = \cos 25°$
$\sin 50° = \cos 40°$ $\sin 45° = \cos 45°$

43. $\sin \theta = \cos(90 - \theta)$ $\cos \theta = \sin(90 - \theta)$

44. $\tan \theta \cdot \cos \theta = \dfrac{\sin \theta}{\cancel{\cos \theta}} \cdot \cancel{\cos \theta} = \sin \theta$

45. $\dfrac{\sin \theta}{\tan \theta} = \dfrac{\sin \theta}{\frac{\sin \theta}{\cos \theta}} = \cancel{\sin \theta} \cdot \dfrac{\cos \theta}{\cancel{\sin \theta}} = \cos \theta$

46. $1 - (\sin \theta)^2 = (\cos \theta)^2$

47. $\dfrac{1}{(\cos \theta)^2} - 1 = \dfrac{1}{(\cos \theta)^2} - \dfrac{(\cos \theta)^2}{(\cos \theta)^2} = \dfrac{1 - (\cos \theta)^2}{(\cos \theta)^2}$
$$= \dfrac{(\sin \theta)^2}{(\cos \theta)^2} = \left(\dfrac{\sin \theta}{\cos \theta}\right)^2 = (\tan \theta)^2$$

48. Since $(\sin \theta)^2 + (\cos \theta)^2 = 1$, then $(\cos \theta)^2 = 1 - (\sin \theta)^2$
and $[(\cos \theta)^2]^2 = [1 - (\sin \theta)^2]^2$
or $(\cos \theta)^4 = 1 - 2 \cdot (\sin \theta)^2 + (\sin \theta)^4$ by using FOIL.

So, $(\sin \theta)^4 - (\cos \theta)^4 = (\sin \theta)^4 - [1 - 2 \cdot (\sin \theta)^2 + (\sin \theta)^4]$
$$= (\sin \theta)^4 - 1 + 2 \cdot (\sin \theta)^2 - (\sin \theta)^4$$
$$= -1 + 2 \cdot (\sin \theta)^2$$
$$= 2(\sin \theta)^2 - 1$$

49. $\sin 72° = \dfrac{h}{20} \Rightarrow h = 20 \cdot \sin 72° \approx 19$ feet

50. Let x be the angle the slide forms with the tower.

Then $\cos x = \dfrac{21}{25}$, so $x = \cos^{-1}\left(\dfrac{21}{25}\right) \approx 33°$.

51. Let θ be the angle the ramp forms with the ground.

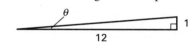

Then $\tan \theta = \dfrac{1}{12}$, so $\theta = \tan^{-1}\left(\dfrac{1}{12}\right) \approx 4.8°$.

PAGE 646, LOOK BACK

52. $y = 7\sqrt{3}$; $z = 2(7) = 14$

53. $x = \dfrac{14}{\sqrt{3}} = \dfrac{14\sqrt{3}}{3}$; $z = 2\left(\dfrac{14\sqrt{3}}{2}\right) = \dfrac{28\sqrt{3}}{3}$

54. $x = \dfrac{13}{2} = 6.5$; $y = 6.5\sqrt{3}$

55. $p = q = 1$; $r = \sqrt{2}$

56. $p = q = 3$; $r = 3\sqrt{2}$

57. $q = \dfrac{16}{\sqrt{2}} = 8\sqrt{2}$; $p = q = 8\sqrt{2}$

58. $\dfrac{x}{360}(40\pi) = 12\pi \Rightarrow x = \dfrac{360}{40\pi}(12\pi) \Rightarrow x = 108°$

59. $\dfrac{x}{360}(200\pi) = 10\pi \Rightarrow x = \dfrac{360}{200\pi}(10\pi) \Rightarrow x = 18°$

60. $\dfrac{x}{360}(50\pi) = 3\pi \Rightarrow x = \dfrac{360}{50\pi}(3\pi) \Rightarrow x = 21.6°$

61. cot θ is the reciprocal of tan θ, sec θ, is the reciprocal of cos θ, and csc θ is the reciprocal of sin θ.

62. a. $\tan\theta = \cot(90 - \theta)$ **b.** $\cot\theta = \tan(90 - \theta)$

    c. $\sec\theta = \csc(90 - \theta)$ **d.** $\csc\theta = \sec(90 - \theta)$

10.3 **PAGE 651, GUIDED SKILLS PRACTICE**

8. $210° - 180° = 30°$. P' has coordinates $\left(-\dfrac{\sqrt{3}}{2}, -\dfrac{1}{2}\right)$.

9. $\sin\theta = 0.6428$
$\cos\theta = 0.7660$

10. Let t be the number of elapsed seconds. Since the wheel turns at one degree per second, after t seconds the point X has rotated through an angle to $t°$. So the coordinates of X are $(\cos t, \sin t)$. The vertical position of point X is $\sin t$. After 10 minutes, $t = 10 \cdot 60 = 600$ seconds, so the vertical position is $\sin 600° \approx -0.866$. After 20 minutes, $t = 20 \cdot 60 = 1200$ seconds, so the vertical position is in $1200° \approx 0.866$. After 1 hour, $t = 60 \cdot 60 = 3600$ seconds, so the vertical position is $\sin 3600° = 0$.

11. In Quadrants I and IV, cosine is positive because the graph of cosine is above the x–axis. In Quadrants II and III, cosine is negative because the graph is below the x–axis.

12. $\sin\theta = 0.7071 \Rightarrow \theta = \sin^{-1}(0.7071) \approx 45°$. Another value of θ for which $\sin\theta = 0.7071$ is $180° - 45° = 135°$.

---

13. 0.9063   **14.** 0.9063   **15.** −0.9063   **16.** −0.9063

17. 0.4226   **18.** −0.4226   **19.** −0.4226   **20.** 0.4226

For problems 21–24, use a 45–45–90 triangle with hypotenuse of length 1 and legs of length $\dfrac{1}{\sqrt{2}}$.

21. $\sin 45° = \dfrac{1}{\sqrt{2}} \approx 0.7071$
$\cos 45° = \dfrac{1}{\sqrt{2}} \approx 0.7071$

22. $\sin 135° = \dfrac{1}{\sqrt{2}} \approx 0.7071$
$\cos 135° = \dfrac{-1}{\sqrt{2}} \approx -0.7071$

23. $\sin 225° = \dfrac{-1}{\sqrt{2}} \approx -0.7071$
$\cos 225° = \dfrac{-1}{\sqrt{2}} \approx -0.7071$

24. $\sin 315° = \dfrac{-1}{\sqrt{2}} \approx -0.7071$
$\cos 315° = \dfrac{1}{\sqrt{2}} \approx 0.7071$

For problems 25–28, use a 30–60–90 triangle with hypotenuse of length 1. The leg across from the 30° angle is of length $\dfrac{1}{2}$ and the leg across from the 60° angle is $\dfrac{\sqrt{3}}{2}$.

25. $\sin 30° = \dfrac{1}{2} = 0.5$
$\cos 30° = \dfrac{\sqrt{3}}{2} \approx 0.8660$

26. $\sin 150° = \dfrac{1}{2} = 0.5$
$\cos 150° = \dfrac{-\sqrt{3}}{2} \approx -0.8660$

27. $\sin 210° = \dfrac{-1}{2} = -0.5$
$\cos 210° = \dfrac{-\sqrt{3}}{2} \approx -0.8660$

28. $\sin 330° = \dfrac{-1}{2} = -0.5$
$\cos 330° = \dfrac{\sqrt{3}}{2} \approx 0.8660$

29. $x = \cos 30° \approx 0.8660$
$y = \sin 30° = 0.5$

30. $x = \cos 60° = 0.5$
$y = \sin 60° \approx 0.8660$

31. $x = \cos 90° = 0$
$y = \sin 90° = 1$

32. $x = \cos 120° = -0.5$
$y = \sin 120° \approx 0.8660$

33. $x = \cos 180° = -1$
$y = \sin 180° = 0$

34. $x = \cos 210° \approx -0.8660$
$y = \sin 210° = -0.5$

35. $x = \cos 300° = 0.5$
$y = \sin 300° \approx -0.8660$

36. $x = \cos 360° = 1$
$y = \sin 360° = 0$

37. $\theta \approx \sin^{-1}(0.7071) \approx 45°$
Another value is $180° - 45° = 135°$.

38. $\theta \approx \sin^{-1}(0.8660) \approx 60°$
Another value is $180° - 60° = 120°$.

39. $\theta \approx \sin^{-1}(0.5000) = 30°$
Another value is $180° - 30° = 150°$.

40. $\theta \approx \sin^{-1}(0.9659) \approx 75°$
Another value is $180° - 75° = 105°$.

41. $\theta \approx \sin^{-1}(0.3217) \approx 19°$
Another value is $180° - 19° = 161°$.

42. $\theta \approx \sin^{-1}(0.9900) \approx 82°$
Another value is $180° - 82° = 98°$.

43. $\theta \approx \sin^{-1}(0.9900) \approx 87°$
Another value is $180° - 87° = 93°$.

44. $\theta \approx \sin^{-1}(0.9999) \approx 89°$
Another value is $180° - 89° = 91°$.

45. $\sin \theta = 0.4756 \Rightarrow \theta \approx \sin^{-1}(0.4756) \approx 28°$.
Since $\sin \theta$ is positive only in Quadrants I and II, the only other value occurs when θ is $180° - 28° = 152°$.

46. $\cos \theta = -0.7500 \Rightarrow \theta \approx \cos^{-1}(-0.7500) \approx 139°$.
Since $\cos \theta$ is negative only in Quadrants II and III, the other value of θ occurs at an angle of $180° - 139° = 41°$ from horizontal in Quadrant III. So the value of θ is $180° + 41° = 221°$.

47. Let t be the number of elapsed hours. Since the satellite orbits at one degree per hour, its position at time t is $(\cos t°, \sin t°)$, with $\cos t°$ representing the horizontal position and $\sin t°$ representing the vertical position. After 2 days, $t = 2 \cdot 24 = 48$ hours and the horizontal position is $\cos 48° \approx 0.6691$. After 5 days, $t = 5 \cdot 24 = 120$ and the horizontal position is $\cos 120° = -0.5$.

PAGE 653, LOOK BACK

48. a

49. a

50. a

51. b

52. $(\cos A)^2$

53. $\tan A$

54. $\cos B$ or $\cos (90° - m\angle A)$

55. $\sin B$ or $\sin (90° - m\angle A)$

PAGE 653, LOOK BEYOND

56. a. • ○ ••
    III   IIIV

b. ○•• • •
  II I   III IV

c. •• •• ○
  IV III   II I

57. I, Io: 1.75 days; II, Europa: 3.5 days; III, Ganymede: 7 days; IV, Callisto: 16 days

58. Sine waves. They are repeatedly traveling around Jupiter in a circular motion that can be modeled by a point rotating around a circle. This creates the sine wave pattern.

5. $\dfrac{\sin 36°}{7} = \dfrac{\sin 58°}{b}$

$b \sin 36° = 7 \sin 58°$

$b = \dfrac{7 \sin 58°}{\sin 36°} \approx 10.1$

$m\angle C = 180° - 36° - 58° = 86°$

$\dfrac{\sin 36°}{7} = \dfrac{\sin 86°}{c}$

$c \sin 36° = 7 \sin 86°$

$c = \dfrac{7 \sin 86°}{\sin 36°} \approx 11.9$

6. $\dfrac{\sin 50°}{22} = \dfrac{\sin S}{9}$

$22 \sin S = 9 \sin 50°$

$\sin S = \dfrac{9 \sin 50°}{22}$

$S = \sin^{-1}\left(\dfrac{9 \sin 50°}{22}\right) \approx 18.26°$

$m\angle Q = 180° - 50° - 18.26°$
$\qquad = 111.74°$

7. $m\angle Y = 180° - 40° - 35° = 105°$

$\dfrac{\sin 40°}{x} = \dfrac{\sin 105°}{10}$

$x \sin 105° = 10 \cdot \sin 40°$

$x = \dfrac{10 \cdot \sin 40°}{\sin 105°} \approx 6.65$

$\dfrac{\sin 105°}{10} = \dfrac{\sin 35°}{z}$

$z \cdot \sin 105° = 10 \sin 35°$

$z = \dfrac{10 \sin 35°}{\sin 105°} \approx 5.94$

The distance from point X to point Y to point Z is $5.94 + 6.65 = 12.59$. It is greater than the distance from point X to point Z by $12.59 - 10 = 2.59$.

8. $m\angle C = 180° - 24° - 56° = 100°$

$\dfrac{\sin 24°}{1.22} = \dfrac{\sin 100°}{c} \Rightarrow (c)(\sin 24°) = 1.22 \sin 100°$

$\Rightarrow c = \dfrac{1.22 \sin 100°}{\sin 24°} \approx 2.954$ cm

9. $\dfrac{\sin A}{3.12} = \dfrac{\sin 29°}{3.28} \Rightarrow \sin A = \dfrac{3.12 \sin 29°}{3.28}$

$\Rightarrow m\angle A = \sin^{-1}\left(\dfrac{3.12 \sin 29°}{3.28}\right) \Rightarrow m\angle A \approx 27.46°$

10. $m\angle C = 180° - 29° - 27.46° = 123.54°$

$\dfrac{\sin 29°}{3.28} = \dfrac{\sin 123.54°}{c} = \Rightarrow (c)(\sin 29°)$

$= 3.28 \sin 123.54° \Rightarrow c = \dfrac{3.28 \sin 123.54°}{\sin 29°} \approx 5.64$ cm

11. $\dfrac{\sin C}{2.13} = \dfrac{\sin 67°}{7.36} \Rightarrow \sin C = \dfrac{2.13 \sin 67°}{7.36}$

$\Rightarrow m\angle C = \sin^{-1}\left(\dfrac{2.13 \sin 67°}{7.36}\right) \Rightarrow m\angle C \approx 15.45°$

12. $m\angle A = 180° - 67° - 15.45° = 97.55°$

$\dfrac{\sin 67°}{7.36} = \dfrac{\sin 97.55°}{a} \Rightarrow (a)(\sin 67°) = 7.36 \sin 97.55°$

$\Rightarrow a = \dfrac{7.36 \sin 97.55°}{\sin 67°} \approx 7.93$ cm

13. $m\angle A = 1800 - 73° - 85° = 22°$

$\dfrac{\sin 22°}{3.14} = \dfrac{\sin 73°}{b} \Rightarrow (b)(\sin 22°) = 3.14 \sin 73°$

$\Rightarrow b = \dfrac{3.14 \sin 73°}{\sin 22°} = 8.02$ cm

14. $m\angle C = 180° - 44° - 35° = 101°$

$\dfrac{\sin 101°}{2.4} = \dfrac{\sin 35°}{a} \Rightarrow (a)(\sin 101°) = 2.4 \sin 35°$

$\Rightarrow a = \dfrac{2.4 \sin 35°}{\sin 101°} = 1.40$ cm

15. $m\angle B = 180° - 72° - 53° = 55°$

$\dfrac{\sin 72°}{2.34} = \dfrac{\sin 55°}{b} \Rightarrow (b)(\sin 72°) = 2.34 \sin 55°$

$\Rightarrow b = \dfrac{2.34 \sin 55°}{\sin 72°} \approx 2.02$ cm

16. $m\angle R = 180° - 30° - 40° = 110°$

$$\frac{\sin 40°}{10} = \frac{\sin 110°}{r}$$

$$r \cdot \sin 40° = 10 \cdot \sin 110°$$

$$r = \frac{10 \cdot \sin 110°}{\sin 40°} \approx 14.62$$

$$\frac{\sin 40°}{10} = \frac{\sin 30°}{p}$$

$$p \cdot \sin 40° = 10 \cdot \sin 30°$$

$$p = \frac{10 \cdot \sin 30°}{\sin 40°} \approx 7.78$$

18. $m\angle Q = 180° - 72° - 36° = 72°$

$$\frac{\sin P}{p} = \frac{\sin Q}{q}$$

$$\frac{\sin 72°}{p} = \frac{\sin 72°}{12}$$

$$12 \cdot \sin 72° = p \cdot \sin 72°$$

$$p = \frac{12 \cdot \sin 72°}{\sin 72°} = 12$$

$$\frac{\sin R}{r} = \frac{\sin P}{p}$$

$$\frac{\sin 36°}{r} = \frac{\sin 72°}{12}$$

$$12 \cdot \sin 36° = r \cdot \sin 72°$$

$$r = \frac{12 \cdot \sin 36°}{\sin 72°} \approx 7.42$$

17. $m\angle P = 180° - 60° - 80° = 40°$

$$\frac{\sin 80°}{7} = \frac{\sin 60°}{q}$$

$$q \cdot \sin 80° = 7 \cdot \sin 60°$$

$$q = \frac{7 \cdot \sin 60°}{\sin 80°} \approx 6.16$$

$$\frac{\sin 80°}{7} = \frac{\sin 40°}{p}$$

$$p \cdot \sin 80° = 7 \cdot \sin 40°$$

$$p = \frac{7 \cdot \sin 40°}{\sin 80°} \approx 4.57$$

19. This may be an ambiguous case.

$$\frac{\sin 60°}{9} = \frac{\sin Q}{7}$$

$$\frac{7 \cdot \sin 60°}{9} = \sin Q$$

$$m\angle Q = \sin^{-1}\left(\frac{7\sin 60°}{9}\right)$$

$$m\angle Q \approx 42.34°$$

Since $180° - 42.34° = 137.66°$ and $137.66° + 60° > 180°$, there is only one triangle possible.

$m\angle R = 180° - 60° - 42.34° = 77.66°$

$$\frac{\sin 60°}{9} = \frac{\sin 77.66°}{r}$$

$$r \sin 60° = 9 \sin 77.66°$$

$$r = \frac{9 \cdot \sin 77.66°}{\sin 60°} \approx 10.15$$

20. This may be an ambiguous case.

$$\frac{\sin 120°}{12} = \frac{\sin R}{8}$$

$$\frac{8 \cdot \sin 120°}{12} = \sin R$$

$$m\angle R = \sin^{-1}\left(\frac{8 \cdot \sin 120°}{12}\right) \approx 35.26°$$

Since $180° - 35.26° = 144.74°$ and $144.74° + 120° > 180°$, there is only one triangle possible.

$m\angle P \approx 180° - 120° - 35.26° = 24.74°$

$$\frac{\sin 24.74°}{p} = \frac{\sin 120°}{12}$$

$$12 \cdot \sin 24.74° = p \cdot \sin 120°$$

$$p = \frac{12 \cdot \sin 24.74°}{\sin 120°}$$

$$p \approx 5.80$$

21. This may be an ambiguous case.

$$\frac{\sin 45°}{8} = \frac{\sin P}{11}$$

$$\frac{11 \cdot \sin 45°}{8} = \sin P$$

$$m\angle P = \sin^{-1}\left(\frac{11 \cdot \sin 45°}{8}\right) \approx 76.48°.$$

The other possible angle is
$180° - 76.48° = 103.52°$.
Since $103.52° + 45° < 180°$, there are two possible triangles.

Case 1: $m\angle P = 76.48°$
$m\angle Q = 180° - 45° - 76.48° = 58.52°$

$$\frac{\sin 45°}{8} = \frac{\sin 58.52°}{q}$$

$$q = \frac{8 \cdot \sin 58.52°}{\sin 45°} \approx 9.65$$

Case 2: $m\angle P = 103.52°$
$m\angle Q \approx 180° - 45° - 103.52° = 31.48°$

$$\frac{\sin 45°}{8} = \frac{\sin 31.48°}{q} \Rightarrow q = \frac{8 \sin 31.48°}{\sin 45°} \Rightarrow q \approx 5.91$$

22. Sample answer: If $a = b \sin A$, the triangle is a right triangle. Since $\angle A$ is fixed and another angle is a right angle, there is only one possibility for the third angle.

23. Sample answer: If $a < b \sin A$, then side a is too short to form any triangle, so no triangle is possible.

24. Sample answer: If $a \geq b$, then there is only 1 triangle possible because side a must "swing" to the right.

25. Sample answer: If $b \sin A < a < b$, there are 2 triangles possible, one by "swinging" side a to the left and the other by "swinging" side a to the right.

26. Sample answer: The only way that side a can "swing" to touch the line in a different point is to the left side of $\angle A$, in which case, the triangle formed does not include $\angle A$, so only one triangle is possible.

27. Sample answer: If $a \le b$ then no triangle is possible since for an obtuse or a right triangle, the side across from the obtuse or right angle must be the longest side of the triangle.

28. Sample answer: Let A be the angle, b the side nearest A and a the side across from A.

Case I: If $m\angle A < 90°$ and

(i) $a < b \sin A$, then no triangle is possible.

(ii) $a \ge b$, then there is 1 triangle possible.

(iii) $b \sin A < a < b$, then there are 2 triangles possible.

Case II: If $m\angle A \ge 90°$ and

(i) $a > b$, there is 1 triangle possible.

(ii) $a \le b$, then no triangle is possible.

29. Since $m\angle A < 90°$ and $a > b$, there is only 1 triangle possible.

30. Since $m\angle A > 90°$ and $a > b$, there is only 1 triangle possible.

31. Since $m\angle A < 90°$ and $b \sin A < a < b$ ($b \sin A = 6 \sin 30° = 3$), there are 2 triangles possible.

32. Since $m\angle A > 90°$ and $a \le b$, no triangle is possible.

33. $\sin A = \frac{a}{c}$; $\sin B = \frac{b}{c}$

34. $\dfrac{\sin A}{a} = \dfrac{\frac{a}{c}}{a} = \dfrac{1}{c}$; $\dfrac{\sin B}{b} = \dfrac{\frac{b}{c}}{b} = \dfrac{1}{c}$; $\dfrac{\sin C}{c} = \dfrac{\sin 90°}{c} = \dfrac{1}{c}$

35. $\dfrac{\sin A}{a} = \dfrac{\sin B}{b} = \dfrac{\sin C}{c}$, since they all are equal to the same quantity.

36. $b \sin A$

37. $a \sin B$

38. Multiplication Property of Equality

39. $a \sin B = b \sin A$

40. Division Property of Equality (both sides divided by ab)

41. $\dfrac{h_2}{a}$

42. Definition of sine

43. $c \sin A$

44. $a \sin C$

45. Multiplication Property of Equality

46. Substitution Property of Equality

47. $\dfrac{\sin A}{a} = \dfrac{\sin C}{c}$

48. $\dfrac{\sin A}{a} = \dfrac{\sin B}{b} = \dfrac{\sin C}{c}$

49. Transitive or Substitution Property of Equality

50. Prove $\dfrac{a - b}{\sin A - \sin B} = \dfrac{a + b}{\sin A + \sin B}$.

Proof:

Since $\dfrac{\sin A}{a} = \dfrac{\sin B}{b}$, by the law of sines, then $b \sin A = a \sin B$.

To complete the proof, show that the left side and the right side are equal to the same quantity.

Left side:

$$\begin{aligned}
\frac{a - b}{\sin A - \sin B} &= \frac{(a - b) \cdot a}{(\sin A - \sin B) \cdot a} && \text{(multiply by } a\text{)} \\
&= \frac{(a - b) \cdot a}{a \cdot \sin A - a \cdot \sin B} && \text{(distribute)} \\
&= \frac{(a - b) \cdot a}{a \cdot \sin A - b \sin A} && \text{(substitution)} \\
&= \frac{(a - b) \cdot a}{(a - b) \cdot \sin A} && \text{(factoring)} \\
&= \frac{a}{\sin A}
\end{aligned}$$

Right side:

$$\begin{aligned}
\frac{a + b}{\sin A + \sin B} &= \frac{(a + b) \cdot a}{(\sin A + \sin B) \cdot a} && \text{(multiply by } a\text{)} \\
&= \frac{(a + b) \cdot a}{a \cdot \sin A + a \cdot \sin B} && \text{(distribute)} \\
&= \frac{(a + b) \cdot a}{a \cdot \sin A + b \cdot \sin A} && \text{(substitution)} \\
&= \frac{(a + b) \cdot a}{(a + b) \cdot \sin A} && \text{(factoring)} \\
&= \frac{a}{\sin A}
\end{aligned}$$

Since the left side and the right side both equal $\dfrac{a}{\sin A}$, the two sides are equal by the Transitive Property.

51. Let A be the angle nearest the polar bear. Then $m\angle A = 180° - 65° - 49° = 66°$.

$$\frac{\sin 66°}{9} = \frac{\sin 49°}{x}$$

$$x = \frac{9 \cdot \sin 49°}{\sin 66°} \approx 7.44 \text{ km}$$

The polar bear is 7.44 km from Station 2.

The shortest distance to the bear from the road is the line segment perpendicular to the road. Its length is y where $\sin 65° = \dfrac{y}{7.44}$.

So $y = (7.44) \sin 65° \approx 6.74$ km.

52. Let A be the angle opposite Oak Street and angle B be the angle opposite 3rd Ave.

$$\frac{\sin A}{42} = \frac{\sin 74°}{48.8} \Rightarrow \sin A = \frac{42 \sin 74°}{48.8}$$
$$\Rightarrow m\angle A = \sin^{-1}\left(\frac{42 \sin 74°}{48.8}\right) \Rightarrow$$
$$m\angle A \approx 55.82°$$
$$m\angle B = 180° - 74° - 55.82° = 50.18°$$

53. Let A be the side across from tower A and b be the side across from tower B.
The third angle is $180° - 70° - 80° = 30°$.

$$\frac{\sin 30°}{5} = \frac{\sin 80°}{a} \Rightarrow a = \frac{5 \cdot \sin 80°}{\sin 30°} \approx 9.85 \text{ mi.}$$
$$\frac{\sin 30°}{5} = \frac{\sin 70°}{b} \Rightarrow b = \frac{5 \cdot \sin 70°}{\sin 30°} \approx 9.40 \text{ mi.}$$

Tower A is closer to the smoke plume. The smoke is 9.40 miles from the tower.

54. The supplement to the 27.5° angle is
$180° - 27.5° = 152.5°$, so $m\angle B$ is
$180° - 11.5° - 152.5° = 16°$.

55. $\dfrac{\sin 11.5°}{BC} = \dfrac{\sin 16°}{5} \Rightarrow BC = \dfrac{5 \cdot \sin 11.5°}{\sin 16°} \approx 3.62$ mi.

56. Let h be the height of the mountain.
Then $\sin 27.5° = \dfrac{h}{BC} = \dfrac{h}{3.62} \Rightarrow h = (3.62) \cdot \sin 27.5°$
$h \approx 1.67$ mi.

PAGE 662, LOOK BACK

57. $S = 4(\pi)(15^2) = 900\pi \approx 2827.4$ m^2
$V = \frac{4}{3}(\pi)(15^3) = 4500\pi \approx 14{,}137.2$ m^3

58. If the radius were tripled, the surface area would be multiplied by $3^2 = 9$. The volume would be multiplied by $3^3 = 27$.

59. Third side of larger triangle $= \sqrt{51^2 - 45^2} = 24$;
$\dfrac{x}{24} = \dfrac{24}{51} \Rightarrow 51x = 576 \Rightarrow x = \dfrac{576}{51} = \dfrac{192}{17}$
≈ 11.3 units

60. $\dfrac{x+2}{2x+1} = \dfrac{2x}{4x-4} \Rightarrow 4x^2 + 2x = 4x^2 + 4x - 8 \Rightarrow$
$2x = 4x - 8 \Rightarrow x = 4$

61. 65.91°

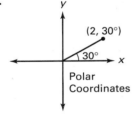

62. 60°

63. 45°

PAGE 662, LOOK BEYOND

64.

Form a right triangle with hypotenuse 2 and one 30° angle. The side across from the 30° angle is $\frac{2}{2} = 1$, and the side across from the 60° angle is $1 \cdot \sqrt{3} = \sqrt{3}$. So the rectangular coordinates are $(\sqrt{3}, 1)$.

65.

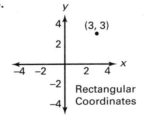

Draw line segments from the point $(3, 3)$ to $(0, 0)$ and from $(3, 3)$ perpendicular to the x–axis, forming a right triangle with the x–axis. Since the legs of the triangle are both of length 3, this is a 45–45–90 triangle. The hypotenuse has length $3\sqrt{2}$, and the angle formed between the hypotenuse and the x–axis is a 45° angle. Therefore, the polar coordinates are $(3\sqrt{2}, 45°)$.

66. $r = \sin \theta$ is a circle of diameter 1, with the bottom edge touching the origin.

$r = \sin (2\theta)$ is a four-petaled flower shape, centered at the origin, with its petals between the coordinate axes.

$r = \sin (3\theta)$ is a three-petaled flower shape, centered at the origin, with one petal on the negative y–axis.

$r = \cos \theta$ is a circle of diameter 1, with the left edge touching the origin.

$r = \cos (2\theta)$ is a four-petaled flower shape, centered at the origin, with its petals on the coordinate axes.

$r = \cos (3\theta)$ is a three-petaled flower shape, centered at the origin, with one petal on the positive x–axis.

10.5 PAGE 666, GUIDED SKILLS PRACTICE

5. $b^2 = 36^2 + 30^2 - 2(36)(30)\cos 50°$
$b^2 = 807.58$
$b \approx 28.42$

$\dfrac{\sin 50°}{28.42} = \dfrac{\sin C}{30} \Rightarrow \sin C = \dfrac{30 \cdot \sin 50°}{28.42} \Rightarrow$
$C = \sin^{-1}\left(\dfrac{30 \cdot \sin 50°}{28.42}\right) \approx 53.96°$

$m\angle A \approx 180° - (50° + c) \approx 76.03°$

6. $45^2 = 24^2 + 28^2 - 2(24)(28) \cdot \cos E$
$2025 = 1360 - 1344 \cos E$
$665 = -1344 \cos E$
$m\angle E = \cos^{-1}\left(\dfrac{665}{-1344}\right) \approx 119.66°$
$\dfrac{\sin 119.66°}{45} = \dfrac{\sin F}{24} \Rightarrow \sin F = \dfrac{24 \cdot \sin 119.66°}{45} \Rightarrow$
$m\angle F = \sin^{-1}\left(\dfrac{24 \cdot \sin 119.66°}{45}\right) \approx 27.61°$
$m\angle D \approx 180° - (119.66° + 27.61°) = 32.73°$

7. $a^2 = 6690^2 + 5750^2 - 2(6690)(5750) \cdot \cos 82°$
$a^2 \approx 67111317.48$
$a \approx 8192.15$ feet
Attendants at sites 2 and 3 cannot communicate directly since their two-way radios have a range of only 7920 feet, and the distance between them is about 8192 feet.

PAGE 667, PRACTICE AND APPLY

8. $b^2 = 12^2 + 17^2 - 2(12)(17) \cos 33° \Rightarrow b^2 \approx 90.822 \Rightarrow b \approx 9.53$ units

9. $c^2 = 2.2^2 + 4.3^2 - 2(2.2)(4.3) \cos 52° \Rightarrow c^2 \approx 11.682 \Rightarrow c \approx 3.42$ units

10. $a^2 = 68.2^2 + 23.6^2 - 2(68.2)(23.6) \cos 87° \Rightarrow a^2 \approx 5039.728 \Rightarrow a \approx 70.99$ units

11. $a^2 = b^2 + c^2 - 2bc \cos A$
$10^2 = 7^2 + 8^2 - 2(7)(8) \cos A$
$100 = 113 - 112 \cos A$
$-13 = -112 \cos A$
$m\angle A = \cos^{-1}\left(\dfrac{-13}{-112}\right) \approx 83.33°$

12. $c^2 = a^2 + b^2 - 2ab \cos C$
$2.5^2 = 3.6^2 + 3.6^2 - 2(3.6)(3.6) \cos C$
$6.25 = 25.92 - 25.92 \cos C$
$-19.67 = -25.92 \cos C$
$m\angle C = \cos^{-1}\left(\dfrac{-19.67}{-25.92}\right) \approx 40.64°$

13. $b^2 = a^2 + c^2 - 2ac \cos B$
$41^2 = 27^2 + 15^2 - 2(27)(15) \cos B$
$1681 = 954 - 810 \cos B$
$727 = -810 \cos B$
$m\angle B = \cos^{-1}\left(\dfrac{727}{-810}\right) \approx 153.84°$

14. $3.7^2 = 3.5^2 + 1.2^2 - 2(3.5)(1.2) \cos F$
$13.69 = 13.69 - 8.4 \cos F$
$0 = -8.4 \cos F$
$m\angle F = \cos^{-1}(0) = 90°$
$\dfrac{\sin 90°}{3.7} = \dfrac{\sin E}{1.2}$
$\sin E = \dfrac{1.2 \sin 90°}{3.7}$
$m\angle E \approx \sin^{-1}\left(\dfrac{1.2}{3.7}\right) \approx 18.92°$
$m\angle D \approx 180° - 90° - 18.92° = 71.08°$

15. $m\angle G = 180° - (72° + 36°) = 72°$

$g = i = 8.8$ by the Converse of the Isosceles Triangle Theorem.

$$\frac{\sin 36°}{h} = \frac{\sin 72°}{8.8}$$

$$h = \frac{8.8 \sin 36°}{\sin 72°} \approx 5.44$$

16. $k^2 = 2^2 + 2^2 - 2(2)(2) \cos 130°$

$k^2 \approx 13.14$

$k \approx 3.63$

$$\frac{\sin 130°}{3.63} = \frac{\sin J}{2}$$

$$\sin J = \frac{2 \cdot \sin 130°}{3.63}$$

$$m\angle J = \sin^{-1}\left(\frac{2 \cdot \sin 130°}{3.63}\right) \approx 24.96°$$

$$m\angle L \approx 180° - 130° - 24.96° = 25.04°$$

17. $$\frac{\sin 40°}{25} = \frac{\sin O}{18}$$

$$\sin O = \frac{18 \cdot \sin 40°}{25}$$

$$m\angle O = \sin^{-1}\left(\frac{18 \cdot \sin 40°}{25}\right) \approx 27.57°$$

Since $180° - 27.57° = 152.43°$ and $152.43° + 40° > 180°$, only one triangle is possible.

$$m\angle M = 180° - 40° - 27.57° = 112.43°$$

$$\frac{\sin 112.43°}{m} = \frac{\sin 40°}{25}$$

$$m = \frac{25 \cdot \sin 112.43°}{\sin 40°} \approx 35.95$$

18. $q^2 = 7^2 + 14^2 - 2(7)(14) \cos 60°$

$q^2 = 147$

$q \approx 12.12$

$$\frac{\sin 60°}{12.12} = \frac{\sin R}{7}$$

$$\sin R = \frac{7 \cdot \sin 60°}{12.12}$$

$$m\angle R = \sin^{-1}\left(\frac{7 \cdot \sin 60°}{12.12}\right) \approx 30.01°$$

$$m\angle P = 180° - 60° - 30.01° = 89.99°$$

19. $36^2 = 30^2 + 28^2 - 2(30)(28) \cos U$

$1296 = 1684 - 1680 \cos U$

$-388 = -1680 \cos U$

$$m\angle U = \cos^{-1}\left(\frac{-388}{-1680}\right) \approx 76.65°$$

$$\frac{\sin 76.65°}{36} = \frac{\sin S}{30}$$

$$\sin S = \frac{30 \cdot \sin 76.65°}{36}$$

$$m\angle S = \sin^{-1}\left(\frac{30 \cdot \sin 76.65°}{36}\right) \approx 54.18°$$

$$m\angle T \approx 180° - 76.65° - 54.18° = 49.17°$$

20. law of sines

21. law of cosines

22. two sides and an angle that is opposite one of the given sides

23. Sample answer:

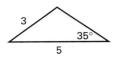

24. Sample answer:

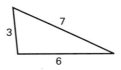

25. law of cosines

26. $\cos \theta = \dfrac{c_1}{b}$

27. $b \cdot \cos \theta = \dfrac{bc_1}{b} \Rightarrow b \cos \theta = c_1$

28. $\theta = 180 - A$,

so $b \cos \theta = c_1 \Rightarrow b \cdot \cos(180 - A) = c_1$

$\Rightarrow -b \cos A = c_1$

29. $a^2 = h^2 + (c_1 + c)^2$

30. $b^2 = h^2 + c_1{}^2$

31. $a^2 = h^2 + (c_1 + c)^2$

$a^2 = h^2 + c_1{}^2 + 2c_1 c + c^2$

$a^2 = b^2 + 2(-b \cos A) \cdot c + c^2$

$a^2 = b^2 - 2bc \cos A + c^2$

$a^2 = b^2 + c^2 - 2bc \cos A$

32. Proof of the Triangle Inequality Theorem:
Given the law of cosines, $c^2 = a^2 + b^2 - 2ab \cos C$ then $c^2 + 2ab \cos C = a^2 + b^2$.
Also, $(a + b)^2 = a^2 + 2ab + b^2 > a^2 + b^2$ since a and b are both positive so $2ab > 0$.
So, using substitution, $(a + b)^2 > c^2 + 2ab \cos C$ and $(a + b)^2 > c^2 + 2ab \cos C > c^2$
since $\cos C$ is positive for all angles C between 0 and 180°. Therefore, $(a + b)^2 > c^2$, and taking the square root of both sides gives $a + b > c$.

Proof that for $a > b, c > a - b$:
Given $c^2 = a^2 + b^2 - 2ab \cos C$, and $a > b$.
Also, $(a - b)^2 = a^2 - 2ab + b^2 < a^2 + b^2 - 2ab \cos C$
since $\cos C < 1$ for all angles C between 0 and 180°.
Therefore, $(a - b)^2 < a^2 + b^2 - 2ab \cos C = c^2$,
so $(a - b)^2 < c^2$ and taking the square root of both sides gives $a - b < c$.

33. Let $a = 60.5$ feet and $b = 90$ feet.
Then let $m\angle C = 45°$ be the angle formed between a and b. The law of cosines gives:
$c^2 = (60.5)^2 + 90^2 - 2(60.5)(90) \cos 45°$
$c^2 = 4059.86$
$c \approx 63.72$ feet

34. Mark walks $2.8 \times 3 = 8.4$ miles and Stephen walks $4.2 \times 3 = 12.6$ miles.
Let $a = 8.4$, $b = 12.6$ and $m\angle C = 72°$.
Then $c^2 = (8.4)^2 + (12.6)^2 - 2(8.4)(12.6) \cos 72°$
$c^2 = 163.91$
$c \approx 12.80$ miles

35. Let $a = 37$ km, $b = 25$ km and $m\angle C = 42°$
Then $c^2 = 37^2 + 25^2 - 2(37)(25) \cos 42°$
$c^2 = 619.18$
$c \approx 24.88$ km

$\dfrac{\sin 42°}{24.88} = \dfrac{\sin B}{25} \Rightarrow \sin B = \left(\dfrac{25 \cdot \sin 42°}{24.88}\right) \Rightarrow B = \sin^{-1}\left(\dfrac{25 \cdot \sin 42°}{24.88}\right) \approx 42.25°$

The road will be 24.88 km long, and the angle formed is 42.25°.

36. Let $a = 300$, $b = 500$ and $m\angle C = 105°$.
Then $c^2 = 300^2 + 500^2 - 2(300)(500) \cos 105°$
$c^2 = 417,645.71$
$c \approx 646.26$ yards

37. Let $a = 10, b = 12$, and $c = 15$. Then
$10^2 = 12^2 + 15^2 - 2(12)(15) \cos A$
$100 = 369 - 360 \cos A$
$-269 = -360 \cos A$
$m\angle A = \cos^{-1}\left(\dfrac{-269}{-360}\right) \approx 41.65°$

$\dfrac{\sin 41.65°}{10} = \dfrac{\sin B}{12} \Rightarrow \sin B = \dfrac{12 \cdot \sin 41.65°}{10}$

$\Rightarrow m\angle B = \sin^{-1}\left(\dfrac{12 \cdot \sin 41.65°}{10}\right) \approx 52.89°$

$m\angle C \approx 180° - 41.65° - 52.89° = 85.46°$

PAGE 699, LOOK BACK

38.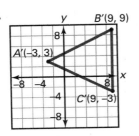

39.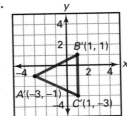

40.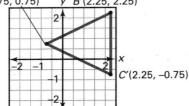
$A'(-0.75, 0.75)$ y $B'(2.25, 2.25)$
$C'(2.25, -0.75)$

41.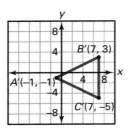
$B'(7, 3)$
$A'(-1, -1)$
$C'(7, -5)$

42. length $= 6\frac{2}{3}$ cm ≈ 6.67 cm;
width $= 33\frac{1}{3}$ cm ≈ 33.33 cm

43. Yes, SAS–similarity

44. No, the congruent angles are not between the corresponding sides in the triangles.

45. $\tan 40° = \frac{PQ}{15} \Rightarrow PQ = 15 \cdot \tan 40° \approx 12.59$

PAGE 669, LOOK BEYOND

46. $1^2 = x^2 + b^2$
$b^2 = 1 - x^2$
$b = \sqrt{1 - x^2}$

47. $\cos \theta = \cos (\sin^{-1} x) = \frac{\sqrt{1 - x^2}}{1} = \sqrt{1 - x^2}$

48. $\tan (\sin^{-1} x) = \tan \theta = \dfrac{x}{\sqrt{1 - x^2}}$

49. $\cos^{-1} x$ implies that the side adjacent to θ is x and the hypotenuse is 1, so the side opposite θ is $\sqrt{1 - x^2}$.
Therefore, $\sin(\cos^{-1} x) = \frac{\sqrt{1 - x^2}}{1} = \sqrt{1 - x^2}$.

50. $\tan^{-1} x$ implies that the side opposite θ is x and the side adjacent to θ is 1. Using the Pythagorean Theorem, $c^2 = x^2 + 1^2 \Rightarrow c = \sqrt{x^2 + 1}$, the hypotenuse. So, $\sin(\tan^{-1} x) = \frac{x}{\sqrt{x^2 - 1}}$.

10.6 PAGE 676, GUIDED SKILLS PRACTICE

6.
$\vec{b}$
$\vec{a} + \vec{b}$
$\vec{a}$

7.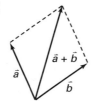
$\vec{a}$
$\vec{a} + \vec{b}$
$\vec{b}$

8.
2 mph
θ
1.5 mph

$c^2 = (1.5)^2 + 2^2 \Rightarrow c^2 = 6.25 \Rightarrow c = 2.5$ mph
$\tan \theta = \frac{2}{1.5}$
$\Rightarrow \theta = \tan^{-1}\left(\frac{2}{1.5}\right)$
$\theta \approx 53.13°$

9.
θ 2.5 mph   15°
2.0 mph

The supplement of the 15° angle is
$180° - 15° = 165°$.
$c^2 = 2^2 + (2.5)^2 - 2(2)(2.5) \cos 165°$
$c^2 \approx 19.91$
$c \approx 4.46$ mph

$\frac{\sin 165°}{4.46} = \frac{\sin \theta}{2.5} \Rightarrow \sin \theta = \frac{2.5 \cdot \sin 165°}{4.46}$
$\Rightarrow \theta = \sin^{-1}\left(\frac{2.5 \cdot \sin 165°}{4.46}\right) \approx 8.34°$

10.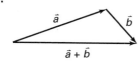

11.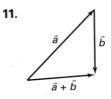

12.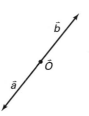

13.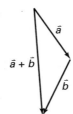

14.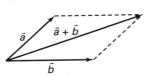

15.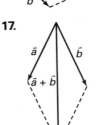

16.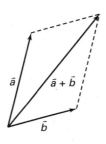

17.

18. $c^2 = 6^2 + 8^2 - 2(6)(8) \cos 140°$
$c^2 \approx 173.54$
$c \approx 13.17 \Rightarrow |\vec{c}| \approx 13.17$
$\frac{\sin 140°}{13.17} = \frac{\sin A}{6} \Rightarrow \sin A = \frac{6 \cdot \sin 140°}{13.17} \Rightarrow m\angle A = \sin^{-1}\left(\frac{6 \cdot \sin 140°}{13.17}\right) \Rightarrow m\angle A \approx 17.02°$

19. $c^2 = 8^2 + (4.5)^2 - 2(4.5)(8) \cos 115°$
$c^2 \approx 114.68$
$c \approx 10.71 \Rightarrow |\vec{c}| \approx 10.71$
$\frac{\sin 115°}{10.71} = \frac{\sin A}{4.5} \Rightarrow \sin A = \frac{(4.5) \cdot \sin 115°}{10.71} \Rightarrow m\angle A = \sin^{-1}\left(\frac{4.5 \sin 115°}{10.71}\right) \Rightarrow m\angle A \approx 22.39°$

20. $c^2 = (3.75)^2 + (7.5)^2 - 2(3.75)(7.5) \cos 120°$
$c^2 \approx 98.44$
$c \approx 9.92 \Rightarrow |\vec{c}| \approx 9.92$
$\frac{\sin 120°}{9.92} = \frac{\sin M}{3.75} \Rightarrow \sin M = \frac{3.75 \sin 120°}{9.92} \Rightarrow m\angle M = \sin^{-1}\left(\frac{3.75 \sin 120°}{9.92}\right) \Rightarrow m\angle M \approx 19.11°$

21. $c^2 = 6^2 + (10.5)^2 - 2(6)(10.5) \cos 110°$
$c^2 \approx 189.34$
$c \approx 13.76$
$\frac{\sin 110°}{13.75} = \frac{\sin M}{6} \Rightarrow \sin M = \frac{6 \sin 110°}{13.76} \Rightarrow m\angle M = \sin^{-1}\left(\frac{6 \sin 110°}{13.76}\right) \Rightarrow m\angle M \approx 24.19$

22. $c^2 = (7.5)^2 + 6^2 - 2(7.5)(6) \cos 60°$
$c^2 = 47.25$
$c \approx 6.87$
$\frac{\sin 60°}{6.87} = \frac{\sin R}{6} \Rightarrow \sin R = \frac{6 \sin 60°}{6.87} \Rightarrow m\angle R = \sin^{-1}\left(\frac{6 \sin 60°}{6.87}\right) \Rightarrow m\angle R \approx 49.11°$

23. $c^2 = (2.5)^2 + (8.5)^2 - 2(2.5)(8.5) \cos 40°$

$c^2 \approx 45.94$

$c \approx 6.78$

$\dfrac{\sin 40°}{6.78} = \dfrac{\sin R}{2.5} \Rightarrow \sin R = \dfrac{2.5 \sin 40°}{6.78} \Rightarrow m\angle R = \sin^{-1}\!\left(\dfrac{2.5 \sin 40°}{6.78}\right) \Rightarrow m\angle R \approx 13.71°$

24.

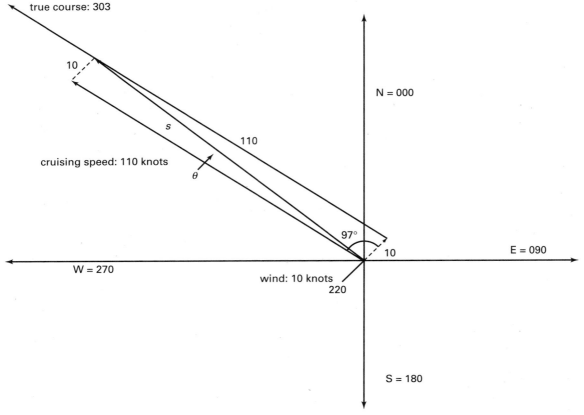

a. The true course is the resultant vector of the wind vector and the vector that represents the heading and speed at which Mina must hold the plane. First, find the wind correction angle: $\dfrac{\sin \theta}{10} = \dfrac{\sin 97°}{110} \Rightarrow \theta = \sin^{-1}\!\left(\dfrac{10 \sin 97°}{110}\right) \approx 5.18°$

Since headings are measured in the clockwise direction, the wind correction angle is $-5.18°$.

Thus, Mina should set out at a heading of $303 - 5.18 = 297.82$.

Next, find Mina's ground speed, s: The angle opposite s in the triangle is supplementary to the angle in the parallelogram measuring $(97 + 5.18)°$, so the angle measures $77.82° \cdot \sin \dfrac{77.82°}{s} = \dfrac{\sin 5.18°}{10} \Rightarrow s = \dfrac{10 \sin 77.82°}{\sin 5.18°} \approx 108.33$ knots.

b. $\dfrac{14}{108.33} \approx 0.13 \text{ hr} = 7.8 \text{ minutes}$

25.

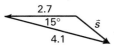

$s^2 = (2.7)^2 + (4.1)^2 - 2(2.7)(4.1) \cos 15° \Rightarrow$

$s^2 \approx 2.714 \Rightarrow s \approx 1.65 \text{ mph}$

26.

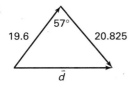

a. $d^2 = 19.6^2 + 20.825^2 - 2(19.6)(20.825) \cos 57°$

$\Rightarrow d^2 \approx 373.23 \Rightarrow \Rightarrow d \approx 19.32 \text{ miles.}$

$\text{time} = \dfrac{19.32}{6} \approx 3.22 \text{ hours}$

b. $\dfrac{\sin \theta}{19.6} = \dfrac{\sin 57°}{19.32} \Rightarrow \sin \theta = \dfrac{19.6 \sin 57°}{19.32}$

$\Rightarrow m\angle \theta = \sin^{-1}\!\left(\dfrac{19.6 \sin 57°}{19.32}\right) \approx 58.32°$

27. $(VA)(VB) = (VC)(VD) \Rightarrow (5)(8) = (6)(VD) \Rightarrow$
$VD = \frac{40}{6} = \frac{20}{3}$. So $CD \approx 6 + \frac{20}{3} = 12\frac{2}{3}$

28. $(7)(x) = (9)(x - 1) \Rightarrow 7x = 9(x - 1)$
$$7x = 9x - 9$$
$$-2x = -9$$
$$x = \frac{9}{2} = 4.5$$

29. $(x - 6)(x) = (3)(9) \Rightarrow$
$$x(x - 6) = 27$$
$$x^2 - 6x - 27 = 0$$
$$(x + 3)(x - 9) = 0$$
$$x = -3 \text{ or } x = 9$$

Since x cannot be negative, $x = 9$.

30. $\frac{VX}{VW} = \frac{VW}{VY} \Rightarrow \frac{4}{VW} = \frac{VW}{9} \Rightarrow (VW)^2 = 4 \cdot 9$
$\Rightarrow VW = \sqrt{36} = 6$

31. $\frac{4}{8} = \frac{8}{VY} \Rightarrow VY = \frac{8 \cdot 8}{4} \Rightarrow VY = 16$

32. $(AY) \cdot (YB) = (YD) \cdot (YC) \Rightarrow (x) \cdot (x - 2) = (6)(4)$
$\Rightarrow \quad x^2 - 2x = 24$
$$x^2 - 2x - 24 = 0$$
$$(x + 4)(x - 6) = 0$$
$$x = -4 \text{ or } x = 6$$
Since x cannot be negative, $x = 6$.

33. $(AY)(YB) = (YD)(YC)$;
Let $x = YB$, and $x + 9 = AY$
$$(x + 9)(x) = 12 \cdot 8$$
$$x^2 + 9x = 96$$
$$x^2 + 9x - 96 = 0$$
$$x = \frac{-9 \pm \sqrt{9^2 - 4(1)(-96)}}{2}$$
$$x = \frac{-9 \pm \sqrt{81 + 384}}{2}$$
$$x = \frac{-9 \pm \sqrt{465}}{2}$$
$$x = \frac{-9 \pm 21.56}{2}$$
$$x \approx 6.28, \ x \approx -15.28$$

Since x must be positive because it represents a length, $x \approx 6.28$.

So, $AY = x + 9 \approx 6.28 + 9 \approx 15.28$

34. Use the parallelogram method:

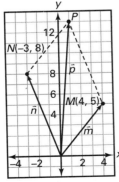

The vector $\vec{m}$ has slope $\frac{5 - 0}{4 - 0} = \frac{5}{4}$. The side
of the parallelogram opposite it has the same slope.
From point $N(-3, 8)$ rise 5 and run 4 to get to the
point $P(1, 13)$. This point is the head of $\vec{p}$.
$$|\vec{p}| = \sqrt{(1 - 0)^2 + (13 - 0)^2} = \sqrt{1 + 169}$$
$$= \sqrt{170} \approx 13.04$$

35. Consider the right triangle with hypotenuse between the vertices $(0, 0)$ and $(1, 13)$.

Then $\tan \theta = \frac{13}{1}$, so $= \theta \tan^{-1}(13) \approx 85.60°$.

36. (1, 13) (see problem 34)

37. The coordinates of the head of $\vec{p}$ are the same as the sums of the x–coordinates and y–coordinates of the vectors $\vec{m}$ and $\vec{n}$. $P = (-3 + 4, 8 + 5) = (1, 13)$.

Rule: To find the coordinates of the head of the vector sum of two given vectors with tails at the origin, add the x–coordinates and add the y–coordinates.

38.

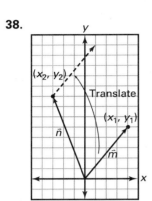

39. *Rule:* If a vector with tail at the origin has coordinates (x_1, y_1), then the angle it makes with the positive x–axis is $\theta = \tan^{-1}\left(\dfrac{y_1}{x_1}\right)$.

NOTE: This is true if the angle θ is between $-90°$ and $90°$.

Let (x, y) be the coordinates of the head of vector $\vec{m}$ and let (x_2, y_2) be the coordinates of the head of vector $\vec{n}$. Vector $\vec{m}$ has slope $\dfrac{y_1 - 0}{x_1 - 0} = \dfrac{y_1}{x_1}$. Translate vector $\vec{m}$ so that the tail of $\vec{m}$ touches the head of $\vec{n}$. Since $\vec{m}$ has a slope of $\dfrac{y_1}{x_1}$, the coordinates of the head after the translation can be found by a vertical translation of y_1 units and a horizontal translation of x_1 units from the point (x_2, y_2). This gives coordinates $(x_2 + x_1, y_2 + y_1)$.

10.7 | **PAGE 683, GUIDED SKILLS PRACTICE**

---

5. $x' = 5 \cos 10° - 4 \sin 10° \approx 4.23$
$y' = 5 \sin 10° + 4 \cos 10° \approx 4.81$

6. $x' = 5 \cos 90° - 4 \sin 90° = -4$
$y' = 5 \sin 90° + 4 \cos 90° = 5$

7. $x' = 5 \cos (-30°) - 4 \sin (-30°) \approx 6.33$
$y' = 5 \sin (-30°) + 4 \cos (-30°) \approx 0.96$

8. $\begin{bmatrix} \cos 45° & -\sin 45° \\ \sin 45° & \cos 45° \end{bmatrix} \begin{bmatrix} 1 & 3 & 2 \\ 1 & 0 & 2 \end{bmatrix}$

$= \begin{bmatrix} 0 & 2.12 & 0 \\ 1.41 & 2.12 & 2.83 \end{bmatrix}$

$A' = (0, 1.41)$, $B' = (2.12, 2.12)$, $C' = (0, 2.83)$

9. $\begin{bmatrix} \cos 72° & -\sin 72° \\ \sin 72° & \cos 72° \end{bmatrix} \begin{bmatrix} 1 & 3 & 2 \\ 1 & 0 & 2 \end{bmatrix} = \begin{bmatrix} -0.64 & 0.93 & -1.28 \\ 1.26 & 2.85 & 2.52 \end{bmatrix}$

$A' = (-0.64, 1.26)$, $B' = (0.93, 2.85)$, $C' = (-1.28, 2.52)$

10. $\begin{bmatrix} \cos (-60°) & -\sin (-60°) \\ \sin (-60°) & \cos (-60°) \end{bmatrix} \begin{bmatrix} 1 & 3 & 2 \\ 1 & 0 & 2 \end{bmatrix} = \begin{bmatrix} 1.37 & 1.5 & 2.73 \\ -0.37 & -2.60 & -0.73 \end{bmatrix}$

$A' = (1.37, -0.37)$, $B' = (1.5, -2.60)$, $C' = (2.73, -0.73)$

11. $(0 \cos 45° - 5 \sin 45°, 0 \sin 45° + 5 \cos 45°) \approx (-3.54, 3.54)$

12. $(-1 \cdot \cos 80° - 1 \cdot \sin 80°, -1 \cdot \sin 80° + 1 \cdot \cos 80°) \approx (-1.16, -0.81)$

13. $(-3 \cos 30° - (-2) \sin 30°, -3 \sin 30° + (-2) \cos 30°) \approx (-1.60, -3.23)$

14. $(2 \cos 140° - 0 \sin 140°, 2 \sin 140° + 0 \cos 140°) \approx (-1.53, 1.29)$

15. $(5 \cos 20° - 5 \sin 20°, 5 \sin 20° + 5 \cos 20°) \approx (2.99, 6.41)$

16. $(2 \cos 400° - 6 \sin 400°, 2 \sin 400° + 6 \cos 400°) \approx (-2.32, 5.88)$

For problems 17 to 22, use the results of the Activity.

17. $90°$ **18.** $180°$ **19.** $270°$ or $-30°$ **20.** $0°$ or $360°$

21. $45°$ **22.** $330°$ or $-30°$

23. $\begin{bmatrix} \cos 45° & -\sin 45° \\ \sin 45° & \cos 45° \end{bmatrix} = \begin{bmatrix} 0.71 & -0.71 \\ 0.71 & 0.71 \end{bmatrix}$

24. $\begin{bmatrix} \cos 30° & -\sin 30° \\ \sin 30° & \cos 30° \end{bmatrix} = \begin{bmatrix} 0.87 & -0.5 \\ 0.5 & 0.87 \end{bmatrix}$

25. $\begin{bmatrix} \cos 120° & -\sin 120° \\ \sin 120° & \cos 120° \end{bmatrix} = \begin{bmatrix} -0.5 & -0.87 \\ 0.87 & -0.5 \end{bmatrix}$

26. $\begin{bmatrix} \cos 320° & -\sin 320° \\ \sin 320° & \cos 320° \end{bmatrix} = \begin{bmatrix} 0.77 & 0.64 \\ -0.64 & 0.77 \end{bmatrix}$

27. $\begin{bmatrix} \cos 90° & -\sin 90° \\ \sin 90° & \cos 90° \end{bmatrix} \begin{bmatrix} 0 & 3 & 2 \\ 1 & 2 & 5 \end{bmatrix}$

$= \begin{bmatrix} -1 & -2 & -5 \\ 0 & 3 & 2 \end{bmatrix}$

28. $\begin{bmatrix} \cos 40° & -\sin 40° \\ \sin 40° & \cos 40° \end{bmatrix} \begin{bmatrix} -2 & 1 & 3 \\ 1 & 2 & -4 \end{bmatrix}$

$= \begin{bmatrix} -2.17 & -0.52 & 4.87 \\ -0.52 & 2.17 & -1.14 \end{bmatrix}$

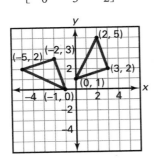

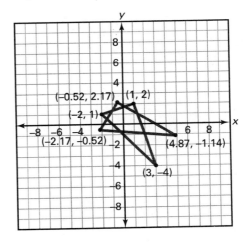

29. $\begin{bmatrix} \cos 225° & -\sin 225° \\ \sin 225° & \cos 225° \end{bmatrix} \begin{bmatrix} 2 & 1 & 2 & 4 \\ 5 & 3 & 1 & 2 \end{bmatrix}$

$= \begin{bmatrix} 2.12 & 1.41 & -0.71 & -1.41 \\ -4.95 & -2.83 & -2.12 & -4.24 \end{bmatrix}$

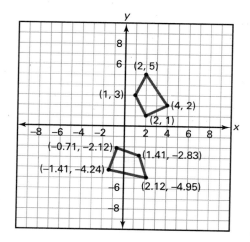

30. $\begin{bmatrix} \cos 70° & -\sin 70° \\ \sin 70° & \cos 70° \end{bmatrix} \begin{bmatrix} 0 & 2 & 4 & 6 \\ 0 & 2 & 1 & 0 \end{bmatrix}$

$= \begin{bmatrix} 0 & -1.20 & 0.43 & 2.05 \\ 0 & 2.56 & 4.10 & 5.64 \end{bmatrix}$

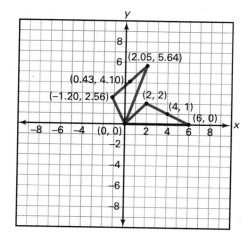

31. $\begin{bmatrix} \cos 30° & -\sin 30° \\ \sin 30° & \cos 30° \end{bmatrix} \begin{bmatrix} \cos 40° & -\sin 40° \\ \sin 40° & \cos 40° \end{bmatrix}$

$= \begin{bmatrix} 0.34 & -0.94 \\ 0.94 & 0.34 \end{bmatrix}$

$\begin{bmatrix} \cos 70° & -\sin 70° \\ \sin 70° & \cos 70° \end{bmatrix}$

$= \begin{bmatrix} 0.34 & -0.94 \\ 0.94 & 0.34 \end{bmatrix}$

A rotation of 30° followed by a rotation of 40° is equivalent to a rotation of 70°.

32.

A graph of a pentagon with vertices labeled (−3.24, 2.35), (1.24, 3.8), (−3.24, −2.35), (1.24, −3.8).

Multiply $\begin{bmatrix} 4 \\ 0 \end{bmatrix}$ repeatedly by the rotation matrix for a

rotation of 72°, which is $\begin{bmatrix} \cos 72° & -\sin 72° \\ \sin 72° & \cos 72° \end{bmatrix} =$

$\begin{bmatrix} 0.309 & -0.951 \\ 0.951 & 0.309 \end{bmatrix}$, until all the vertices are found.

The other vertices are $\begin{bmatrix} 1.24 \\ 3.80 \end{bmatrix}; \begin{bmatrix} -3.24 \\ 2.35 \end{bmatrix}; \begin{bmatrix} -3.24 \\ -2.35 \end{bmatrix};$

$\begin{bmatrix} 1.24 \\ -3.80 \end{bmatrix}$

33. $[R_0] = \begin{bmatrix} \cos 0° & -\sin 0° \\ \sin 0° & \cos 0° \end{bmatrix} = \begin{bmatrix} 1 & 0 \\ 0 & 1 \end{bmatrix}$

The product of any rotation matrix and the identity matrix gives the rotation matrix.

34. $[R_{60}] = \begin{bmatrix} \cos 60° & -\sin 60° \\ \sin 60° & \cos 60° \end{bmatrix} = \begin{bmatrix} 0.5 & -0.87 \\ 0.87 & 0.5 \end{bmatrix}$

$[R_{60}] \times [R_{60}] \times [R_{60}] \times [R_{60}] \times [R_{60}] \times [R_{60}] = \begin{bmatrix} 1 & 0 \\ 0 & 1 \end{bmatrix}$

The product is $[R_0] = \begin{bmatrix} 1 & 0 \\ 0 & 1 \end{bmatrix}$ because 6 60° rotations gives a 360° rotation, which is

the same as a 0° rotation.

35. $[R_{35}] = \begin{bmatrix} \cos 35° & -\sin 35° \\ \sin 35° & \cos 35° \end{bmatrix} = \begin{bmatrix} 0.82 & -0.57 \\ 0.57 & 0.82 \end{bmatrix}$

$[R_{-35}] = \begin{bmatrix} \cos(-35°) & -\sin(-35°) \\ \sin(-35°) & \cos(-35°) \end{bmatrix} = \begin{bmatrix} 0.82 & 0.57 \\ -0.57 & 0.82 \end{bmatrix}$

$[R_{35}] \times [R_{-35}] = \begin{bmatrix} 1 & 0 \\ 0 & 1 \end{bmatrix}$

The product of the two matrices gives the identity matrix.

36. Let $A = (x, y)$. Since $O = (0, 0)$, $(OA)^2 = \left(\sqrt{(x-0) + (y-0)^2} \right)^2 = x^2 + y^2$.
$A' = (x \cdot \cos\theta - y\sin\theta, x\sin\theta + y\cos\theta)$,

so $(OA')^2 = \left(\sqrt{(x\cos\theta - y\sin\theta - 0)^2 + (x\sin\theta + y\cos\theta - 0)^2} \right)^2$
$\qquad = x^2(\cos\theta)^2 - 2x\cos\theta\, y\sin\theta + y^2(\sin\theta)^2 + x^2(\sin\theta)^2$
$\qquad\quad + 2x\sin\theta\, y\cos\theta + y^2(\cos\theta)^2$
$\qquad = x^2(\cos\theta)^2 + y^2(\sin\theta)^2 + x^2(\sin\theta)^2 + y^2(\cos\theta)^2$
$\qquad = x^2[(\cos\theta)^2 + (\sin\theta)^2] + y^2[(\cos\theta)^2 + (\sin\theta)^2]$
$\qquad = x^2(1) + y^2(1) = x^2 + y^2$.

Since $(OA)^2 = x^2 + y^2 = (OA')^2$, the distances are equal.

37. a. After 20 minutes, the restaurant has rotated for $\frac{20}{60} = \frac{1}{3}$ of an hour, so it has rotated through an angle of $\frac{1}{3} \cdot 360° = 120°$.

$\begin{bmatrix} \cos 120° & -\sin 120° \\ \sin 120° & \cos 120° \end{bmatrix} \begin{bmatrix} 30 \\ -42 \end{bmatrix} = \begin{bmatrix} 21.37 \\ 46.98 \end{bmatrix}$

The coordinates are $(21.37, 46.98)$ after rotating for 20 minutes.

b. After t minutes, the restaurant has rotated for $\frac{t}{60}$ of an hour, so it has rotated through an angle of $\frac{t}{60} \cdot 360° = 6t°$.

$\begin{bmatrix} \cos(6t°) & -\sin(6t°) \\ \sin(6t°) & \cos(6t°) \end{bmatrix} \begin{bmatrix} 30 \\ -42 \end{bmatrix} = \begin{bmatrix} 30\cos(6t°) + 42\sin(6t°) \\ 30\sin(6t°) - 42\cos(6t°) \end{bmatrix}$

The coordinates are $(30\cos(6t°) + 42\sin(6t°), 30\sin(6t°) - 42\cos(6t°))$ after rotating for t minutes.

PAGE 685, LOOK BACK

38.

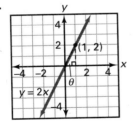

The point $(1, 2)$ is on the line $y = 2x$, so form a right triangle with the leg opposite θ of length 2 and the leg adjacent to θ of length 1.

Then $\tan\theta = \frac{2}{1} \Rightarrow \theta = \tan^{-1}(2) \approx 63.43°$.

39.

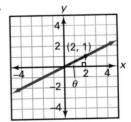

The point $(2, 1)$ is on the line $y = 0.5x$, so form a right triangle with the leg opposite θ of length 1 and the leg adjacent to θ of length 2.

Then $\tan\theta = \frac{1}{2} \Rightarrow \theta = \tan^{-1}\left(\frac{1}{2}\right) \approx 26.57°$.

40.

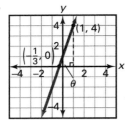

The points $\left(\frac{-1}{3}, 0\right)$ and $(1, 4)$ are on the line $y = 3x + 1$, so form a right triangle with the leg opposite θ of length 4 and the leg adjacent to θ of length $1 + \frac{1}{3} = \frac{4}{3}$. Then $\tan \theta = \frac{4}{\frac{4}{3}} = 3 \Rightarrow \theta \tan^{-1}(3) \approx 71.57°$.

41. $\sin 37° \approx 0.60$; $\cos 37° \approx 0.80$; $\tan 37° \approx 0.75$ **42.** $\sin 90° = 1$; $\cos 90° = 0$; $\tan 90°$ is undefined

43. $\sin 250° \approx -0.94$; $\cos 250° \approx -0.34$; $\tan 250° \approx 2.75$

44. $\sin(-10°) \approx -0.17$; $\cos(-10°) \approx 0.98$; $\tan(-10°) \approx -0.18$

45. $\theta = \tan^{-1}(2.75) \approx 70°$ **46.** $\theta = \cos^{-1}(0.36) \approx 69°$ **47.** $\theta = \cos^{-1}(0.81) \approx 36°$

48. $\theta = \sin^{-1}(0.77) \approx 50°$ **49.** $\tan \theta = \frac{6}{10} \Rightarrow \theta = \tan^{-1}\left(\frac{6}{10}\right) \approx 30.96°$
The slope is too steep.

PAGE 685, LOOK BEYOND

50. $C = 2\pi r$

The number of radians is $360°$ is $\frac{2\pi r}{r} = 2\pi$.

51. $\frac{\pi}{180°} \times 180° = \pi$ radians

52. $\frac{\pi}{180°} \times 90° = \frac{\pi}{2}$ radians

53. $\frac{\pi}{180°} \times 60° = \frac{\pi}{3}$ radians

CHAPTER REVIEW AND ASSESSMENT

1. $\frac{3}{7}$ **2.** $\frac{7}{3}$ **3.** $m\angle A = \tan^{-1}\left(\frac{3}{7}\right) \approx 23.20°$

4. $m\angle B = \tan^{-1}\left(\frac{7}{3}\right) \approx 66.80°$ **5.** $\sin A = \frac{55}{73}$; $\cos A = \frac{48}{73}$ **6.** $\sin B = \frac{48}{73}$; $\cos B = \frac{55}{73}$

7. $m\angle A = \sin^{-1}\left(\frac{55}{73}\right) \approx 48.89°$ or $\cos^{-1}\left(\frac{48}{73}\right) \approx 48.89°$ **8.** $m\angle B = \sin^{-1}\left(\frac{48}{73}\right) \approx 41.11°$ or $\cos^{-1}\left(\frac{55}{73}\right) \approx 41.11°$

9. $(\cos 70°, \sin 70°) \approx (0.34, 0.94)$ **10.** $(\cos 130°, \sin 130°) \approx (-0.64, 0.77)$

11. $\sin \theta = 0.2 \Rightarrow \theta = \sin^{-1}(0.2) \approx 11.54°$.
The other angle is $180° - 11.54° = 168.46°$.

12. $\cos \theta = 0.8 \Rightarrow \theta = \cos^{-1}(0.8) \approx 36.87°$.
The other angle is $-36.87°$ because cosine ratios are positive in the first and fourth quadrants.
Since $0° < \theta < 360°$, we can express $-36.87°$ as $-36.87° + 360° = 323.13°$

13. $\frac{\sin 115°}{40} = \frac{\sin R}{16} \Rightarrow m\angle R = \sin^{-1}\left(\frac{16 \cdot \sin 115°}{40}\right) \approx 21.26°$
Since $\angle P$ is obtuse, $\angle R$ must be acute, so the triangle is not ambiguous.
$m\angle Q \approx 180° - 115° - 21.26° = 43.74°$.

14. $\frac{\sin 115°}{40} = \frac{\sin 43.74°}{q} \Rightarrow q = \frac{40 \cdot \sin 43.74°}{\sin 115°} \approx 30.52$.

15. $m\angle U = 180° - 25° - 48° = 107°.$

16. $\dfrac{\sin 107°}{18} = \dfrac{\sin 48°}{t} \Rightarrow t = \dfrac{18 \cdot \sin 48°}{\sin 107°} \approx 13.99$

$\dfrac{\sin 107°}{18} = \dfrac{\sin 25°}{s} \Rightarrow s = \dfrac{18 \cdot \sin 25°}{\sin 107°} \approx 7.95$

17. $d^2 = 20^2 + 28^2 - 2(20)(28) \cos 40° \approx 326.03°$

$d^2 \approx 326.03$

$d \approx 18.06$

18. $\dfrac{\sin 40°}{18.06} = \dfrac{\sin F}{20} \Rightarrow m\angle F = \sin^{-1}\left(\dfrac{20 \cdot \sin 40°}{18.06}\right) \approx 45.38°$

$m\angle E = 180° - 40° - 45.38° = 94.62°$

19. $34^2 = 24^2 + 18^2 - 2(24)(18) \cos I$

$1156 = 900 - 864 \cos I$

$\cos I = \dfrac{256}{-864} \Rightarrow m\angle I = \cos^{-1}\left(\dfrac{256}{-864}\right) \approx 107.24°$

20. $\dfrac{\sin 107.24°}{34} = \dfrac{\sin H}{24} \Rightarrow m\angle H = \sin^{-1}\left(\dfrac{24 \cdot \sin 107.24}{34}\right)$

$\approx 42.39°$

$m\angle G \approx 180° - 107.24° - 42.39° = 30.37°$

21.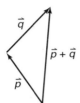

22.

23.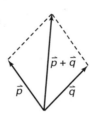

24.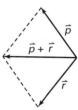

25. $(1 \cdot \cos 270° - 5 \sin 270°, 1 \sin 270° + 5 \cos 270° = (5, -1)$

26. $(4 \cos 45° - (-3) \sin 45°, 4 \sin 45° + (-3) \cos 45°) \approx (4.95, 0.71)$

27. $\begin{bmatrix} \cos 30° & -\sin 30° \\ \sin 30° & \cos 30° \end{bmatrix} \begin{bmatrix} 0 & 3 & 3 \\ 0 & 1 & 2 \end{bmatrix} = \begin{bmatrix} 0 & 2.10 & 1.60 \\ 0 & 2.37 & 3.23 \end{bmatrix}$

The image $\Delta D' E' F'$ has vertices $D'(0, 0)$, $E'(2.10, 2.37)$, and $F'(1.60, 3.23)$.

28. $\begin{bmatrix} \cos 135° & -\sin 135° \\ \sin 135° & \cos 135° \end{bmatrix} \begin{bmatrix} 1 & -1 & 1 \\ 1 & 1 & -1 \end{bmatrix} = \begin{bmatrix} -1.41 & 0 & 0 \\ 0 & -1.41 & 1.41 \end{bmatrix}$

The image $\Delta G' H' I'$ has vertices $G'(-1.41, 0)$, $H'(0, -1.41)$, and $I'(0, 1.41)$.

29. The supplement of the 50° angle is 130°. The third angle in the obtuse triangle is
$180° - 130° - 30° = 20°.$

$\dfrac{\sin 20°}{35} = \dfrac{\sin 30°}{a} \Rightarrow a = \dfrac{35 \cdot \sin 30°}{\sin 20°} \approx 51.17$, where a is the hypotenuse of the

right triangle.

$\sin 50° = \dfrac{h}{51.17} \Rightarrow h = 51.17 \cdot \sin 50° \approx 39.20$ m

30. After 40 minutes:

$100 \text{ mph} \times \dfrac{2}{3} = 66.67$ miles traveled

$115 \text{ mph} \times \dfrac{2}{3} = 76.67$ miles traveled

The angle measurement between the vectors is:

$150 - 45 = 105°$

Using the Law of Cosines:

$d^2 = (66.67)^2 + (76.67)^2 - 2(66.67)(76.67) \cos 105°$

$d^2 \approx 12,967.93$

$d \approx 113.88$ miles apart

31.

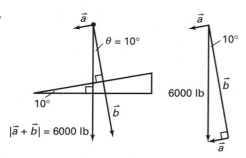

The triangles are similar by AA, so $\theta = 10°$. We have a right triangle, so

$$\sin 10° = \frac{|\vec{a}|}{6000 \text{ lb}} \Rightarrow |\vec{a}| = 6000 \cdot \sin 10° \approx 1041.89 \text{ lb}$$

A force of 1041.89 lb is required to push the truck uphill.

32.
$$\begin{aligned}
(186{,}000{,}000)^2 &= d^2 + d^2 - 2(d)(d) \cos (0.0004222°) \\
3.4596 \times 10^{16} &= 2d^2 - 2d^2(\cos(0.0004222°)) \\
3.4596 \times 10^{16} &= 2d^2(1 - \cos(0.0004222°))
\end{aligned}$$

$$\frac{3.4596 \times 10^{16}}{2(1 - \cos(0.0004222°))} = d^2$$

$$6.3713 \times 10^{26} = d^2$$
$$d = \sqrt{6.3713 \times 10^{26}} \approx 2.52 \times 10^{13} \text{ ft}$$

CHAPTER TEST

1. $\tan A = \frac{8}{5}$

2. $\tan B = \frac{5}{8}$

3. $m\angle A = \tan^{-1}\frac{8}{5} \approx 57.99°$

4. $m\angle B = \tan^{-1}\frac{5}{8} \approx 32.01°$

5. $\sin A = \frac{30}{34} = \frac{15}{17}$; $\cos A = \frac{16}{34} = \frac{8}{17}$

6. $\sin B = \frac{16}{34} = \frac{8}{17}$; $\cos B = \frac{30}{34} = \frac{15}{17}$

7. $m\angle A = \sin^{-1}\frac{15}{17} = \cos^{-1}\frac{8}{17} \approx 61.93°$

8. $m\angle B = \sin^{-1}\frac{8}{17} = \cos^{-1}\frac{15}{17} \approx 28.07°$

9. $(x, y) = (\cos 36°, \sin 36°) \approx (0.81, 0.59)$

10. $(x, y) = (\cos 115°, \sin 115°) \approx (-0.42, 0.91)$

11. $\sin^{-1} 0.6 \approx 36.87°$ or $180° - 36.87° = 143.13°$

12. $\cos^{-1} 0.3 \approx 72.54°$ or $360° - 72.54° = 287.46°$

13. $m\angle K = 180° - 102° - 50° = 28°$

14. $\dfrac{\sin 102°}{l} = \dfrac{\sin 28°}{10}$

$l = \dfrac{10 \sin 102°}{\sin 28°} \approx 20.84$ units

$\dfrac{\sin 50°}{j} = \dfrac{\sin 28°}{10}$

$j = \dfrac{10 \sin 50°}{\sin 28°} \approx 16.32$ units

15. $\dfrac{\sin D}{15} = \dfrac{\sin 106°}{25}$

$m\angle D = \sin^{-1}\left(\dfrac{15 \sin 106°}{25}\right) \approx 35.22°$

$m\angle F = 180° - 106° - 35.22° = 38.78°$

16. $\dfrac{\sin 38.78°}{DE} = \dfrac{\sin 106°}{25}$

$DE = \dfrac{25 \sin 38.78°}{\sin 106°} \approx 16.29$ units

17. $SR^2 = 12^2 + 15^2 - 2(12)(15)\cos 110°$

$SR = \sqrt{12^2 + 15^2 - 2(12)(15)\cos 110°} \approx 22.18$

18. $\dfrac{\sin S}{15} = \dfrac{\sin 110°}{22.18}$

$m\angle S = \sin^{-1}\left(\dfrac{15 \sin 110°}{22.18}\right) \approx 39.46°$

$m\angle R = 180° - 110° - 39.46° = 30.54°$

19. $d^2 = 7.2^2 + 8.2^2 - 2(7.2)(8.2)\cos 38°$

$d = \sqrt{7.2^2 + 8.2^2 - 2(7.2)(8.2)\cos 38°}$

$d \approx 5.1$

About 5.1 miles

20.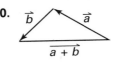

21.

22. $s^2 = 1.8^2 + 2^2$ $\quad \vartheta = \tan^{-1}\dfrac{2}{1.8}$

$s = \sqrt{1.8^2 + 2^2}$ $\quad \vartheta \approx 48.01°$

$s \approx 2.7$ mph $\quad \theta = 90° - 48.01° = 41.99°$

23. $\begin{bmatrix} \cos 90° & -\sin 90° \\ \sin 90° & \cos 90° \end{bmatrix}\begin{bmatrix} 3 \\ -1 \end{bmatrix} = \begin{bmatrix} 3(0) + 1 \\ 3(1) - 0 \end{bmatrix} = \begin{bmatrix} 1 \\ 3 \end{bmatrix}; (1, 3)$

24. $\begin{bmatrix} \cos 150° & -\sin 150° \\ \sin 150° & \cos 150° \end{bmatrix} = \begin{bmatrix} -\dfrac{\sqrt{3}}{2} & -\dfrac{1}{2} \\ \dfrac{1}{2} & -\dfrac{\sqrt{3}}{2} \end{bmatrix}$

CHAPTERS 1–10 CUMULATIVE ASSESSMENT

1. Area of shaded region: $\pi(5\sqrt{2})^2 = 50\pi$

Area of nonshaded region: $\pi(10)^2 - 50\pi = 50\pi$

Area of the large circle : $\pi(10)^2 = 100\pi$

The probability that the point is chosen inside the shaded area is $\dfrac{50\pi}{100\pi} = \dfrac{1}{2}$.

The probability that the point is chosen outside the shaded area is $\dfrac{50\pi}{100\pi} = \dfrac{1}{2}$.

The answer is C.

2. $V = (24)(26)(15) = 9360$ in^3

$SA = 2(24)(26) + 2(15)(26) + 2(15)(24) = 2748$ in^2

Since $9360 > 2748$, the answer is A.

3. The answer is C, since the triangles are similar and $\angle E$ corresponds to $\angle H$.

4. The answer is D, since no information is known about the triangle.

5. The answer is b, since $T(x, y) = (-x, y)$ reflects an image across the y-axis.

6. The answer is c, since $T(x, y) = (-x, -y)$ results in a rotation about the origin.

7. The answer is b, since no angle measures are known.

8. c

9. a

10. The answer is d, since the product of the slopes is $2\left(-\dfrac{1}{2}\right) = -1$.

11. $(3\cos 45° - 5\sin 45°, 3\sin 45° + 5\cos 45°) \approx (-1.41, 5.66)$

12. $10 + 6 = 16$

13. Vertices = 10, faces = 7, edges = 15

14. $\dfrac{24}{36} = \dfrac{36}{x} \Rightarrow 24x = 36^2 \Rightarrow x = \dfrac{36^2}{24} = 54$

15. $m\angle L = 180° - 36° - 50° = 94°$

$\dfrac{\sin 94°}{12} = \dfrac{\sin 36°}{k} \Rightarrow k = \dfrac{12 \cdot \sin 36°}{\sin 94°} \approx 7.07$

16. Area of square = $5^2 = 25$.

Area of a circle = $\pi r^2 = 25 \Rightarrow r^2 = \dfrac{25}{\pi} \Rightarrow r \approx 2.82$

17. Volume of cube $= 5^3 = 125$

Volume of a sphere $= \frac{4}{3}\pi r^3 = 125 \Rightarrow r^3 = \frac{125}{\pi} \cdot \frac{3}{4} \Rightarrow r \approx 3.10$

18. When $\theta = 135°$, $\sin\theta \approx 0.71$ and $-\cos\theta \approx -(-0.71) = 0.71$

11.1 | PAGE 702, GUIDED SKILLS PRACTICE

5. $\frac{\ell}{s} \approx 1.618 \Rightarrow \frac{\ell}{10} \approx 1.618 \Rightarrow \ell \approx 16.18$

6. $\frac{\ell}{s} \approx 1.618 \Rightarrow \frac{9}{s} \approx 1.618 \Rightarrow s \approx \frac{9}{1.618} \Rightarrow s \approx 5.56$

7. Sample answer:
They intersect at right angles, that is $\overline{AC} \perp \overline{DE}$. The intersection seems to be the center of the spiral.

8. Check students' drawings. The length of the longer side is approximately $1\frac{3}{5}$ inches.

PAGES 702–705, PRACTICE AND APPLY

9. $\frac{\ell}{3} = \frac{1 + \sqrt{5}}{2} \Rightarrow \ell = 3\left(\frac{1 + \sqrt{5}}{2}\right) \approx 4.85$

10. $\frac{8}{s} = \frac{1 + \sqrt{5}}{2} \Rightarrow 16 = s(1 + \sqrt{5})$
$\Rightarrow s = \frac{16}{1 + \sqrt{5}} \approx 4.94$

11. $\ell = \left(\frac{1 + \sqrt{5}}{2}\right)s$

$\sqrt{s^2 + \left(\frac{1 + \sqrt{5}}{2}\right)^2 s^2} = 10$
$\Rightarrow \sqrt{3.618034}\, s = 10$
$\Rightarrow s \approx 5.26$

12. $\frac{5}{s} = \frac{1 + \sqrt{5}}{2} \Rightarrow 10 = s(1 + \sqrt{5}) \Rightarrow$
$s = \frac{10}{1 + \sqrt{5}} \approx 3.09$

13. $\frac{\phi}{1} = \frac{1}{\phi - 1}$

14. $\phi^2 \approx 2.618033989$
ϕ^2 is 1 more than ϕ. This is found by cross multiplying $\frac{\phi}{1} = \frac{1}{\phi - 1} \Rightarrow \phi(\phi - 1) = 1$
$\Rightarrow \phi^2 - \phi = 1$
$\Rightarrow \phi^2 = \phi + 1$

15. $\phi^{-1} \approx 0.6180339887$
$\phi^{-1} = \phi - 1$
This is found from the ratio in Exercise 13.
$\frac{\phi}{1} = \frac{1}{\phi - 1} \Rightarrow \frac{1}{\phi} = \phi - 1$

16.

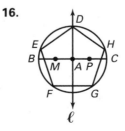

17. Sample answer: In a circle of radius of 3 cm, $DE \approx 3.8$ and $EH \approx 6.1$

18. $\frac{EH}{DE} \approx 1.605$
The difference is about 0.01.

19. $m\angle D = 180° - \frac{360°}{5} = 108°$

20. $m\angle DEH = m\angle DHE = 36°$

21. $\dfrac{\sin 108°}{EH} = \dfrac{\sin 36°}{DE} \Rightarrow \dfrac{EH}{DE} = \dfrac{\sin 108°}{\sin 36°} \approx 1.618033989 \approx \phi$

22.

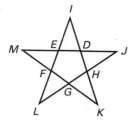

23. $m\angle EDH = 108°$
$m\angle IDE = m\angle IED = 180° - 108° = 72°$
$m\angle I = 180° - (72° + 72°) = 36°$

24. $\dfrac{EI}{\sin 72°} = \dfrac{DE}{\sin 36°} \Rightarrow \dfrac{EI}{DE} = \dfrac{\sin 72°}{\sin 36°} \approx 1.618033989 \approx \phi$

25. Sample answer:
If the original pentagon was inscribed in a circle of radius 3 cm, then $IL \approx 14.9$ and $IF \approx 9.2$.
$\dfrac{IL}{IF} \approx 1.6$

26. Sample answer:
If the original pentagon was inscribed in a circle of radius 3 cm, then $IF \approx 9.2$ and $IE \approx 5.7$.
$\dfrac{IF}{IE} \approx 1.6$

27. 34, 55, 89, 144, 233

28.

Term	1	1	2	3	5	8	13	21	34	55	89	144	233
Ratio	–	$\frac{1}{1}$	$\frac{2}{1}$	$\frac{3}{2}$	$\frac{5}{3}$	$\frac{8}{5}$	$\frac{13}{8}$	$\frac{21}{13}$	$\frac{34}{21}$	$\frac{55}{34}$	$\frac{89}{55}$	$\frac{144}{89}$	$\frac{233}{144}$
Value	–	1	2	1.5	1.67	1.6	1.625	1.615	1.619	1.618	1.618	1.618	1.618

29. The ratios seem to approach ϕ. **30.** Yes; $\dfrac{514,229}{317,811} \approx 1.618033989$ **31.** $F_{20} = \dfrac{\phi^{20} - (\phi')^{20}}{\sqrt{5}} = 6765$

32. $\dfrac{F^n}{F^{n-1}} = \left(\dfrac{\phi^n - (\phi')^n}{\sqrt{5}}\right) \div \left(\dfrac{\phi^{n-1} - (\phi')^{n-1}}{\sqrt{5}}\right)$

$= \left(\dfrac{\phi^n - (\phi')^n}{\sqrt{5}}\right) \cdot \dfrac{\sqrt{5}}{\phi^{n-1} - (\phi')^{n-1}}$

$= \dfrac{\phi^n - (\phi')^n}{\phi^{n-1} - (\phi')^{n-1}}$

As n increases, $(\phi')^n$ gets very close to 0, so $\dfrac{F^n}{F^{n-1}}$ approaches $\dfrac{\phi^n}{\phi^{n-1}} = \phi$.

33. $\dfrac{5}{3} \approx 1.67$
$\dfrac{7}{5} = 1.4$
$\dfrac{10}{8} = 1.25$
$\dfrac{14}{11} \approx 1.27$
$\dfrac{20}{16} = 1.25$

3×5 is closest to the golden ratio.

34. $\dfrac{\ell}{s} = \phi$ $\ell s = 104 \text{ in}^2$
$\ell = s\phi \Rightarrow s^2\phi = 104$
$s^2 = \dfrac{104}{\phi}$
≈ 64.276
$\Rightarrow s \approx 8 \text{ in.}$
$\Rightarrow \ell = \dfrac{104}{8} = 13 \text{ in.}$

35. $\dfrac{\ell}{s} = \phi$ so $\ell = s\phi$.
$2\ell + 2s = 36 \text{ in.}$
$\Rightarrow 2s\phi + 2s = 36$
$\Rightarrow 2s(\phi + 1) = 36$
$\Rightarrow s = \dfrac{18}{\phi + 1}$
$\Rightarrow s \approx 6.88$
$\Rightarrow s \approx 7 \text{ in.}$
$\ell = \dfrac{1}{2}(36 - 2 \cdot 7) = 11 \text{ in.}$

PAGE 705, LOOK BACK

36. $d = \sqrt{(4-0)^2 + (7-0)^2} = \sqrt{16 + 49} = \sqrt{65} \approx 8.1$

37. $d = \sqrt{(4-1)^2 + (2-3)^2} = \sqrt{9 + 1} = \sqrt{10} \approx 3.2$

38. $d = \sqrt{(5-(-6))^2 + (-2-0)^2} = \sqrt{121 + 4} = \sqrt{125} \approx 11.2$

39. $d = \sqrt{(-1-3)^2 + (-3-(-5))^2} = \sqrt{16+4} = \sqrt{20} \approx 4.5$

40. $(x-2)^2 + (y-1)^2 = 3^2$ or
$(x-2)^2 + (y-1)^2 = 9$

41. $(x', y') = (\cos 50°, \sin 50°)$
$= (0.6428, 0.7660)$

42. The point is in quadrant 2, hence the angle is
$\angle A = 180° - \sin^{-1}(0.766) \approx 180° - 50°$
$\angle A \approx 130°$

43.

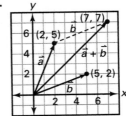

$\vec{a} + \vec{b}$ points from the origin to $(7, 7)$.

PAGE 705, LOOK BEYOND

44. 34 and 55; They are in the Fibonacci sequence.

45. Answers will vary.

11.2 PAGE 709, GUIDED SKILLS PRACTICE

6. $AB = |0-2| + |1-3| = 2 + 2 = 4$

7. $BC = |2-(-2)| + |3-1| = 4 + 2 = 6$

8. $AC = |0-(-2)| + |1-1| = 2 + 0 = 2$

9.

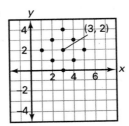

10.

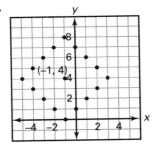

PAGES 709–710, PRACTICE AND APPLY

11. $|0-7| + |0-3| = 7 + 3 = 10$

12. $|5-(-2)| + |-3-4| = 7 + 7 = 14$

13. $|1-(-2)| + |5-(-3)| = 3 + 8 = 11$

14. $|-9-(-3)| + |-3-(-1)| = 6 + 2 = 8$

15. $|-11-(-3)| + |4-9| = 8 + 5 = 13$

16. $|-129-152| + |43-236| = 281 + 193 = 474$

17. $4(2) = 8$

18. $4(4) = 16$

19. $4(12) = 48$

20. $8(1) = 8$

21. $8(5) = 40$

22. $8(10) = 80$

23.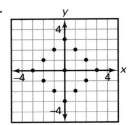

24.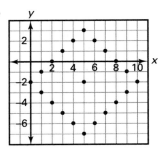

25.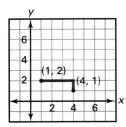

26. $(0, 0)$ and $(0, 4)$ have one path between them. $(0, 0)$ and $(1, 3)$ have several paths.

27.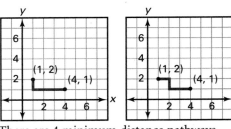

There are 4 minimum distance pathways.

28.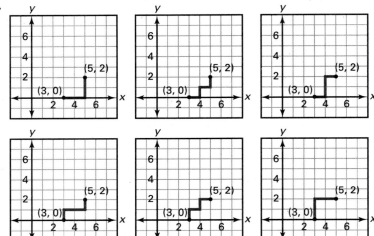

There are 6 minimum distance pathways.

29. a. $(1, 2)(2, 1), (3, 0)$

 b. $(3, -1), (1, 3)$

 c.

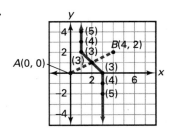

30. The perpendicular bisector is defined in taxicab geometry in the same way as it was defined in Euclidian geometry. Graphically, there appears to be a "bend" in the bisector.

31.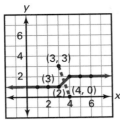

The bisector has a bend in it. However, the bend consists of only two points, rather than the three formed in Exercise 29.

32.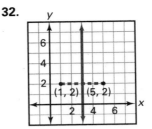

Any point with coordinates $(3, y)$ lies on the bisector. This bisector is the same line obtained when using Euclidian geometry.

33. Because the taxidistance between $(2, 0)$ and $(3, 2)$ is odd, a perpendicular bisector cannot be found.

34. A taxicab circle with radius r and center (h, k) has the general formula
$$|x - h| + |y - k| = r$$

35. The points are:

$(-1, 0), (-1, 1), (-1, 2), (0, -1), (0, 3), (1, -1), (1, 3)$
$(2, -1), (2, 3), (3, 0), (3, 1), (3, 2)$

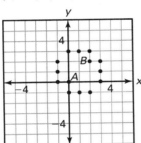

Sample answer:
The points form two sets of parallel line segments of length 3; or a square with side length 4 which has its corner points deleted.

36.

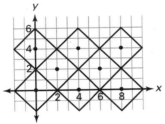

Sample answer:
If one corner of the city grid is $(0, 0)$, then put call boxes at $(4m, 4n)$ and $(2 + 4m, 2 + 4n)$ where m and n are non-negative integers.

37. The set of points (x, y) will have the property
$$|x| + |y| < 10$$
$$|x - 5| + |y - 4| < 8$$

That is, the interior of the intersection of the circle of radius 10 centered about $(0, 0)$ and the circle of radius 8 centered about $(5, 4)$.

Jenny should look for places within the region shown, including the edge points and sides.

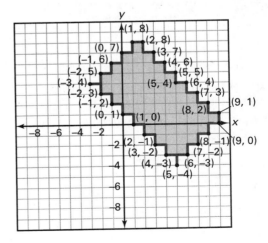

38. $10 \cdot 4 = 6y \Rightarrow y = \frac{40}{6} \approx 6.67$

39. $4 \cdot 15 = 7y \Rightarrow y = \frac{60}{7} \approx 8.57$

40. $AC \cdot BC = EC \cdot DC$

$\Rightarrow 22 \cdot 5 = EC \cdot 6 \Rightarrow EC = \frac{110}{6}$

$\Rightarrow DE = EC - CD = \frac{110}{6} - 6 = \frac{37}{3} \approx 12.33$

41. $MP \cdot MN = (MR)^2$

$\Rightarrow 18 \cdot 7 = (MR)^2$

$\Rightarrow MR = \sqrt{126} \approx 11.22$

42. $\sqrt{53^2 - 28^2} = 45$

$\tan \theta = \frac{28}{45} \approx 0.62$

43. $\cos \theta = \frac{45}{53} \approx 0.85$

44. $\cos \theta = \frac{45}{53} \Rightarrow \theta \approx 31.89°$

45. c. Sample answer:
The pieces do not quite fit together to form the rectangle. Note that in the rectangle the piece labeled *B* has a base of 5 units, while in the original square, the base is shorter than 5 units.

5. Yes, only two vertices are odd.

6. No, all 4 vertices are odd.

7. Yes, only two vertices are odd.

8.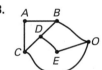

The graph does not contain an Euler path since vertices *B*, *C*, *D*, and *O* are odd.

9. Yes. Since the driver must cover both sides of every street, every street contributes two edges to the graph.

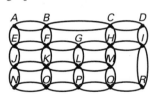

Note that every vertex is even, hence an Euler circuit exists.

10. There is an Euler path. However there are two odd vertices so there is no Euler circuit.

11. Each of the four vertices are odd, so there are no Euler paths or circuits.

12. Each of the four vertices are odd, so there are no Euler paths or circuits.

13. Each vertex is even, so there is an Euler path and circuit.

14. There is an Euler path, but no Euler circuit since there are two odd vertices.

15. Each vertex is even, so there is an Euler path and circuit.

16.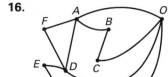

Yes. Each vertex is even, so there is an Euler circuit.

17.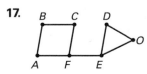

Vertices *E* and *F* are odd, so there is no Euler circuit.

18.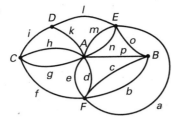

19. There is an Euler path but no Euler circuit since D and E are odd vertices.

20. Sample answer:

Note that the path must start at vertex D or E.

$(E)\ a - b - c - d - e - f - g - h - i - k - p - o - n - m - \ell(D)$

21. 5

22. $3 + 3 + 4 = 10$

23. 6

24. $3 + 3 + 3 + 3 = 12$

25. 8

26. $3 + 3 + 3 + 3 + 4 = 16$

27. The sum of the degrees of the vertices of a graph is twice the number of edges.

28. Each edge has two endpoints, therefore it is counted twice when adding the degrees of the vertices, so the sum must be a multiple of 2.

29. Let S_e be the sum of the degrees of the even vertices, S_o be the sum of the degrees of the odd vertices, and E be the number of edges. Then $S_e + S_o = 2E$.
Note that S_e is even since the sum of even numbers is even. Then $S_o = 2E - S_e$ is even, since the difference of two even numbers is even. Therefore S_o is an even number, that is, the sum of degrees of odd vertices is an even number. But the only way a sum of odd numbers is even is if there is an even number of those odd numbers. Therefore the number of odd vertices must be even.

30. Yes; note that each intersection has even degree, so there is an Euler path.

31. Yes; note that each intersection has even degree, so there is an Euler path.

32. Yes. One solution is shown.

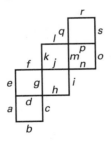

Travel the edges in alphabetical order.

33. Sample answer:

It is not possible. Consider each box and the outside to be a vertex and every segment between two boxes as an edge.

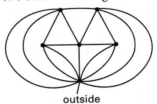

outside

The vertices representing the three larger boxes will have degree 5, so there can be no Euler circuit.

34.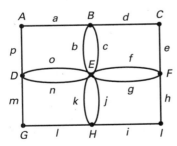

35. Yes; all vertices in the graph are even.

36. No.

The associated graph has 4 odd vertices, so there is no Euler path or circuit.

37. Graph for Jo

In Jo's graph, every vertex is even, so there is an Euler path.

Graph for Tamara

In Tamara's graph, there are two odd vertices, so there is an Euler path.

PAGE 719, LOOK BACK

38.

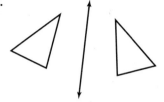

39.

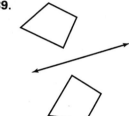

40. If people live in California, then they live in the United States.
Converse: If people live in the United States, then they live in California.

41. If S is a square, then S is a rectangle.
Converse: If S is a rectangle, then S is a square.

42. If S is a square, then S is a parallelogram with four congruent sides and four congruent angles.
Converse: If S is a parallelogram with four congruent sides and four congruent angles, then S is a square.

43. If S is a dodecagon, then S is a polygon with 12 sides.
Converse: If S is a polygon with 12 sides, then S is a dodecagon.

44. If θ is an angle with $\theta < 45°$, then $\sin \theta < \cos \theta$.
Converse: If θ is an angle with $\sin \theta < \cos \theta$, then $\theta < 45°$.

45. $V = Bh$
$B = \frac{1}{2}(4)(6) = 12$
$100 = 12 \cdot h \Rightarrow h = \frac{100}{12} \approx 8.3$

46. $\ell wh = 100$
$4 \cdot 5h = 100 \Rightarrow 20h = 100 \Rightarrow h = 5$

47. $\frac{1}{3}\pi r^2 h = 100$
$\frac{1}{3}\pi r^2 5 = 100$
$\frac{5\pi}{3}r^2 = 100 \Rightarrow r^2 = \frac{300}{5\pi} \Rightarrow r \approx 4.4$

48. $\frac{4}{3}\pi r^3 = 100 \Rightarrow r^3 = \frac{300}{4\pi} \Rightarrow r = \sqrt[3]{\frac{75}{\pi}} \approx 2.9$

49. Edge $\overline{CE}$ is a bridge.

50.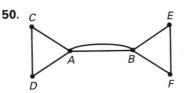

No edge is a bridge.
A graph that contains an Euler circuit cannot contain a bridge because every vertex has even degree. Removing an edge cannot disconnect the graph because there is still some path between any two vertices.

51. Sample answer:
ABCDEFECBGHCHDEHGBFGAFA

11.4 **PAGE 724, GUIDED SKILLS PRACTICE**

6. **a, b, d**

7. **b, c**

8. $V - E + F = 8 - 12 + 6 = 2$

9. $V - E + F = 6 - 12 + 8 = 2$

PAGES 725–727, PRACTICE AND APPLY

10. Inside

11. Outside

12. Answers may vary from 1 to 5.

13. The number of times the ray from P crosses the curve is always odd.

14. The number of times the ray from Q crosses the curve is always even.

15. odd; even

16. A ray from the point intersects the curve an odd number of times, thus, the point is inside the curve.

17.–18. Sample answer:

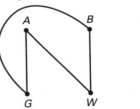

19. If one were inside and one were outside, there would be no way to connect C to E without intersecting the simple closed curve.

20.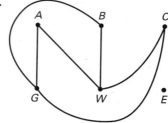

Sample answer:
Note that the curve *BGCW* is a simple closed curve with *A* on the inside and *E* on the outside. By the Jordan Curve Theorem, every curve connecting *A* to *E* intersects an already connected utility line. If the edges are redrawn so that *E* is on the inside of curve *BGCW*, then another simple closed curve, *AGCW*, is formed with *E* on the inside and *B* on the outside. Thus, either *A* or *B* cannot be connected to *E* without any intersecting edges.

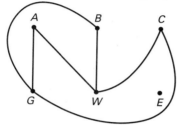

21.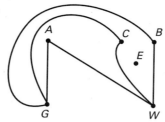

Sample answer:
The diagram above has both *C* and *E* inside the simple closed curve *ABGW*. Note that the curve *BGCW* is a simple closed curve with *E* on the inside and *A* on the outside. By the Jordan Curve Theorem, every curve connecting *A* to *E* intersects an already connected utility line.
If the edges are redrawn so that *E* is on the outside of curve *BGCW*, then another simple closed curve, *AGCW*, is formed with *E* on the inside and *B* on the outside. Thus, either *A* or *B* cannot be connected to *E* without any intersecting edges.

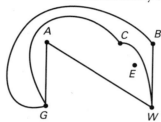

It is impossible to connect all three houses to two utilities without forming a simple closed curve with one house inside and the third utility outside or the utility inside and the house outside. Thus, the three houses cannot be connected to the three utilities without any intersecting lines.

22. $V = 8, E = 12, F = 6$
$V - E + F = 8 - 12 + 6 = 2$

23. $V = 20, E = 30, F = 12$
$V - E + F = 20 - 30 + 12 = 2$
The polyhedron has 12 faces, so it is a dodecahedron.

24. Sample answer:

$V = 6, E = 12, F = 8$
$6 - 12 + 8 = 2$

25. The path covers both sides of the Möbius strip.

26. You must traverse the strip twice in order to get back to the starting point.

27. The result is a loop of paper with four half-twists.

28. The result is two loops of paper, linked together, one large and one small. The small loop is a Möbius strip, and the large loop has four half-twists.

29. Yes.

30. The result is one loop of paper with four half-twists.

31. Yes.

32. The result is two Möbius strips.

33. Cut the bottle "diagonally". Start at any point on the surface and return to it without crossing itself. By avoiding the joined edges entirely this produces a surface without dividing it into more than one piece. Then cut around the self-intersection.

34. Sample answer:
The length of the belt's life will be longer since both sides of the belt get used.

PAGES 727–728, LOOK BACK

35. $27x - 21 = 10x + 30$
$17x = 51$
$x = 3$

36. Sample answer:
$\angle 1 \cong \angle 5$ and $\angle 2 \cong \angle 7$ by the Alternate Interior Angles Theorem

37. Answers may vary. Sample answer:
$\angle 1$ and $\angle 4$, $\angle 3$ and $\angle 7$, $\angle 6$ and $\angle 7$

38.

Statement	Reason
$m\angle 1 + m\angle 2 + m\angle 3 = 180°$	Linear Pair Property
$m\angle 5 = m\angle 1$ $m\angle 6 = m\angle 3$	Alternate Interior Angles Theorem
$m\angle 5 + m\angle 2 + m\angle 6 = 180°$	Substitution Property

39. $VG = \frac{4}{3}\pi r^3 = \frac{4}{3}\pi(3^3) = 36\pi$
$VO = \frac{4}{3}\pi r^3 = \frac{4}{3}\pi(2^3) = \frac{32}{3}\pi$
$\frac{VG}{VO} = \frac{36\pi}{\frac{32\pi}{3}} = 3\frac{3}{8}$, so you will need $3\frac{3}{8}$ oranges.

40. $a^2 = b^2 + c^2 - 2bc \cos A$
$a^2 = 25^2 + 20^2 - 2(25)(20) \cos 55° \approx 451.42356$
$a \approx 21.2$

41. $b^2 = a^2 + c^2 - 2ac \cos B$
$b^2 = (104)^2 + (47)^2 - 2(104)(47) \cos (92°) \approx 13{,}366.177$
$b \approx 115.6$

PAGE 728, LOOK BEYOND

42. a. move 1

b. move 2

c. move 3

43. Sample answer:

44. Sample answer:

5. If *A*, *B* and *C* are angles of a triangle, m∠*A* + m∠*B* + m∠*C* > 180° in spherical geometry and m∠*A* + m∠*B* + m∠*C* < 180° in hyperbolic geometry.

6. Yes

7. No

8. No

9. Yes

10. Yes

11. No

12. Yes

13. No

PAGES 734–736, PRACTICE AND APPLY

14. 90° + 90° + 90 = 270°

15. Sample answer:

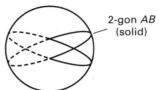

2-gon *AB* (solid)

The vertices must be opposite each other on the sphere.

16. The sum of the measures of the angles of a 2-gon is greater than 0° and less than 360° since each angle will measure between 0° and 180°.

17. Sample answer:

The diagonal is $\overline{BD}$.

18. The sum of the angles of the quadrilateral is greater than 360° because the 4-gon can be divided into two triangles by the diagonal. In spherical geometry the sum of the angles in a triangle is greater than 180°, so the sum of the angles of two triangles would be greater than 360°.

19.

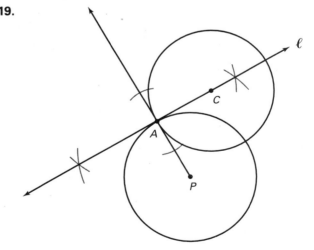

20. As the distance from *A* increases, the radius of the arc increases. The other endpoint of the arc on the circle moves farther from point *A*. If *C* were infinitely far from *A* we would obtain the diameter $\overline{AB}$.

21. If an arc is orthogonal to a circle at a point, then the tangent of the arc is perpendicular to the tangent of the circle at that point. By the Tangent Theorem, the tangent of the arc is also perpendicular to a radius of the circle drawn to the point of tangency. The tangent of the circle at the given point must contain the radius of the arc, and so it must contain the center of the arc.

22.

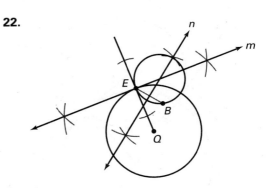

23. As E gets closer to the line $\overline{BQ}$, the radius of the arc gets very large.

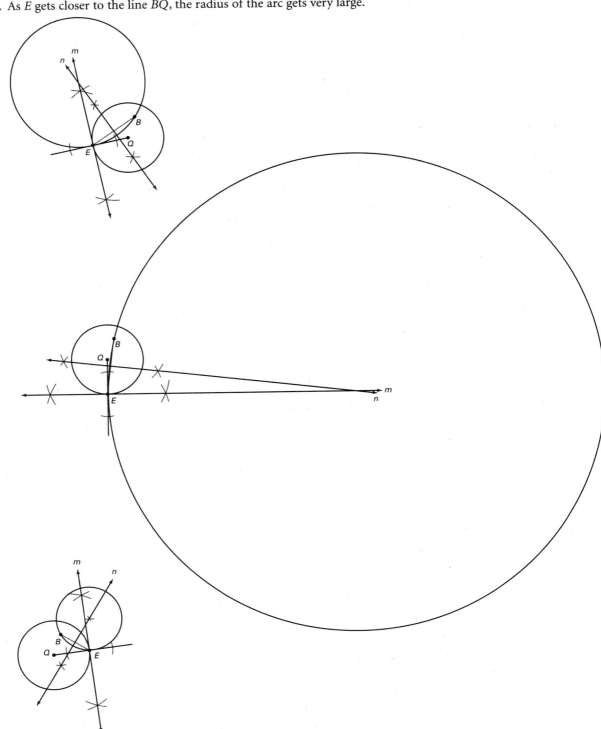

24.

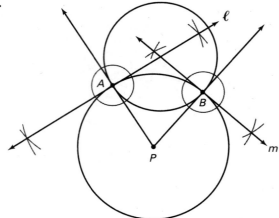

25. If A and B are endpoints of a diameter, then $\ell \parallel m$ and they never intersect. In this case the line through A and B is the diameter $\overline{AB}$.

26.

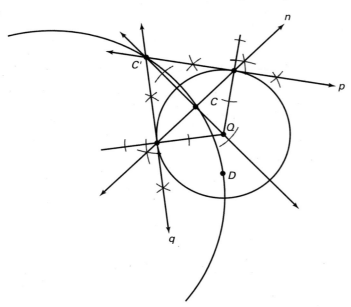

27. Infinitely many lines can be drawn through A that do not intersect ℓ. This shows that the Parallel Postulate does not hold in Poincaré's system.

28. Sample answer:

$m\angle A \approx 30°$, $m\angle B \approx 30°$, $m\angle C \approx 90°$

29. The sum of the measures of the angles in the triangle is less than 180°. Measure the angles with a protractor to get a fairly accurate reading. In Euclidean geometry the sum of the angle measures in a triangle is 180°, while in spherical geometry it would be greater than 180°.

30.

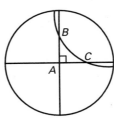

The sum of the acute angles in Poincaré's model is less than 90°, since the angle sum of the triangles is less than 180°.

31. Sample answer:

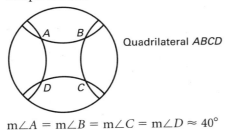

Quadrilateral *ABCD*

$m\angle A = m\angle B = m\angle C = m\angle D \approx 40°$

32. The sum of the measures of the angles in the The sum of the measures of the angles in the quadrilateral is less than 360° because there are two triangles in a quadrilateral (divided by a diagonal). In Exercise 29 it was shown that the sum of the angle measures of a triangle is less than 180°, so if you added the sum of the angles of the two triangles, it would be less than 360°.

33. Euclidian geometry
Sum of the angles of an *n*-gon $= (n - 2)180°$
Hyperbolic geometry
Sum of the angles of an *n*-gon $< (n - 2)180°$

34. Sample answer: The model is a convex surface (or concave, depending on how it is turned).

35. Sample answer: The triangles form a flat surface.

36. Sample answer: The model will not lie flat and it is neither concave nor convex overall. The surface is "wavy," going up and down.

37. Sample answer: The models are all composed of a number of equilateral triangles about a given vertex. In the first model the angle measures of the triangles around the vertex add to less than 360°, so the model will not lie flat: it is convex. In the second model the sum of the angle measures equals 360°, so the model will lie flat. In the third model the sum of the angle measures is greater than 360, so it will not lie flat: it is "wavy."

38. Sample answer: The surface cannot be extended to infinitely many triangles; it will come back on itself to form a closed surface like a sphere. (In fact, an icosohedron will be formed, which is a polyhedral approximation of a sphere.)

39. Sample answer: A Euclidean plane will be formed. This is called a tessellation of the plane by equilateral triangles and will extend to infinitely many triangles.

40. Sample answer: A surface will be formed that has no overall tendency to be either concave or convex and will extend to infinitely many triangles.

PAGE 737, LOOK BACK

41.–42. Check students' constructions.

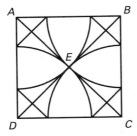

43.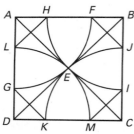

ABCD is a square, so $AB = BC = CD = DA$ and $\angle A \cong \angle B \cong \angle C \cong \angle D$. Also, the same compass setting was used to draw all four arcs, so $AE = AF = AG = BH = BI = CJ = CK = DL = DM$. By the Overlapping Segments Theorem, $AH = FB = BJ = IC = CM = KD = DG = LA$. By the Segment Addition Postulate, $HF = JI = MK = GL$. By SAS, $\triangle AHL \cong \triangle BJF \cong \triangle CMI \cong \triangle DGK$, so $HL = FJ = IM = KG$ because CPCTC. Let $AB = 1$ unit. Since $\triangle ABE$ is a 45-45-90 triangle, $AE = AF = BH = \frac{\sqrt{2}}{2}$. By the Segment Addition Postulate, $AF + BH - HF = AB = 1$, so $HF = 1 - \frac{\sqrt{2}}{2} - \frac{\sqrt{2}}{2} = 1 - \sqrt{2}$. Also by the Segment Addition Postulate, $AH = 1 - \frac{\sqrt{2}}{2}$. $\triangle AHL$ is a 45-45-90 triangle, so $HL = \sqrt{2}\left(1 - \frac{\sqrt{2}}{2}\right) = 1 - \sqrt{2}$, so $AH = HL$. Thus, all sides of $HFJIMKGL$ are congruent. Also, because CPCTC, $\angle AHL \approx \angle ALH \approx \angle BFJ \approx \angle BJF \approx \angle CIM \approx \angle CMI \approx \angle DKG \approx \angle DGK$. By the Linear Pair of Property, $\angle LHF \approx \angle HFJ \approx \angle FJI \approx \angle JIM \approx \angle IMK \approx \angle MKG \approx \angle KGL \approx \angle GLH$. Since all of the sides and angles are congruent, $HFJIMKGL$ is a regular octagon.

44. Sample answer:

45. Sample answer:
The graph has four odd vertices, so no Euler path or circuit exists.

46. C, G, I, J, L, M, N, S, U, V, W, Z are all topologically equivalent to the letter Z.

47. SIDE and CLOT are topologically equivalent. LAST and COZY are not topologically equivalent since A is not topologically equivalent to O.

PAGE 737, LOOK BEYOND

48. $\frac{1}{8}$ of the surface area is covered by this triangle.

$$A = \pi r^2 \left(\frac{90° + 90° + 90°}{180°} - 1\right)$$
$$= \pi r^2\left(\frac{1}{2}\right)$$
$$= \frac{1}{8}(4\pi r^2), \text{ which is } \frac{1}{8} \text{ of the surface area of a sphere.}$$

49. $\frac{m\angle A + m\angle B + m\angle C}{180°}$ is very close to 1, so

$$A = \pi r^2 \left(\frac{m\angle A + m\angle B + m\angle C}{180°} - 1\right)$$

is very close to 0.

50. $\pi r^2\left(\dfrac{m\angle A + m\angle B + m\angle C}{180°} - 1\right) < 4\pi r^2$

$\dfrac{m\angle A + m\angle B + m\angle C}{180°} - 1 < 4$

$\dfrac{m\angle A + m\angle B + m\angle C}{180°} < 5$

$m\angle A + m\angle B + m\angle C < 5(180°) = 900°$

11.6 PAGE 742, GUIDED SKILLS PRACTICE

5. The points where segments connect to the base form the Cantor dust.

6. If the two boxes above a box are both shaded or both unshaded, then the box is left unshaded. However, if one of the boxes is shaded and one is unshaded, the box beneath them is shaded.

7.

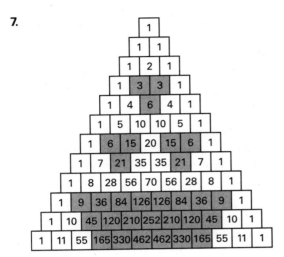

The shaded areas are inverted pyramids with increasingly longer bases arranged in a self–similar pattern, as in Activity 2.

PAGES 743–744, PRACTICE AND APPLY

8. $\frac{1}{2}(6)(8) = 24$

9. $3 \cdot \left[\frac{1}{2} \cdot 3 \cdot 4\right] = 18$

10. $9 \cdot \left[\frac{1}{2} \cdot \frac{3}{2} \cdot 2\right] = 13.5$

11. $27\left[\frac{1}{2} \cdot \frac{3}{4} \cdot 1\right] = 10.125$

12. $6 + 8 + 10 = 24$

13. $3(3 + 4 + 5) = 3(12) = 36$

14. $9(1.5 + 2 + 2.5) = 9(6) = 54$

15. $27(.75 + 1 + 1.25) = 27(3) = 81$

16. $\frac{3}{4}$

17. The area decreases and its limit is 0.

18. The perimeter increases without bound.

19. Check students' constructions.

20. Step 2:

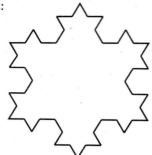

Step 3:

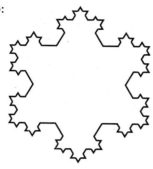

21. Step 0: $P = 3 \times 18 = 54$
Step 1: $P = 12 \times 6 = 72$
Step 2: $P = 48 \times 2 = 96$

22. The perimeter is increasing. It increases by a factor of $\frac{4}{3}$ each time. The perimeter increases by a larger amount each time, so that means the perimeter will continue to increase without bound.

23. The area of the snowflake is always increasing. The area never becomes infinite because the snowflake can always be enclosed in a circle of radius 10.4 cm.

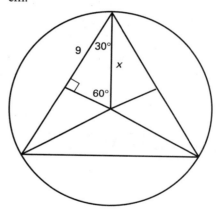

24. Sample answer:

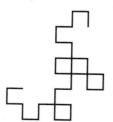

25. Sample answers:

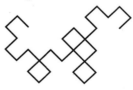

4 iterations

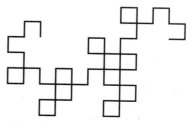

5 iterations

26. The length after the first iteration is $4\left(\frac{\sqrt{2}}{2}\right)$ $= 2\sqrt{2}$, since each line segment is a leg of an isosceles right triangle with hypotenuse of length 1. After two iterations the length is $8\left(\frac{\sqrt{2}}{2}\right)\left(\frac{\sqrt{2}}{2}\right) = 8\left(\frac{2}{4}\right) = 4$. After three iterations the length is $16\left(\frac{\sqrt{2}}{2}\right)\left(\frac{1}{2}\right) = 4\sqrt{2}$. The length of a dragon curve after $n+1$ iterations is $\sqrt{2}$ times the length after n iterations.

27. The area increases as iterations increase, but not without bound, because the amount it increases by is always less than half of the increase of the previous iteration.

28. If the volume of the cube is 1 cubic unit, each side has length 1 unit. In each iteration, each cube is divided into 27 equal cubes, and 7 are removed (one from each face and one from the center of the cube) so the volume is $\frac{20}{27}$ of the volume after the previous iteration. After 1 iteration, the volume is $\frac{20}{27}$ units^3. After 2 iterations, the volume is $\left(\frac{20}{27}\right)^2 = \frac{400}{729}$ units^3.

29. For the second iteration there are $4(4) = 16$ tetrahedrons.
For the third iteration there are $4(16) = 64$ tetrahedrons.

PAGE 737, LOOK BACK

30. $d = \sqrt{(4-2)^2 + (-2+1)^2} = \sqrt{2 + (-1)} = \sqrt{5} \approx 2.2$

31. $d = \sqrt{(5-(-2))^2 + (-10-3)^2} = \sqrt{7^2 + 13^2} = \sqrt{49 + 169} = \sqrt{218} \approx 14.8$

32. $d = \sqrt{(15-(-6))^2 + (2-5)^2} = \sqrt{21^2 + 3^2} = \sqrt{450} \approx 21.2$

33. $d = \sqrt{(4-2)^2 + (3-(-3))^2 + (2-5)^2} = \sqrt{2^2 + 6^2 + 3^2}$
$= \sqrt{4 + 36 + 9} = \sqrt{49} = 7$

34. $d = \sqrt{(18-0)^2 + (1-(-1))^2 + (0-5)^2} = \sqrt{324 + 4 + 25} = \sqrt{353} \approx 18.8$

35. $d = \sqrt{(5-2)^2 + (1-(-12))^2 + (-5-0)^2} = \sqrt{3^2 + 13^2 + 5^2}$
$= \sqrt{9 + 169 + 25} = \sqrt{203} = 14.2$

36. The point opposite the corner is $(8, 8, 8)$. Therefore the measure is
$d = \sqrt{(0-8)^2 + (0-8)^2 + (0-8)^2} = \sqrt{192} \approx 13.9$

PAGE 745, LOOK BEYOND

37. Check students' work.

38. The entire paper would appear shaded.

11.7 PAGE 752, GUIDED SKILLS PRACTICE

5. Sample answer:

6. The resulting figure is a rectangle with vertices $(0, 0)$, $(0, 3)$, $(4, 3)$, and $(4, 0)$.

7. Yes

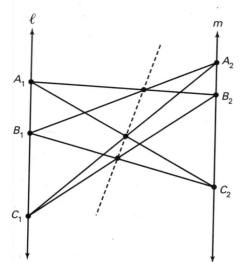

8. The points should be collinear.

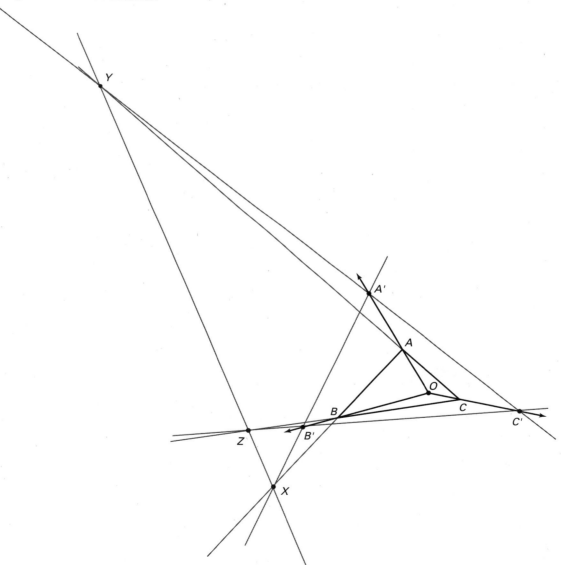

9.

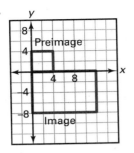

10.

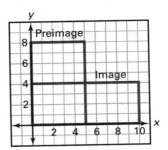

11.

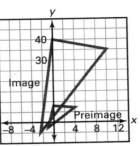

12.

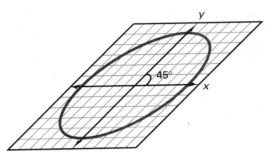

13.

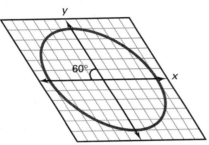

14. a. P
b. $\overrightarrow{PR}, \overrightarrow{PS}, \overrightarrow{PT}$

15. a. N
b. $\overrightarrow{NJ}, \overrightarrow{NK}, \overrightarrow{NL}$

16. a. M
b. $\overrightarrow{MG}, \overrightarrow{MH}, \overrightarrow{MI}$

17.

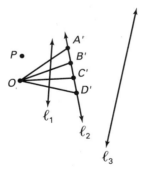

18.

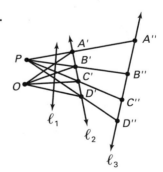

19. $C; H; K$

20. $B; A$

21. $B; D$ or E

22.–27., 29.

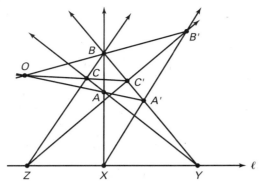

28. One triangle is a projection of the other.

PAGES 754–755, LOOK BACK

30. If the wind blows, the then the trees shake.
If the tree shakes, then the apple falls.
If the apple falls, then the worm squirms.
⇒ If the wind blows, the worm squirms.

31. Deductive

32. Inductive

33. Deductive

34. Sample answer: The golden ratio is the ratio of the sides of a golden rectangle. A golden rectangle is a rectangle with short side of length s and long side of length ℓ where $\frac{\ell}{s} = \frac{s}{\ell - s}$.

35. $d = |4 - 2| + |3 - 1| = 2 + 2 = 4$

36. $d = |-3 - 1| + |2 - 1| = 4 + 1 = 5$

37. $d = |1 - 5| + |3 - 5| = 4 + 2 = 6$

38. Every vertex of the figure is even, therefore it contains an Euler path and an Euler circuit.

39. There are no parallel lines in spherical geometry.

PAGE 755, LOOK BEYOND

40. Arrange the coins as follows:

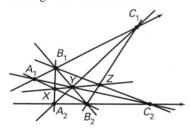

Note that points C_1 and C_2 in the Theorem of Pappus were chosen so that B_1, B_2, and Y are collinear.

CHAPTER REVIEW AND ASSESSMENT

1. $\frac{\ell}{s} \approx 1.168 \Rightarrow \ell \approx 1.618s \Rightarrow \ell \approx 1.618(3) \approx 4.85$

2. $\frac{\ell}{s} \approx 1.618 \Rightarrow s \approx \frac{\ell}{1.618} \Rightarrow s \approx \frac{9}{1.618} \approx 5.56$

3.

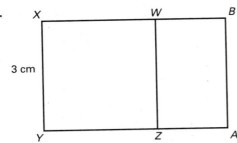

4. *AZWB* is also a golden rectangle.

5. $d = |6 - 4| + |1 - 5| = 2 + 4 = 6$

6. $d = |7 - (-1)| + |2 - 0| = 8 + 2 = 10$

7.

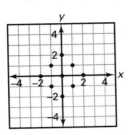

8.

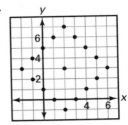

9. Euler path; no Euler circuit

10. Euler path; no Euler circuit

11. Neither

12. Euler path; no Euler circuit

13. Yes

14. No; the curve on the left intersects itself once while the curve on the right intersects itself twice.

15. The figure is topologically equivalent to a figure with 40 vertices, 80 edges and 36 faces (similar to the one pictured in the text).
$V - E + F = 40 - 80 + 36 = -4$

16. A sphere has Euler characteristic 2. This is a topological invariant, so any surface with a different Euler characteristic is not topologically equivalent to a sphere.

17. The two curves that intersect at the endpoints of a diameter of the ball are both lines in spherical geomtry.

18. The curve that wraps around the ball, intersecting each of the other curves in two places, is not a line in spherical geometry.

19. $\overleftrightarrow{KL}$, $\overleftrightarrow{KN}$, and $\overline{MN}$ are lines in hyperbolic geometry.

20.

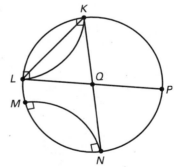

Drawing the chord from *K* to *L* on the circle gives us a curve which is not a line, since it does not intersect the outer circle at right angles.

21.

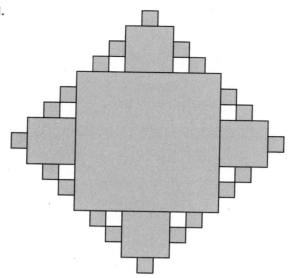

22.

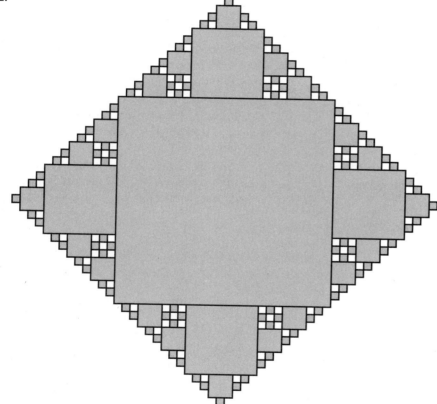

23. The area increases by $4 \cdot \left(\frac{1}{3}s\right)^2 = \frac{4}{9}s^2$ after the first iteration. The area of the nth iteration is then $\left(\frac{4}{9}\right)^n s^2$ more than the area of the $(n-1)$st iteration. The area of the fractal increases, but is bounded.

24. The perimeter of the fractal increases by $\frac{2}{3}s$ for each square added in the first iteration. Thus it increases by $\frac{8}{3}s$ in the first iteration. Therefore the perimeter increases without bound as $n \to \infty$.

25.

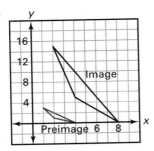

26.

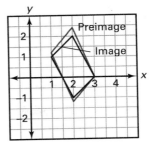

27.

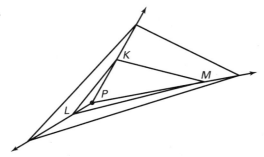

28.

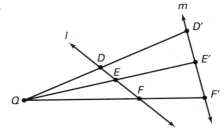

29. $\frac{\ell}{s} \approx 1.618$

$s = 45 \Rightarrow \ell \approx 1.618(45) \approx 72.81 \approx 73$ ft

$\ell = 45 \Rightarrow s = \frac{45}{1.618} \approx 27.8 \approx 28$ ft

30. The pilot should be flying along a great circle around the planet. The path goes above eastern Russia, into the Arctic Circle, then south above Norway.

31. They were measuring from different heights. Alex's photos have more detail, so his measurements may include small inlets or peninsulas that are not visible in the satellite photo.

CHAPTER TEST

1. $\frac{l}{15} = \frac{15}{l - 15}$; $l(l - 15) = 15^2$; $l^2 - 15l - 225 = 0$;

$l = \frac{15 + \sqrt{(-15)^2 - 4(1)(-225)}}{2(1)} =$

$\frac{15 + \sqrt{1125}}{2} \approx 24.27$ units

2. $\frac{8}{s} = \frac{s}{8 - s}$; $s^2 = 64 - 8s$; $s^2 + 8s - 64 = 0$;

$s = \frac{-8 + \sqrt{8^2 - 4(1)(-64)}}{2(1)} = \frac{-8 + \sqrt{320}}{2} \approx$

4.94 units

4. $\frac{l}{21} = \frac{21}{l - 21}$; $l(l - 21) = 21^2$; $l^2 - 21l - 441 = 0$;

$l = \frac{-(-21) + \sqrt{(-21)^2 - 4(1)(-441)}}{2(1)} =$

$\frac{21 + \sqrt{2205}}{2} \approx 34$ inches

3.

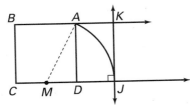

5. $4 - 1 + 6 - 5 = 4$ units

6. $8 - (-3) + 2 - (-1) = 14$ units

7. Check students' graphs.

8. Check students' graphs.

9. The graph contains an Euler path because there are 2 odd vertices, but not an Euler circuit because not all vertices are even.

10. The graph contains an Euler path and an Euler circuit because each vertex is even.

11. Yes, because there are 2 odd vertices, so there is an Euler path, but not an Euler circuit.

12. A rectangular prism has 8 vertices, 12 edges, and 6 faces: $8 - 12 + 6 = 2$.

13. The torus is topologically equivalent to a rectangular prism with a smaller rectangular prism removed from it, so it has 16 vertices, 32 edges, and 16 faces: $16 - 32 + 16 = 0$.

14. Yes, both figures contain one "twist" of a loop.

15. No, the figure on the left contains one loop with a "twist," but the figure on the left contains two loops.

16. $\overline{LM}$ is not a line in spherical geometry because it is not a great circle.

17. $\overleftrightarrow{JK}$ is a line in spherical geometry because it is a great circle.

18. Yes, because in spherical geometry the sum of the angle measures in a triangle are greater than $180°$.

19.

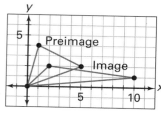

20–21.

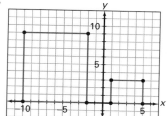

22. The area increases.

23. The perimeter increases.

24.

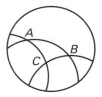

25.

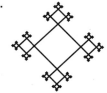

CHAPTERS 1–11 CUMULATIVE ASSESSMENT

1. $\angle 1$ and $\angle 2$ are corresponding angles. Thus they are congruent. The answer is C.

2. The hypotenuse is the longest side of a right triangle. The answer is A.

3. Area $= 6\left(\frac{1}{2}\right)(2)(\sqrt{3}) = 6\sqrt{3} \approx 10.4$
Perimeter $= 6 \cdot 2 = 12$
The answer is B.

4. $AG > BE$, since $AG = \sqrt{\ell^2 + w^2 + h^2}$ and $BE = \sqrt{w^2 + h^2}$, and $\ell > 0$. The answer is A.

5. The x-intercept must have y- and z-coordinates that are 0. The only choice where this is true is $(0, 0, 0)$. Then it must be that $4t - 8 = 0$, $2t - 3 = 0$, and $t + 4 = 0$. But then $4t - 8 = 2t - 3$, so $t = \frac{5}{2}$. But $\frac{5}{2} + 4 \neq 0$, so this is impossible. Therefore, the answer is d.

6. $SA = 6s^2$ and $V = s^3$, so as s increases, $\frac{6s^2}{s^3} = \frac{6}{s}$ decreases. Choose d.

7. $V = \frac{4}{3}\pi r^3$, so if the radius is doubled then the volume is increased by a factor of $2^3 = 8$. Choose d.

8. The answer is c, by definition of sine.

9. Sample answer:
$\frac{AB}{ZY} = \frac{BC}{YX} = \frac{CD}{XW} = \frac{DA}{WZ}$

10. There are two odd vertices; therefore the graph contains an Euler path but no Euler circuit.

11. There are five axes of symmetry.

12. $m = \frac{4 - (-1)}{2 - 3} = \frac{5}{-1} = -5$

13. $15x + 12 = 20x - 1$
$13 = 5x$
$x = \frac{13}{5}$ or 2.6

14. $\frac{\ell}{s} \approx 1.618 \Rightarrow s = \frac{\ell}{1.618}$
$s = \frac{12}{1.618} \approx 7.42$

15. $d = |1 - 7| + |4 - 10| = 6 + 6 = 12$

CHAPTER 12

A Closer Look at Proof and Logic

6. Argument form: *modus ponens* (valid)
If p then q

p	Premises
Therefore, q	Conclusion

p: weather takes a turn for the worse
q: local farmers will suffer a loss of income

7. Argument form: *modus tollens* (valid)
If p then q

$\sim q$	Premises
Therefore, $\sim p$	Conclusion

p: Stokes was in top form
q: Stokes won the competition

8. Invalid form: Affirming the Consequent
If p, then q

q	Premises
Therefore, p	Conclusion

p: the car is a Dusenberg
q: the car is a classic

9. Invalid form: Denying the Antecedent
If p then q

$\sim p$	Premises
Therefore, $\sim q$	Conclusion

p: Sean stuffed himself at lunch
q: Sean is feeling sleepy now

10. By the *modus tollens* argument form:
Therefore, the team did not win on Saturday.

11. By the *modus ponens* argument form:
Therefore, Sabrina did a stupendous amount of work at the last minute.

12. If Samantha is ill, then she is absent.
Samantha is absent.
Therefore, Samantha is ill.
Argument form: Affirming the Consequent
Invalid

13. If Sims is a man of good moral character, then he is innocent.
Sims is not a man of good moral character.
Therefore, Sims is not innocent.
Argument form: Denying the Antecedent
Invalid

14. If hedgehogs are tone deaf, they will seldom be seen at symphony concerts.
Hedgehogs are tone deaf.
Therefore, hedgehogs will seldom be seen at symphony concerts.
Argument form: *modus ponens*
Valid

15. If the plan was foolproof there were no unpleasant surprises.
There were unpleasant surprises.
Therefore, the plan was not foolproof.
Argument form: *modus tollens*
Valid

16. Sample answer: If the building was of sound construction, then it survived the storm.
The building was of sound construction.
Therefore, it survived the storm.

17. Sample answer: If the building was of sound construction, then it survived the storm.
The building did not survive the storm.
Therefore, it was not of sound construction.

18. Sample answer: If the building was of sound construction, then it survived the storm.
The building survived the storm.
Therefore, it was of sound construction.
Invalid

19. Sample answer: If the building was of sound construction, then it survived the storm.
The building was not of sound construction.
Therefore, it did not survive the storm.
Invalid

20. Valid, by *modus ponens*

21. Invalid, since the premises follow the *modus ponens* form, so the valid conclusion is "Eleanor will succeed."

22. Invalid, since the argument follows neither the *modus ponens* nor the *modus tollens* form.

23. Invalid, since the argument has an invalid form (denying the antecedent).

24. Invalid, since the argument has an invalid form (affirming the consequent)

25. Invalid, since the argument follows neither the *modus ponens* nor the *modus tollens* form.

26. Invalid, since the premises follow the *modus tollens* form, so the valid conclusion is "Mary does not study."

27. Valid, by *modus tollens*

28. Valid, by *modus ponens*

29. False. The diagonals of a parallelogram are congruent only if it is a rectangle.

30. The conclusion is a true statement without regard to the rest of the argument, based on the definitions of a rectangle and a square.

31. False. If *PQRS* is a parallelogram and the diagonals are congruent, then by Theorem 4.6.5, *PQRS* is a rectangle.

32. Not necessarily. If the premises of an argument are false, then there is no guarantee that the conclusion is true, even though the argument might be valid.

33. Yes, since a valid argument guarantees that its conclusion is true if its premises are true.

34. *r* might be false if one or more of the premises is false.

35. Sample answer:
In football, if a team does not move the ball 10 yards in 4 downs, then they lose possession of the ball. The Mammoths did not lose possession of the ball. Therefore, they must have moved the ball at least 10 yards in 4 downs. (*modus tollens*)

36. Sample answer:
If tulips are not planted in the fall, then they will not flower in the spring.
Nina's tulips flowered in the spring.
Therefore, Nina planted them in the fall.
(*modus tollens*)

PAGES 774-775, LOOK BACK

37. Yes, by using the SSS Postulate

38. No, since none of the congruence postulates apply to the triangles

39. Yes, by using the SAS Postulate

40. By the Inscribed Angle Theorem, the measure of $\overset{\frown}{QR}$ is $2 \cdot \text{m}\angle 1 = 2 \cdot 20° = 40°$.

41. By Therorem 9.4.3 and Exercise 40, since $\text{m}\angle 2 = 35°$ the measure of arc *PS* is $2(35) + 40 = 110°$.

42. Since the measure of arc *PS* is 110°, the measure of arc *SR* is 80°, and the measure of arc *RQ* is 40°, so the measure of arc *PQ* must be $360 - (110 + 80 + 40) = 130°$.

43. $\tan \theta = \sqrt{3}$
$\theta = 60°$
$\sin \theta = \frac{\sqrt{3}}{2} \approx 0.866$
$\cos \theta = 0.5$

44. An angle in the second quadrant cannot have the same tangent because the tangent is negative in the second quadrant.

45. 240° has the same tangent as θ.

46. Not *y*, from *modus tollens*, and also not *x*, from *modus tollens* again.

47. *m*, *q*, and *r* can be concluded by the If-Then Transitive Property

12.2 **PAGE 779, GUIDED SKILLS PRACTICE**

5. False, because only one statment is true in the conjunction.

6. True, because one of the statments is true in the disjunction.

7. True, because both statments are true in the conjunction.

8. False, because both statements are false in the disjunction.

9. ~(*r* AND *s*)

10. ~*t* OR ~*u*

PAGES 779–781, PRACTICE AND APPLY

11. A carrot is a vegetable and Florida is a state. The conjuncion is true because both statements are true.

12. A ray has only one endpoint and kangaroos can sing. The conjunction is false because one statement (kangaroos can sing) is false.

13. The sum of the measures of the angles of a triangle is 180° and two points determine a line. The conjunction is true because both statements are true.

14. Triangles are circles or squares are parallelograms. The disjunction is true because one of the statments (squares are parallelograms) is true.

15. Points in a plane equidistant from a given point form a circle or the sides of an equilateral triangle are congruent. The disjunction is true because both of the statements are true.

16. An orange is a fruit or cows have kittens. The disjunction is true because one of the statements (an orange is a fruit) is true.

17. The figure is not a rectangle.

18. My client is guilty.

19. Rain does not make the road slippery.

20. Triangles do not have six sides.

21. a.

p	~*p*	~(~*p*)
T	F	T
F	T	F

b. ~(~*p*) is logically equivalent to *p* because they both have the same truth values.

22. △*ABC* is not isosceles.

23. △*ABC* has two equal angles or △*ABC* is isosceles

24. △*ABC* is isosceles and △*ABC* has two equal angles.

25. △*ABC* does not have two equal angles.

26. ∠1 and ∠2 are not (both) acute angles.

27. ∠1 and ∠2 are adjacent or ∠1 and ∠2 are acute angles.

28. ∠1 and ∠2 are acute angles and ∠1 and ∠2 are not adjacent.

29. △*ABC* has two equal angles or ∠1 and ∠2 are not (both) acute angles.

30.

p	q	r	p AND q	(p AND q) AND r
T	T	T	T	T
T	T	F	T	F
T	F	T	F	F
T	F	F	F	F
F	T	T	F	F
F	T	F	F	F
F	F	T	F	F
F	F	F	F	F

All three of the statements must be true in order for (p AND q) AND r to be true. The statement is false otherwise.

31.

p	q	r	p OR q	(p OR q) OR r
T	T	T	T	T
T	T	F	T	T
T	F	T	T	T
T	F	F	T	T
F	T	T	T	T
F	T	F	T	T
F	F	T	F	T
F	F	F	F	F

All three statements must be false in order for (p OR q) OR r to be false.

32.

p	q	r	s	p AND q	r AND s	(p AND q) OR (r AND s)
T	T	T	T	T	T	T
T	T	T	F	T	F	T
T	T	F	T	T	F	T
T	T	F	F	T	F	T
T	F	T	T	F	T	T
T	F	T	F	F	F	F
T	F	F	T	F	F	F
T	F	F	F	F	F	F
F	T	T	T	F	T	T
F	T	T	F	F	F	F
F	T	F	T	F	F	F
F	T	F	F	F	F	F
F	F	T	T	F	T	T
F	F	T	F	F	F	F
F	F	F	T	F	F	F
F	F	F	F	F	F	F

(p AND q) OR (r AND s) is false if and only if both of the statements in the disjunction are false. This happens when one or both of p or q is false and when one or both of r or s is false.

33. Any of the following will make the statement true:
Flora will cook; Vernon will vacuum.
Flora will cook; Vernon will wash the windows.
Flora will cook; Vernon will vacuum and wash the windows.
Flora will wash the dishes; Vernon will vacuum.
Flora will wash the dishes; Vernon will wash the windows.
Flora will wash the dishes; Vernon will vacuum and wash the windows.
Flora will cook and wash the dishes; Vernon will vacuum.
Flora will cook and wash the dishes; Vernon will wash the windows.
Flora will cook and wash the dishes; Vernon will vacuum and wash the windows.

34. No; the advertisement has the form "if p then q," where p represents "one bulb burns out" and q represents "not all bulbs go out." The statement being compared is of the form "If p then r," where r represents "all the remaining bulbs continue burning." r is not the same as q, since if q is true it is possible that some bulbs are not burning.

35. The records in which the individual was born after 1950 AND the annual income is greater than $30,000 are:

Last name	First name	State	Year of birth	Annual income
Mallo	Elizabeth	TX	1956	$50,000
Brookshier	Mary	OH	1960	$62,000
Lamb	Charles	TX	1951	$41,000
Raemsch	Martin	OK	1965	$32,000

The records in which the individual lives in Texas OR the annual income is less than $30,000 are:

Last name	First name	State	Year of birth	Annual income
Craighead	Alicia	TX	1955	$25,000
Tuggle	Lawrence	LA	1972	$20,000
Mallo	Elizabeth	TX	1956	$50,000
Tong	Jun	TX	1952	$18,000
Lamb	Charles	TX	1951	$41,000

PAGE 781, LOOK BACK

36. They are not similar because not all the corresponding sides are proportional: $\frac{18}{6} = \frac{9}{3} \neq \frac{15}{4}$

37. Yes, by the AA Similarity Postulate

38. Yes, by the SAS Similarity Theorem

39. $\sin A = \frac{\text{opposite}}{\text{hypotenuse}} = \frac{12}{15} = \frac{4}{5}$

40. $\cos B = \frac{\text{adjacent}}{\text{hypotenuse}} = \frac{12}{15} = \frac{4}{5}$

41. $\tan B = \frac{\text{opposite}}{\text{adjacent}} = \frac{9}{12} = \frac{3}{4}$

42. $\cos A = \frac{\text{adjacent}}{\text{hypotenuse}} = \frac{9}{15} = \frac{3}{5}$

PAGE 781, LOOK BEYOND

43. The sentence cannot be true, because that would contradict the sentence's statement. But the sentence cannot be false, because then the sentence's statement would be true. Therefore the sentence is neither true nor false.

44. The woman should ask one of the people what the other one would say if he or she were asked the correct direction. Then take the opposite of the answer. If the woman asked the truthful person, the person would give the incorrect answer, because the person would truthfully say what the liar would say. If the woman asked the liar, the person would give the incorrect answer because the person would lie about what the truthful person would say.

6.

p	q	$p \Rightarrow q$
T	T	T
T	F	F
F	T	T
F	F	T

7.

p	q	$q \Rightarrow p$
T	T	T
T	F	T
F	T	F
F	F	T

8.

p	q	$\sim p$	$\sim q$	$\sim p \Rightarrow \sim q$
T	T	F	F	T
T	F	F	T	T
F	T	T	F	F
F	F	T	T	T

9.

p	q	$\sim q$	$\sim p$	$\sim q \Rightarrow \sim p$
T	T	F	F	T
T	F	T	F	F
F	T	F	T	T
F	F	T	T	T

PAGES 788-790, PRACTICE AND APPLY

10. Conditional: True, because squares have four 90° angles.
Converse: If a figure is a rectangle, then it is a square.
False; rectangles do not necessarily have four congruent sides.
Inverse: If a figure is not a square, then it is not a rectangle.
False; the figure can still be a rectangle if it is not a square—it could still have congruent opposite sides.
Contrapositive: If a figure is not a rectangle, then it is not a square.
True; if the figure is not a rectangle, it does not have four 90° angles, so it cannot be a square.

11. Conditional: True, since squaring both sides of an equation preserves equality.
Converse: If $a^2 = b^2$, then $a = b$.
False: if $a = -3$ and $b = 3$, then $(-3)^2 = (3)^2$ but a and b are not equal.
Inverse: If $a \neq b$, then $a^2 \neq b^2$.
False; if $a = -3$ and $b = 3$, they are not equal but $(-3)^2 = (3)^2$.
Contrapositive: If $a^2 \neq b^2$, then $a \neq b$.
True; if $a^2 \neq b^2$ then taking the square root of each side will not yield the same number, so $a \neq b$.

12. Conditional: False, since squaring both sides does not necessarily maintain inequality; it can reverse it.
Converse: If $a^2 < b^2$ then $a < b$.
False; if $b = -4$ and $a = 3$ then $a^2 < b^2$ but a is not less than b.
Inverse: If $a \geq b$ then $a^2 \geq b^2$
False; if $b = -4$ and $a = 3$ then $a \geq b$ but b^2 is not less than a^2.
Contrapositive: If $a^2 \geq b^2$ then $a \geq b$.
False; if $a = -4$ and $b = 3$ then $a^2 \geq b^2$ but a is not greater than or equal to b.

13. Conditional: False, since AAA does not guarantee triangle congruence.
Converse: If two triangles are congruent, then the three angles of one triangle are congruent to the three angles of the other triangle.
True; given that two triangles are congruent, their corresponding angles are congruent by the Polygon Congruence Postulate.
Inverse: If the three angles of one triangle are not congruent to the three angles of another triangle, then the triangles are not congruent.
True; if all the angles of one triangle are not congruent to the angles of the other triangle, then there is no way to set up a correspondence between the sides and angles so that corresponding angles are congruent.
Contrapositive: If two triangles are not congruent, then the three angles of one triangle are not congruent to the three angles of the other triangle.
False; If the triangles are not congruent they may still have congruent corresponding angles. They could be similar.

14. Conditional: True, because the sum of two even numbers is always even.
Converse: If $p + q$ is an even number, then p and q are even numbers.
False: p and q can be odd and their sum even ($3 + 7 = 10$).
Inverse: If p and q are not even numbers, then $p + q$ is not an even number.
False; If p and q are both odd, then their sum is even.
Contrapositive: If $p + q$ is not an even number, then p and q are not even numbers.
True; if $p + q$ is odd, then either p or q must be odd (but not both).

15. Conditional: True, by the laws of physics.
Converse: If the water temperature is less than or equal to 32°F, then it will freeze at normal atmospheric pressure.
True, by the laws of physics.
Inverse: If water does not freeze at normal atmospheric pressure, then its temperature is greater than 32°F.
True, by the laws of physics.
Contrapositive: If water's temperature is greater than 32°F, then it will not freeze at normal atmospheric pressure.
True, by the laws of physics.

16. Contrapositive: If $\sim q$ then $\sim p$
To form the contrapositive switch q and p and negate both. Contrapositive of the contrapositive: If $\sim(\sim p)$ then $\sim(\sim q)$, or If p then q.
The result of taking the contrapositive of a contrapositive is the original conditional.

17. **b** is true because it is the contrapositive of the original conditional. Since the statement is true, the contrapositive is also true.
The other two could be false because school could be canceled for other reasons besides snow.

18. Pythagorean Theorem: If a triangle is a right triangle, then the square of the length of the hypotenuse is equal to the sum of the squares of the lengths of the legs.
Converse: If the square of the length of the hypotenuse of a triangle is equal to the sum of the squares of the lengths of the legs, then the triangle is a right triangle.
True, this is proven in Lesson 5.4.
Inverse: If a triangle is not a right triangle, then the square of the length of the hypotenuse is not equal to the sum of the squares of the lengths of the legs.
True; if a triangle is not a right triangle, then the square of the length of the hypotenuse is greater than or less than the sum of the squares of the lengths of the legs.
Contrapositive: If the square of the length of the hypotenuse of a triangle is not equal to the sum of the squares of the lengths of the legs, then the triangle is not a right triangle.
True; The contrapositive is logically equivalent to the original conditional, which is the Pythagorean Theorem.

19. Sample answer: Theorem 3.3.3: If two lines cut by a transversal are parallel, then alternate interior angles are congruent.
Converse: If two lines cut by a transversal have alternate interior angles that are congruent, then the lines are parallel.
The converse of the theorem is Theorem 3.4.3, so it is true.
Inverse: If two lines cut by a transversal are not parallel, then alternate interior angles are not congruent.
The inverse is logically equivalent to the converse, so it is true.
Contrapositive: If two lines cut by a transversal do not have alternate interior angles that are congruent, then the lines are not parallel.
The contrapositive is logically equivalent to the original theorem, so it is true.

20. Sample Answer: Theorem 9.3.3: If two inscribed angles intercept the same arc, then they have the same measure.
Converse: If two inscribed angles have the same measure, then they intercept the same arc.
False; inscribed angles may have the same measure and not intercept the same arc.
Inverse: If two inscribed angles do not intercept the same arc, then they do not have the same measure.
False; if two inscribed angles do not intercept the same arc they can still have the same measure.
Contrapositive: If two inscribed angles do not have the same measure, then they do not intercept the same arc.
True; the contrapositive is logically equivalent to the original theorem.

21. If you are a senior, then you must report to the auditorium.

22. If a point is on the perpendicular bisector of a segment, then it is equidistant from the endpoints of the segment.

23. If she is going to be late, then she will call me.

24. If you do mathematics homework every night, then you will improve your grade in mathematics.

25. All of them are true.
r if and only if s can be written as $r \Leftrightarrow s$
In other words, r implies s and s implies r.
So if the statement r if and only if s is true then:

a. If r is true then s must also be true, because r implies s.

b. The statement is true because s implies r.

c. The statement is true because it is the contrapositive of **a,** and since **a** is true, then this statement is also true.

d. The statement is true because it is the contrapositive of **b,** and **b** is true.

26. Many of the theorems in the book can be rewritten in biconditional form.
Examples: Theorem 4.4.1—The Isosceles Triangle Theorem:
Two sides of a triangle are congruent if and only if the angles opposite those sides are congruent.
The converse is true, the inverse is true, and the conrapositive is true so an "if and only if" statement can be substituted in for the IF-THEN statement.
This also works with Theorem 9.2.2, the Tangent Theorem:
A line is perpendicular to a radius of a circle at its endpoint if and only if the line is tangent to the circle.
Any theorem in the book will work if its converse can be proven true, because the inverse has the same truth value as the converse, and the contrapositive has the same truth value as the original statement.

27. a. Same meaning; implies that getting at least a B in mathematics is a requirement to make the honor roll.

    b. Different meaning; this is the converse of the original statement.

    c. Different meaning; this is the inverse of the original statement.

    d. Same meaning; implies that getting at least a B in math is a requirement to make the honor roll.

PAGE 790, LOOK BACK

28. First find $m\angle C = 180° - 37° - 50° = 93°$

then, $\dfrac{\sin 37°}{100} = \dfrac{\sin 93°}{c}$

$c \sin 37° \approx 99.86$

$c \approx \dfrac{99.86}{\sin 37°} \approx 165.94$ units

29. First find $m\angle B = 180° - 65 - 47° = 68°$

then, $\dfrac{\sin 65°}{3.45} = \dfrac{\sin 68°}{b}$

$b \sin 65° \approx 3.20$

$b \approx \dfrac{3.20}{\sin 65°} \approx 3.53$ units

30.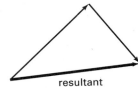

31.

32.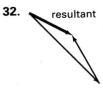

PAGE 790, LOOK BEYOND

33. p is given
q is to be proved
Once $p \Rightarrow q$ is proved to be true,
then given p, q follows by *modus ponens*.

34. p is given
q is to be proved.
Once $\sim q \Rightarrow \sim p$ is proved, then its contrapositive
($p \Rightarrow q$) is true and since p is given then q follows.
However then q is true, so $\sim q$ is false.

12.4 PAGE 794, GUIDED SKILLS PRACTICE

7. (Lines ℓ and m are parallel) AND (lines ℓ and m are not parallel)

8. ($\triangle ABC$ is isoceles) AND ($\triangle ABC$ is not isoceles)

9. (All squares are rectangles) AND (all squares are not rectangles)

10. ($ABCD$ is a square) AND ($ABCD$ is not a square)   **11.** Lines ℓ and m are not parallel

12. The same-side interior angles are supplementary

PAGES 794–795, PRACTICE AND APPLY

13. Yes; the proof starts by assuming the negation of the statement to be proved. Then a contradiction results after logical arguments. Lastly, the opposite of the assumption is stated to be true because the assumption must be false.

14. Yes; the proof starts by assuming the negation of the statement to be proved. Then a contradiction results after logical arguments. Lastly, the opposite of the assumption is stated to be true because the assumption must be false.

15. No; the proof does not start by assuming the negation of the statement to be proved.

16. No; the proof does not start by assuming the negation of the statement to be proved.

17. Yes; the proof starts by assuming the negation of the statement to be proved. Then a contradiction results after logical arguments. Lastly, the opposite of the assumption is stated to be true because the assumption must be false.

18. $\overline{AB} \not\cong \overline{AC}$

19. $\overline{DB} \cong \overline{AC}$

20. $\overline{BC}$

21. $\angle ACB$

22. $\triangle DCB$

23. SAS

24. $\overline{AB} \cong \overline{AC}$

25. $\overline{JK} \cong \overline{JL}$

26. isosceles

27. $\angle K \cong \angle L$

28. $\angle KJM \cong \angle LJM$

29. $\triangle KJM \cong \triangle LJM$

30. ASA

31. $\overline{KM} \cong \overline{LM}$

32. $\overline{JM}$ is not a median of $\triangle JKL$

33. $\overline{JK} \not\cong \overline{JL}$

34. Suppose that there is a largest integer m.
But if m is an integer, then $m + 1$ is integer, and $m + 1$ is larger than m.
So m is not the largest integer.
This is a contradiction.
Therefore there is no largest integer.

35. Suppose that there is a smallest positive real number x.

Then $\frac{1}{2}x$ is a real number, and $\frac{1}{2}x < x$, so x is not the smallest positive real number.

This is a contradiction.
Therefore there is no smallest positive real number.

36. Given: m and n are integers
and m^2 does not divide n^2 with no remainder.

Suppose m does divide n with no remainder (assume the opposite of what is to be proven).
Then n can be factored into $k \cdot m$ for some integer k.

$n = k \cdot m$
$n^2 = k^2 m^2$
$\frac{n^2}{m^2} = k^2$

Thus m^2 divides n^2 with no remainder, since k^2 is an integer.
This is a contradiction.
Therefore, for two integers m and n, if m^2 does not divide n^2 with no remainder, then m does not divide n with no remainder.

37. Given: a fraction $\frac{x}{y}$ in lowest terms with a decimal expansion that terminates after n places.

Suppose the denominator y does not divide any power of 10 with no remainder.
Then y does not divide 10^n. But $10^n\left(\frac{x}{y}\right) = m$, an integer, and so $10^n x = my$. Then y divides my so y divides $10^n x$. But y does not divide x, so y divides 10^n, which is a contradiction. Therefore y divides 10^n for some n.

38. Sample answer:
This contradicts the evidence that my client had a broken leg, and thus was unable to perform that task. Therefore, then assumption that my client is guilty is false, so my client is innocent.

39. $7x - 10 + 5x + 10 + 3x = 180$
$$15x = 180$$
$$x = 12$$

40. $a^2 + b^2 = 10^2 + 12^2$
$$= 100 + 144$$
$$= 244$$
$$c^2 = 14^2 = 196$$
Since $c^2 < a^2 + b^2$, the triangle is acute.

41. $a^2 + b^2 = 7^2 + 9^2$
$$= 49 + 81$$
$$= 130$$
$$c^2 = 15^2 = 225$$
Since $a^2 + b^2 < c^2$, the triangle is obtuse.

42. Since $4 + 7 < 12$, this is not a triangle.

43. $a^2 + b^2 = 8^2 + 12^2$
$$= 64 + 144$$
$$= 208$$
$$c^2 = 17^2 = 289$$
Since $a^2 + b^2 < c^2$, the triangle is obtuse.

44. $\sin 72° \approx 0.951$; $\cos 72° \approx 0.309$; $\tan 72° \approx 3.078$

45. $\sin 45° \approx 0.707$; $\cos 45° \approx 0.707$; $\tan 45° = 1$

46. $\sin 140° \approx 0.643$; $\cos 140° \approx -0.766$; $\tan 140° \approx -0.839$

47. $\sin 5° \approx 0.087$; $\cos 5° \approx 0.996$; $\tan 5° \approx 0.087$

48. Let h be the height of the tower and d the distance from point A to the tower.
$d = 90 + x$, where x is the distance from point B to the tower.
The following equations can be written to estimate the height of the tower:
$$\tan 55° = \frac{h}{90 + x}$$
$$\tan 68° = \frac{h}{x}$$
Solve each equation for h:
$$h = (90 + x)(\tan 55°)$$
$$= 90 \cdot \tan 55° + x \tan 55°$$
$$h = x \tan 68°$$
Then
$$x \tan 68° = 90 \cdot \tan 55° + x \tan 55°$$
$$x \tan 68° - x \tan 55° = 90 \cdot \tan 55°$$
$$x(\tan 68° - \tan 55°) = 90 \cdot \tan 55°$$
$$x = \frac{90 \cdot \tan 55°}{\tan 68° - \tan 55°}$$
$$x \approx 122.77$$
Then $h = x \tan 68°$
$$\approx 122.77 \tan 68°$$
$$\approx 303.87$$
So the tower is about 304 ft tall.

49. $n = 1$
$5^n - 1 = 5 - 1 = 4$ is divisible by 4.

50. Assume that $5^n - 1$ is divisible by 4.
It means that $5^n - 1$ can be written as $4k$ for some positive integer k. Then
$$5^n - 1 = 4k$$
$$5^n = 4k + 1$$
$$5 \cdot 5^n = 5(4k + 1)$$
$$5^{n+1} = 20k + 5$$
$$5^{n+1} - 1 = 20k + 4$$
$$5^{n+1} - 1 = 4(5k + 1)$$
so $5^{n+1} - 1$ is divisible by 4.

51. The statement is true for $n = 1$ and if it is true for n, then it is true for $n + 1$. Then according to the Principle of Mathematical Induction, the statement is true for all values of n.

12.5 **PAGE 801–802, GUIDED SKILLS PRACTICE**

5.

Input-Output Table			
Input		Output	
p	q	NOT p	NOT p OR q
1	1	0	1
1	0	0	0
0	1	1	1
0	0	1	1

6.

Input-Output Table			
Input		Output	
p	q	p AND q	NOT (p AND q)
1	1	1	0
1	0	0	1
0	1	0	1
0	0	0	1

7. Read from left to right, one branch at a time.
Combine the results of the branches when they flow together.
1. NOT q appears first: NOT q
2. The AND gate takes p and the output from step 1:
p AND (NOT q)

8. Read from left to right, one branch at a time.
Combine the results of the branches when they flow together.
1. p and q appear first: p AND q.
2. The OR gate takes r and the output from step 1:
(p AND q) OR r.

PAGES 802–804, PRACTICE AND APPLY

9.

10. ─[AND]─

11. ─)OR)─

12. If $p = 1$, the output is 0.

13. If $p = 1$, the output is 1.

14. If $p = 1$ and $q = 0$, the output is 0.

15. If $p = 1$ and $q = 0$, the output is 1.

	p	q	NOT p	NOT q	(NOT p) OR (NOT q)
16.	1	1	0	0	0
17.	1	0	0	1	1
18.	0	1	1	0	1
19.	0	0	1	1	1

	p	q	r	NOT p	(NOT p) AND q	NOT r	((NOT p) AND q) OR (NOT r)
20.	1	1	1	0	0	0	0
21.	1	1	0	0	0	1	1
22.	1	0	1	0	0	0	0
23.	1	0	0	0	0	1	1
24.	0	1	1	1	1	0	1
25.	0	1	0	1	1	1	1
26.	0	0	1	1	0	0	0
27.	0	0	0	1	0	1	1

28. Read from left to right, one branch at a time. Combine the results of the branches when they flow together.
1. NOT p appears first: NOT p
2. The OR gate takes q and the output from step 1:
   (NOT p) OR q
3. The NOT gate gives NOT ((NOT p) OR q)

29. Read from left to right, one branch at a time. Combine the results of the branches when they flow together.
1. NOT p
2. NOT q
3. The AND gate takes the input from steps 1 and 2:
   (NOT p) AND (NOT q)

30. Read from left to right, one branch at a time. Combine the results of the branches when they flow together.
1. p OR q
2. The OR gate takes r and the input from step 1.
   (p OR q) OR r

31. Read from left to right, one branch at a time. Combine the results of the branches when they flow together.
1. NOT p
2. q AND r
3. The OR gate takes the inputs from step 1 and 2:
   (NOT p) OR (q and r)
4. The NOT gate gives NOT((NOT p) OR (q AND r))

32.

33.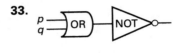

34.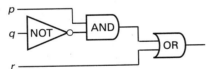

35.

36.

p	q	p AND q	NOT (p and q)	NOT p	NOT q	(NOT p) AND (NOT q)	(NOT p) OR (NOT q)
1	1	1	0	0	0	0	0
1	0	0	1	0	1	0	1
0	1	0	1	1	0	0	1
0	0	0	1	1	1	1	1

a NOT(p AND q) and **c** (NOT p) OR (NOT q) are functionally equivalent.

37.

p	q	NOT q	p OR (NOT q)	p OR q	NOT (p OR q)	NOT p	(NOT p) AND (NOT q)
1	1	0	1	1	0	0	0
1	0	1	1	1	0	0	0
0	1	0	0	1	0	1	0
0	0	1	1	0	1	1	1

b NOT (p OR q) and **c** (NOT p) AND (NOT q) are functionally equivalent.

38. Sample answer:

NOT q	(NOT q) OR r	p OR ((NOT q) OR r)
0	1	1
0	0	1
1	1	1
1	1	1
0	1	1
0	0	0
1	1	1
1	1	1

p OR ((NOT q) OR r)

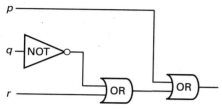

39. a. The arrangement corresponds to the logic function AND.
Both A and B have to be closed for the bulb to burn.

   b. The arrangement corresponds to the logic function OR. Either A or B need to be closed for the bulb to burn.

PAGE 804, LOOK BACK

---

40. Sample answer:
About five of the grids will cover the spill. (There is perspective distortion, but the grid is at middle distance.) $\approx 6{,}000$ ft

41. $\frac{5 \text{ ft}}{50 \text{ ft}} = \frac{1}{10}$ or 0.1

42. $\dfrac{\frac{1}{2}(3)(3)}{} = \dfrac{9}{50}$ or 0.18

43. False: Example: $\frac{2}{3} = \frac{4}{6}$ but $\frac{(2+1)}{3} \neq \frac{(4+1)}{6}$ because $\frac{3}{3} \neq \frac{5}{6}$

44. False: Example: $\frac{2}{3} = \frac{4}{6}$ but $\frac{2}{3} \neq \frac{(2+4+1)}{3+6+1}$ because $\frac{2}{3} \neq \frac{7}{10}$

45. $\frac{18}{25}$ or $\frac{25}{18} = $ the ratio of the perimeters
$\frac{18}{25}$ or $\frac{25}{18} = $ the ratio of the sides

The ratio of the perimeters is equal to the ratio of the sides in similar triangles.

46.

p	q	If p then q
1	1	1
1	0	0
0	1	1
0	0	1

47.

p	q	NOT p	(NOT p) AND q
1	1	0	0
1	0	0	0
0	1	1	1
0	0	1	0

p	q	NOT q	p OR (NOT q)
1	1	0	1
1	0	1	1
0	1	0	0
0	0	1	1

p	q	NOT p	(NOT p) OR q
1	1	0	1
1	0	0	0
0	1	1	1
0	0	1	1

48. The network whose logical expression is (NOT p) OR q is functionally equivalent to "If p then q."

CHAPTER REVIEW AND ASSESSMENT

1. Valid, by *modus ponens*

2. Invalid, since the argument is of the form "Affirming the Consequent."

3. Therefore, a cat is not a rodent.

4. Therefore, 25 is not divisible by 6.

5. John is my brother and 17 is prime.

6. John is my brother or 17 is prime.

7.

p	q	p AND q
T	T	T
T	F	F
F	T	F
F	F	F

8.

p	q	p OR q
T	T	T
T	F	T
F	T	T
F	F	F

9. Converse: If the ground is wet, then it is raining.
Inverse: If it is not raining, then the ground is not wet.
Contrepositive: If the ground is not wet, then it is not raining.

10.

p	q	$q \Rightarrow p$
T	T	T
T	F	T
F	T	F
F	F	T

11.

p	q	$\sim p \Rightarrow \sim q$
T	T	T
T	F	T
F	T	F
F	F	T

12.

p	q	$\sim q \Rightarrow \sim p$
T	T	T
T	F	F
F	T	T
F	F	T

13. $\overrightarrow{AD} \perp \overline{BC}$

14. $\triangle ADC \cong \triangle ADB$

15. $\overline{AC} \cong \overline{AB}$

16. the given fact that $\triangle ABC$ is scalene

17. p OR (NOT q)

18.

p	q	NOT q	p OR (NOT q)
1	1	0	1
1	0	1	1
0	1	0	0
0	0	1	1

19. 1. p OR q
2. NOT r
3. (p OR q) AND (NOT r)

20.

p	q	r	p OR q	NOT r	(p OR q) AND (NOT r)
1	1	1	1	0	0
1	1	0	1	1	1
1	0	1	1	0	0
1	0	0	1	1	1
0	1	1	1	0	0
0	1	0	1	1	1
0	0	1	0	0	0
0	0	0	0	1	0

21. It is a valid argument. It has *modus tollens* form.

22.

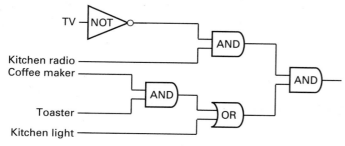

CHAPTER TEST

1. Valid by *modus ponens*

2. Not valid (affirming the consequent)

3. Not valid (denying the antecedent)

4. If it is not a rectangle, then it is not a square. A rhombus is not a rectangle. Therefore, a rhombus is not a square.

5. If a number is not a multiple of 4, then it is not a multiple of 8. The number 50 is not a multiple of 4. Therefore, the number 50 is not a multiple of 8.

6. A dolphin is a mammal and circles are polygons.

7. The conjunction is false because circles are not polygons.

8. A dolphin is a mammal or circles are polygons.

9. The disjunction is true because only one of the two statements must be true, and it is true that a dolphin is a mammal.

10.

p	q	p or q
T	T	T
T	F	T
F	T	T
F	F	F

11. Converse: If $QRST$ is a quadrilateral, then $QRST$ is a trapezoid.
Inverse: If $QRST$ is not a trapezoid, then $QRST$ is not a quadrilateral.
Contrapositive: If $QRST$ is not a quadrilateral, then $QRST$ is not a trapezoid.

12.

p	q	$q \Rightarrow p$
T	T	T
T	F	T
F	T	F
F	F	T

13.

p	q	$\sim p$	$\sim q$	$\sim p \Rightarrow \sim q$
T	T	F	F	T
T	F	F	T	T
F	T	T	F	F
F	F	T	T	T

14.

p	q	$\sim p$	$\sim q$	$\sim q \Rightarrow \sim p$
T	T	F	F	T
T	F	F	T	F
F	T	T	F	T
F	F	T	T	T

15. Converse: If water's temperature is greater than or equal to 100°, then it boils at normal atmospheric pressure. True

Inverse: If water does not boil at normal atmospheric pressure, then its temperature is not greater than or equal to 100°. True

Contrapositive: If water's temperature is not greater than or equal to 100°, then it does not boil at normal atmospheric pressure. True

16. $\overline{AB} \cong \overline{AC}$

17. isosceles

18. $\angle B \cong \angle C$

19. $\angle BAD \cong \angle CAD$

20. $\triangle BAD \cong \triangle CAD$

21. ASA

22. $\overline{BD} \cong \overline{CD}$

23. $\overline{AD}$ is not a median of $\triangle ABC$.

24. $\overline{AB}$ is not $\cong \overline{AC}$

25. $\sim(p \text{ OR } q)$

26.

p	q	p OR q	$\sim(p$ OR $q)$
1	1	1	0
1	0	1	0
0	1	1	0
0	0	0	1

27. $(p \text{ AND } \sim q) \text{ OR } r$

28.

p	q	r	$\sim q$	p AND $\sim q$	$(p$ AND $\sim q)$ OR r
1	1	1	0	0	1
1	1	0	0	0	0
1	0	1	1	1	1
1	0	0	1	1	1
0	1	1	0	0	1
0	1	0	0	0	0
0	0	0	1	0	0
0	0	1	1	0	1

1. The two lengths are equal.

   Length of $a = \sqrt{(-9 - (-3))^2 + (14 - 3)^2} = \sqrt{157}$

   Length of $b = \sqrt{(7 - 1)^2 + (-12 - (-1))^2} = \sqrt{157}$

   The answer is C.

2. Since BE is a length, it has a positive value. Then $\frac{1}{2}BE < BE$, so the answer is A.

3. $\overline{DF}$ is the hypotenuse. It is longer than $\overline{DE}$. The answer is B.

4. **b.** It can be rewritten as "the shortest path between two points is a line segment" and still be true.

5. Slope of $\overline{AF} = \frac{(20 - 15)}{(9 - 13)} = \frac{5}{-4}$

   $\frac{(-3 - (-13))}{(4 - 12)} = \frac{10}{-8} = \frac{5}{-4}$

   The answer is d.

6. $EA \times EB = EC \times ED$

   $EA \times EB = EC$

   ED

   b. $EB \times \frac{EA}{ED}$

7. **c.** The sum of the angles of a triangle is less than 180°.

8. **b.** The fog has not lifted. (*modus tollens*)

9. The boat can leave the harbor.

10. If the boat cannot leave the harbor, then the fog has not lifted.

11. **c.** $\left(\frac{a}{2}, \frac{a\sqrt{3}}{2}\right)$

12. $\frac{360}{13} \approx 27.7°$

13. $\frac{\left(\frac{1}{2}\right)(4)(4) + \frac{1}{2}(4)(5)}{(9)(9)} = \frac{18}{81} = \frac{2}{9}$

14. $MN = \frac{10}{\tan 50°} + 14 + \frac{10}{\tan 35°} \approx 36.67$

15. $\frac{\sin 72°}{22} = \frac{\sin 68°}{AB}$

    $(\sin 72°)(AB) = 22(\sin 68°)$

    $AB = \frac{22(\sin 68°)}{\sin 72°} \approx 21.4 \text{ mm}$